T0145241

Studies in Fuzziness and Soft Computing

Volume 422

Series Editor

Janusz Kacprzyk, Systems Research Institute, Polish Academy of Sciences, Warsaw, Poland

The series "Studies in Fuzziness and Soft Computing" contains publications on various topics in the area of soft computing, which include fuzzy sets, rough sets, neural networks, evolutionary computation, probabilistic and evidential reasoning, multi-valued logic, and related fields. The publications within "Studies in Fuzziness and Soft Computing" are primarily monographs and edited volumes. They cover significant recent developments in the field, both of a foundational and applicable character. An important feature of the series is its short publication time and world-wide distribution. This permits a rapid and broad dissemination of research results.

Indexed by SCOPUS, DBLP, WTI Frankfurt eG, zbMATH, SCImago.

All books published in the series are submitted for consideration in Web of Science.

Shahnaz N. Shahbazova · Ali M. Abbasov ·
Vladik Kreinovich · Janusz Kacprzyk ·
Ildar Z. Batyrshin
Editors

Recent Developments and the New Directions of Research, Foundations, and Applications

Selected Papers of the 8th World Conference on Soft Computing, February 03–05, 2022, Baku, Azerbaijan, Vol. I

 Springer

Editors
Shahnaz N. Shahbazova
Institute of Control Systems, Ministry
of Science and Education of the Republic
of Azerbaijan
Azerbaijan State University of Economics
UNEC
Baku, Azerbaijan

Vladik Kreinovich
Department of Computer Science
University of Texas at El Paso
El Paso, TX, USA

Ildar Z. Batyrshin
Centro de Investigacion en Computacion
Instituto Politecnico Nacional
Mexico City, Mexico

Ali M. Abbasov
Institute of Control Systems, Ministry
of Science and Education of the Republic
of Azerbaijan
Azerbaijan State University of Economics
UNEC
Baku, Azerbaijan

Janusz Kacprzyk
Systems Research Institute
Polish Academy of Science
Warsaw, Poland

ISSN 1434-9922   ISSN 1860-0808 (electronic)
Studies in Fuzziness and Soft Computing
ISBN 978-3-031-20155-4   ISBN 978-3-031-20153-0 (eBook)
https://doi.org/10.1007/978-3-031-20153-0

This Springer imprint is published by the registered company Springer Nature Switzerland AG
The registered company address is: Gewerbestrasse 11, 6330 Cham, Switzerland

The 8th World Conference on Soft Computing is dedicated to the Research Heritage and the 100th Anniversary Birthday of professor Lotfi A. Zadeh.

Preface

These two volumes constitute the Selected papers from the 8th World Conference on Soft Computing or WConSC 2022, held on February 03–05, 2022, in Baku, Azerbaijan. The conference was organized by the Institute of Control System, Ministry of Science and Education of the Republic of Azerbaijan, University of California, Berkeley, California, USA, and Azerbaijan State University of Economics UNEC, Azerbaijan.

In 1991, Zadeh introduced the concept of Soft Computing—a consortium of methodologies that collectively provide a foundation for the conception, design, and utilization of intelligent systems. One of the principal components of Soft Computing is Fuzzy Logic. Today, the concept of Soft Computing is growing rapidly in visibility and importance. Many papers are written in the eighties and early nineties were concerned, for the most part, with applications of fuzzy logic to knowledge representation and commonsense reasoning. Soft Computing facilitates the use of fuzzy logic, neurocomputing, evolutionary computing, and probabilistic computing in combination, leading to the concept of hybrid intelligent systems.

Combining these intelligent systems tools has already led to a large number of applications, which shows the great potential of Soft Computing in many application domains.

The volumes cover a broad spectrum of Soft Computing techniques, including theoretical results and practical applications that help find solutions for industrial, economic, medical, and other problems.

The conference papers included in these proceedings were grouped into the following research areas:

- Soft Computing and Fuzzy Logic
- Fuzzy Applications and Fuzzy Control
- Fuzzy Logic and Its Applications
- Soft Computing and Logic Aggregation
- Fuzzy Model and Z-Information
- Fuzzy Neural Networks
- Artificial Intelligence and Expert Systems

- Probabilistic Uncertainty
- Intelligent Methods and Fuzzy Approach
- Fuzzy Methods and Applications in Medicine
- Fuzzy Control Systems
- Fuzzy Logic and Fuzzy Sets
- Fuzzy Simulation
- Fuzzy Recognition
- Fuzzy Applications and Measurement Systems

During the WConSC 2022 conference, we had two panels. The first panel titled **Lotfi Zadeh Legacy: Inspirations for Mathematics, Systems Theory, Decision, Control** featured presentations by Prof. Janusz Kacprzyk, Prof. Oscar Castillo, Prof. Ildar Z. Batyrshin, Prof. Nadezhda Yarushkina, Prof. Valentina E. Balas, Prof. Vadim Stefanuk, Prof. Jozo Dujmovic, and Prof. Shahnaz N. Shahbazova. The second panel titled **Fuzzy Logic and Soft Computing, Computational Intelligence, and Intelligent Technologies** featured Prof. Vladik Kreinovich, Prof. Oscar Castillo, Prof. Takahiro Yamanoi, Prof. Aminaga Sadiqov, Prof. Averkin Alexey, Prof. Ildar Z. Batyrshin, Prof. Marius Balas, Prof. Nishchal Kumar Verma, and Prof. Shahnaz N. Shahbazova. Many thanks to all the panel speakers for their very valuable memories of Prof. Lotfi A. Zadeh and for their interesting talks.

At the WConSC 2022 conference, we had 16 keynote speakers: Prof. Janusz Kacprzyk (Poland), Prof. Vladik Kreinovich (Texas, USA), Prof. Oscar Castillo (Mexico), Prof. Ildar Z. Batyrshin (Mexico), Prof. Nadezhda Yarushkina (Russia), Prof. Valentina E. Balas (Romania), Prof. Vadim Stefanuk (Russia), Prof. Alexey Averkin (Russia), Prof. Marius Balas (Romania), Prof. Nishchal Kumar Verma (India), Prof. Imre Rudas (Hungary), Prof. Elizabet Chang (Austria), Prof. Asaf Hajiyev (Azerbaijan), Prof. Praveen K. Khosla (India), Prof. Takahiro Yamnoi (Japan) and Prof. Shahnaz N. Shahbazova (Azerbaijan). Summaries of their talks are included in this book.

The editors of this book would like to acknowledge all the authors for their contributions that kept the quality of the WConSC 2022 conference at a high level. We have received invaluable help from the members of the International Program Committee and from the Conference. Thanks to all of them we had remarkably interesting presentations and stimulating discussions.

We express our gratitude to the Institute of Control Systems of the Ministry of Science and Education of the Republic of Azerbaijan, the Director General of the Institute, Prof. Ali M. Abbasov, for assistance in organizing the conference and publications of the 8th World Conference on Soft Computing.

Our special thanks go to Janusz Kacprzyk, Editor-in-Chief of the Springer book series Studies in Fuzziness and Soft Computing for the opportunity to organize this guest-edited volumes.

We are grateful to the Springer staff, especially to Dr. Thomas Ditzinger (Senior Editor, Applied Sciences and Engineering, Springer-Verlag), for their excellent collaboration, patience, and help during the preparation of these volumes.

We hope that the volumes will provide useful information to professors, researchers, and graduated students in the area of Soft Computing techniques and applications and that all of them will find this collection of papers inspiring, informative and useful. We also hope to see you all at future World Conferences on Soft Computing.

Baku, Azerbaijan   Prof. Shahnaz N. Shahbazova
Baku, Azerbaijan   Prof. Ali M. Abbasov
El Paso, USA   Prof. Vladik Kreinovich
Warsaw, Poland   Prof. Janusz Kacprzyk
Mexico City, Mexico   Prof. Ildar Z. Batyrshin

Contents

Fuzzy Logic and Its Applications

Soft Computing and Logic Aggregation

About the Editors

Dr. Shahnaz N. Shahbazova received the academic degree of Ph.D. in 1995, the academic title of associate professor in 1996, the academic degree of Doctor of Sciences in Engineering 2015. Since 2002, she has been an academician, and since 2014, Vice-President of the Lotfi A. Zadeh International Academy of Sciences. Since 2011, she has held the position of General Chairman and organizer of the World Conference on Soft Computing (WConSC), dedicated to the preservation and development of the scientific heritage of Prof. Lotfi A. Zadeh. She is an honorary professor at the University of Obuda in Hungary and the "Aurel Vlaicu" University of Arad in Romania, an Honorary Doctor of Philosophy of Technical Sciences of the UNESCO International Personnel Academy.

She is a member of the Board of Directors of the North American Society for Fuzzy Information Processing (NAFIPS); moderator of the Berkeley Soft Computing Initiative (BISC) group; member of the council 3338.01 "System Analysis, Management and Information Processing" for the defense of Ph.D. and doctoral dissertations at the Institute of Control Systems, Ministry of Science and Education of the Republic of Azerbaijan.

Dr. Shahbazova participated in many international conferences as Organizer, Honorary Chair, Session Chair, a member in Steering, Advisory, or International Program Committees and Keynote Speaker. She is a president of the Fuzzy System Association in Baku.

She awarded: India—three months (1998), Germany (DAAD)—three months (1999, 2003, 2010), USA, California, Berkeley University (Fulbright)—(2007, 2008, 2012, 2015, 2016, 2017). She is the author of more than 202 scientific articles, twelve methodological manuals, nine textbooks, three monographs, and seven Springer publications. Her research interests include Artificial Intelligence, Soft Computing, Intelligent Systems, Machine Learning Methods for Decision-Making and Fuzzy Neural Network.

Academician Ali M. Abbasov is the Director General of the Institute of Control Systems of the National Academy of Sciences of Azerbaijan. He received the Ph.D. of Technical Sciences in 1981, doctorate degree in technical sciences in 1993, and the academic title of professor in 1996. He was appointed to the State post of Minister of Information Technology and Communications from February 20, 2004, where he worked until November 2015. For four years (2000–2004), he served as Rector of the Azerbaijan State University of Economics. He earned an international reputation as a politician, being a member of the National Parliament of the Republic of Azerbaijan (2000–2004) and a member of the Parliamentary Assembly of the Council of Europe (2001–2004). He is an Honorary Doctor of the University of Pannonia, Moscow Power Engineering University, Odessa National Academy of Communications and Belarusian State University of Informatics and Radio electronics.

He is the full member of the International Academy of Sciences (Azerbaijan Branch) since 2000, a full member of the National Academy of Sciences of the Republic of Azerbaijan since 2001. He is the chairman of the Council 3338.01 "System Analysis, Management and Information Processing" for the defense of Ph.D. and doctoral dissertations from Institute of Control Systems of the National Academy Sciences of Azerbaijan.

He is the author of more than 240 scientific articles, six methodological manuals, five textbooks, six monographs and four Springer publications and nine certificates of authorship, and patents. He was the supervisor of 14 doctors of sciences and 25 doctors of philosophy. His research interests include Artificial Intelligence, Soft Computing, Speech Recognition, Machine Learning Methods for Decision-Making and Fuzzy Neural Network.

Vladik Kreinovich received his MS in Mathematics and Computer Science from St. Petersburg University, Russia, in 1974, and Ph.D. from the Institute of Mathematics, Soviet Academy of Sciences, Novosibirsk, in 1979. From 1975 to 1980, he worked with the Soviet Academy of Sciences; during this time, he worked with the Special Astrophysical Observatory (focusing on the representation and processing of uncertainty in radioastronomy). For most of the 1980s, he worked on error estimation and intelligent information processing for the National Institute for Electrical Measuring Instruments, Russia. In 1989, he was a visiting scholar at Stanford University. Since 1990, he has worked in the Department of Computer Science at the University of Texas at El Paso. In addition, he has served as an invited professor in Paris (University of Paris VI), France; Hannover, Germany; Hong Kong; St. Petersburg and Kazan, Russia; and Brazil.

His main interests are the representation and processing of uncertainty, especially interval computations and intelligent control. He has published 12 books, 39 edited books, and more than 1800 papers. Vladik is a member of the editorial board of the international journal *Reliable Computing (formerly Interval Computations)* and several other journals. In addition, he is the co-maintainer of the international Web site on interval computations http://www.cs.utep.edu/interval-comp.

Vladik is Vice President of the International Fuzzy Systems Association (IFSA), Vice President of the European Society for Fuzzy Logic and Technology (EUSFLAT), Fellow of International Fuzzy Systems Association (IFSA), Fellow of Mexican Society for Artificial Intelligence (SMIA), Fellow of the Russian Association for Fuzzy Systems and Soft Computing, Treasurer of the IEEE Systems, Man, and Cybernetics Society; he served as Vice President for Publications of IEEE Systems, Man, and Cybernetics Society

2015–2018, and as President of the North American Fuzzy Information Processing Society 2012–2014; is a foreign member of the Russian Academy of Metrological Sciences; was the recipient of the 2003 El Paso Energy Foundation Faculty Achievement Award for Research awarded by the University of Texas at El Paso; and was a co-recipient of the 2005 Star Award from the University of Texas System.

Janusz Kacprzyk is Professor of Computer Science at the Systems Research Institute, Polish Academy of Sciences, WIT—Warsaw School of Information Technology, and Chongqing Three Gorges University, Wanzhou, Chongqing, China, and Professor of Automatic Control at PIAP—Industrial Institute of Automation and Measurements in Warsaw, Poland. He is Honorary Foreign Professor at the Department of Mathematics, Yli Normal University, Xinjiang, China. He is Full Member of the Polish Academy of Sciences, Member of Academia Europaea, European Academy of Sciences and Arts, European Academy of Sciences, Foreign Member of the: Bulgarian Academy of Sciences, Spanish Royal Academy of Economic and Financial Sciences (RACEF), Finnish Society of Sciences and Letters, Flemish Royal Academy of Belgium of Sciences and the Arts (KVAB), National Academy of Sciences of Ukraine and Lithuanian Academy of Sciences. He was awarded with six honorary doctorates. He is Fellow of IEEE, IET, IFSA, EurAI, IFIP, AAIA, I2CICC, and SMIA.

His main research interests include the use of modern computation computational and artificial intelligence tools, notably fuzzy logic, in systems science, decision making, optimization, control, data analysis and data mining, with applications in mobile robotics, systems modeling, ICT etc.

He authored seven books, (co) edited more than 150 volumes, (co) authored more than 650 papers, including ca. 150 in journals indexed by the WoS. He is listed in 2020 and 2021 "World's 2% Top Scientists" by Stanford University, Elsevier (Scopus) and ScieTech Strategies and published in PLOS Biology Journal.

He is the editor in chief of eight book series at Springer, and of two journals, and is on the editorial boards of ca. 40 journals. He is President of the Polish Operational and Systems Research Society and Past President of International Fuzzy Systems Association.

Ildar Z. Batyrshin graduated from the Moscow Physical-Technical Institute. He received Ph.D. from the Moscow Power Engineering Institute and Dr. Sc. from the Higher Attestation Committee of the Russian Federation. He occupied professor and research positions in Applied Mathematics and Computer Science at the National Research Technological University, Kazan, Russia, the Institute of Problems of Informatics of Academy of Sciences of the Republic of Tatarstan, Russia, and Mexican Petroleum Institute (IMP). Since 2014 he with the Center for Computing Research (CIC) of National Polytechnic Institute (IPN), Mexico, as a Titular Professor C.

He is the President of the Mexican Society for Artificial Intelligence (SMIA), Past-President of the Russian Association for Fuzzy Systems and Soft Computing (RAFSSoftCom), a member of the NAFIPS Board of Directors, a member of BoG of IEEE SMCS, and Senior Member of IEEE.

His awards: Fellow of IFSA, SMIA and RAFSSoftCom; Regular Member of the Mexican Academy of Sciences; Level three (highest) Researcher of the National System of Researchers of Mexico; Honorary Professor of Obuda University, Budapest, Hungary; Honorary Researcher of the Republic of Tatarstan, Russia; 1st Prize for the Best Research of IMP for the Development of Expert System in Diagnostics of Water Production (SMART-Agua) in 2007; State Research Fellowship of the Presidium of Russian Academy of Sciences for Distinguished Researchers (1997–2003).

He is an associate editor of several scientific journals. He is a co-author and co-editor of more than 20 books and special volumes of journals.

He served as a Co-chair of Program or Organizing Committees of more than 10 International Conferences on Soft Computing, Artificial Intelligence, Computational Intelligence, and Data Mining.

Soft Computing and Fuzzy Logic

Lotfi A. Zadeh's Life and His Scientific Legacy

Shahnaz N. Shahbazova⬛

Abstract The paper describes in detail the life and scientific heritage of Professor Lotfi A. Zadeh. The paper covers the period from the place of his birth, the city of Baku, to the end of his life. It also describes Lotfi's information about education (doctoral studies), work, and family in the United States. Here, we also describes Zadeh's papers and the impact research results of his scientific contribution, as well as new knowledge discoveries (the history and features of the fuzzy theory, and how it originated). His scientific collaboration with other scientists (his mentors were Norbert Wiener, John R. Ragazzini. Rudolf Kalman, Charles A. Desoer and others) is described in detail and we also describe their scientific life by years up to the last stage of his life. The paper also provides information about his visit to his homeland, and participation in the international exhibition of telecommunications and information technologies BakuTel 2008. Following his will, Lotfi A. Zadeh was buried in his homeland.

Keywords Fuzzy logic · Fuzzy sets · Soft computing · Computing with words (CWW)

1 Lotfi A. Zadeh's Life

Prof. Lotfi A. Zadeh was born on February 4, 1921, in Baku, the capital of Azerbaijan— then a part of the Soviet Union. His parents were Iranian citizens. Lotfi's father was a foreign correspondent for the newspaper, Iran, a leading newspaper in Iran at that time. In addition, he was a businessman. His mother studied medicine. Foreign

Sh. N. Shahbazova (✉)
Ministry of Science and Education of the Republic of Azerbaijan, Institute of Control Systems, 68 Bakhtiyar Bahabzadeh str., Baku AZ1141, Azerbaijan
e-mail: shahbazova@cyber.az; shahnaz_shahbazova@unec.edu.az

Department of Digital Technologies and Applied Informatics, Azerbaijan State University of Economics (UNEC), Baku AZ1001, Azerbaijan

correspondents had a privileged position because it was important for the government to present to the outside world a favorable view of what was going on inside. He was the only child and a pampered child. Lotfi always had a maid and he had a governess.

Lotfi was enrolled in the Elementary School Number 16 in Baku. For him it was a very good school and he had fond memories of his student days. As was the rule in the Soviet era, what was extolled was science, scientists, and engineers. Under this influence, Lotfi decided at an early age that what he wanted was to become an engineer. This decision became the core of his outlook on life.

Although Lotfi's parents had no problems maintaining a high standard of living, the situation around them in the late twenties was getting grim. He remembered very long lines at bakeries, with bread a hard-to-get commodity. A food crisis was in the making. In 1931, Lotfi's parents decided to return to Iran, taking him with them. In Baku, Lotfi completed three years of elementary school. In Tehran, Lotfi's parents enrolled him in a missionary school, the American College. In relation to School Number 16 in Baku, the American College was in a different world. In School Number 16, believing in God was viewed as a criminal offense. At the American College he had a chapel at 10 am every day, singing hymns and reciting the Bible. When he was enrolled in the American College, through an administrative mistake he was placed in the eighth grade, not knowing Farsi and not knowing English. Miraculously he survived, but what happened is that the Ministry of Education issued a regulation requiring that all students in foreign schools should have completed at least six years in an Iranian school. At the end of his eighth year, he was forced to transfer to sixth grade in an Iranian school. Miraculously, he survived again. After completing his sixth grade in the Iranian school and passing a nation-wide examination, he returned to the American College as a student in the ninth grade. He did not quite understand how he managed to jump from the third grade to the eighth grade, from the eighth grade to the sixth grade and back to the ninth grade.

The last four years at the American College for him presented no problems. Most of his teachers were Presbyterian missionaries from the Midwest. For him, they were role models—dedicated, principled, well-educated, and fair-minded. At a distance, he fell in love with the United States and American values. During Lotfi's student days at the American College, he met Fay, his wife, who at that time was a student at the Women's branch of the American College. They were acquaintances but not close friends.

After graduating from the American College, he decided to apply for admission to the University of Tehran. He had to pass a stiff nationwide entrance examination. He passed the examination, ranking number 3. At that time, rankings were very important and highly ranked students got a lot of publicity. In Iran, knowledge and culture has always been put on a pedestal.

His years at the University of Tehran were happy years of his life. In Tehran, he had many friends, especially within the Russian-speaking community. As in Baku, they lived very well. His mother was an MD and his father was a partner in a building supply company. They had a maid and he had a car and a valet. At the American College, the ambient culture was Anglo-Saxon Protestant. At the University of Tehran, the ambient culture was Iranian and French. Most of his professors were graduates of elite French universities. The standards of instruction were very high. At that time, the regime in Iran was anti-clerical. Some of his professors had communist sympathies. During his student days and thereafter, his outlook on the world was not influenced by the prevailing views. In a photo, he is shown in his study with a sign above the desk saying in Russian ОДИН, meaning Alone. This was a proclamation of his commitment to nonconformity—a commitment that he adhered to throughout his life.

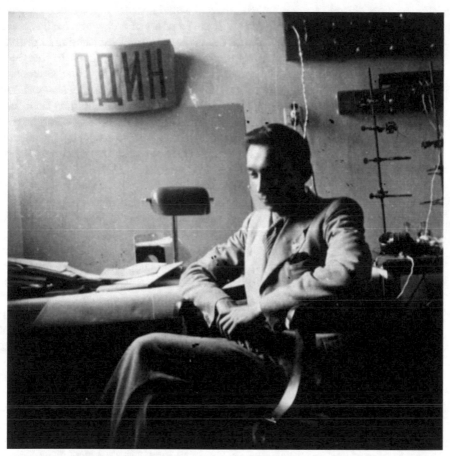

In Europe, Germany was preparing to invade other countries. War seemed to be imminent. He graduated from the University of Tehran in 1942 with a degree in Electrical Engineering. At that time, Iran was occupied by the Allies with Americans controlling the center, including Tehran. Lotfi's father ran a company that supplied construction materials. Thanks to his knowledge of English, Lotfi served as an intermediary between him and the Persian Gulf Command. As an intermediary, he had a high income. He could have stayed in Tehran, acquiring wealth and living in high style, but this was not what he wanted. His calling was to be an engineer working on the frontiers of science. To this end, he decided to leave behind his comfortable life in Tehran and emigrate to the United States in pursuit of a career in the academic world.

Lotfi applied for admission to MIT and was admitted because at that time MIT did not have many students. He left Tehran early in 1944, traveling to the United States by air and sea. Lotfi arrived in New York in July 1944 and moved to Cambridge after spending the summer months working at the International Electronics Corporation. This was his only stint in the industry.

To him, MIT was a new world. Throughout the war, MIT was the Mecca of research in electronics, communications, and what later developed into information technology. He had a feeling that he is at the frontiers of a new era. Many subjects were new to him. At the University of Tehran, he got excellent training in the basic sciences but his exposure to the world of electronics was not there. In Iran, at the time he was a student at the University, there was no industry and no research. He took courses in armature winding of motors but nothing related to electronic circuits. At MIT, it did not take him a long time to absorb the new ideas. In fact, Lotfi found that MIT was less demanding than the University of Tehran. In most cases, he could complete the exams in half the time. His Master's thesis dealt with helical antennas but he was inspired, above all, by Professor Guillemin's lectures on circuit theory.

Lotfi graduated from MIT with a Masters's degree in electrical engineering in February 1946. His parents came to the United States shortly thereafter and settled in New York. He did not want to be away from them. Lotfi decided not to continue his graduate studies at MIT toward a Ph.D. degree, and moved to New York. Professor Guillemin tried to persuade him not to do so. Here is what he said in his letter of recommendation. "I regard Mr. Zadeh as one of the most brilliant students that it has recently been my pleasure to know. You will be fortunate indeed if he decides to join your staff. I say this will feeling for I had hoped to be able to get Mr. Zadeh to take part in research activities here at MIT and shall very definitely regret losing him."

Lotfi was very fortunate in finding a position as an instructor in electrical engineering at Columbia University—a leading university in the United States. Lotfi shuddered at the thought of where he would have been if he had to take a job at a teaching university in the New York area, with no research and no opportunity to move upward. In March of 1946, Lotfi married Fay, the love of his life. His move to Columbia University and marrying Fay were decisive events in Lotfi's life.

After spending three years as an instructor, he received his Ph.D. degree in 1949 under the supervision of Professor John R. Ragazzini. His thesis was concerned with the frequency analysis of time-varying networks. But Lotfi's interests began to shift to systems analysis and information systems. In 1950, Lotfi published a significant paper "An extension of Wiener's theory of prediction," co-authored with Professor Ragazzini. In 1952, he co-authored with Professor Ragazzini a paper on what has come to be known as the Z-transform method—a method that is in wide use today in digital signal processing. During the next few years, he was brimming with ideas, publishing about ten papers per year. In 1954, Lotfi was promoted to the rank of Associate Professor, and was promoted to the rank of Professor in 1957. In 1956–57, he was a visiting member of the Institute for Advanced Study in Princeton, New Jersey, where Lotfi became acquainted with some of the leading intellects at that time. Of particular importance was his auditing a course on logic taught by Stephen Kleene. Kleene's lectures were inspirational. He became Lotfi's mentor in logic.

In the fifties, he became very interested in probability theory and its application to decision analysis. Lotfi's best friend was Herbert Robbins, a brilliant mathematician and Chair of the Department of Mathematical Statistics. Another very close friend was Richard Bellman, the father of dynamic programming. Robbins and Bellman had a profound influence on Lotfi's intellectual outlook.

In the fifties, Lotfi's work made him a visible figure in the world of systems analysis and information systems. His position at Columbia University was secure, but clouds began to appear in the blue sky. A conflict developed between Lotfi's Department and the Electronics Research Laboratory—a laboratory supported by the Air Force—over giving tenure appointments in the Department to employees of the Laboratory. This was equivalent to selling such appointments—this is what he refer to today as money-centricity. He was an opponent of money-centricity then, as he was all his life. Professor John Whinnery, who was Chair of Electrical Engineering at UC Berkeley, heard about the conflict. He was passing through New York on a Sunday in January 1959. He called Lotfi at home to inquire if he would be interested in moving to Berkeley. There were pros and cons. Eventually, Lotfi decided to leave his comfortable position at Columbia University and move to Berkeley, knowing that at UC he would have to work much harder than he had to at Columbia. In July of 1959, Fay, Stella, Norman, and Lotfi started on a long journey by car to Berkeley. Moving to Berkeley was a decisive event for Lotfi and his family.

At Berkeley, there were many new challenges. As anticipated, he had to work much harder at UC than he had to at Columbia, but the challenges and pressures were good for Lotfi. Lotfi coauthored a book with Professor Charles DeSoer in which a novel state-based approach to linear system theory was described. In 1963, he was on semi-sabbatical leave at MIT. While he was there, he received a message from John Whinnery, who was Dean of Engineering at that time, saying that he would like Lotfi to serve as Chair of Electrical Engineering. Although administration was not his cup of tea, Lotfi felt obliged to respond affirmatively to his request. This was an important event in his career.

In 1965, Lotfi was invited to attend a conference in the Soviet Union which was held on board of Admiral Nakhimov—a ship that years later, sank in the Black Sea. On the way to the conference, Fay and Lotfi stopped in Baku for one day. It was, for him, a memorable visit after leaving Baku in 1931. Lotfi met many leading figures at that time, including Professor Ibraghimov, whom he met again during his visit to Baku in 2008. Lotfi and Fay were overwhelmed by the warm hospitality. Lotfi's visit to Baku brought back memories of his childhood.

As Chair of EE, Lotfi found himself involved in a bitter conflict with Computer Science over whether Computer Science should be within EE or on the outside. A preemptive move on his part was to request that the name of the Department be changed to EECS. This request was approved by the University in 1967. Eventually, almost all EE departments in the United States have changed their name, mostly to Department of Electrical and Computer Engineering. Changing the name of the Department and building up a program in Computer Science was important achievements, in recognition of which Lotfi was awarded the IEEE Education Medal.

While Lotfi was serving as Chair, he continued to do a lot of thinking about basic issues in systems analysis, especially the issue of the unsharpness of class boundaries. In July of 1964, he was attending a conference in New York and was staying at the home of his parents. They were away. He had a dinner engagement but it had to be canceled. Lotfi was alone in the apartment. His thoughts turned to the unsharpness of class boundaries. It was at that point that the simple concept of a fuzzy set occurred to him. It did not take him long to put his thoughts together and write a paper on the subject. This was the genesis of fuzzy set theory. He knew that the word "fuzzy" would make the theory controversial. Knowing how the real world functions, he submitted his paper to Information and Control because he was a member of the Editorial Board. There was just one review—which was very lukewarm. Lotfi believed that his paper would have been rejected if he were not on the Editorial Board. Today, with over 256099 Google Scholar citations, "Fuzzy sets" is by far the highest cited paper in Information and Control.

Lotfi's paper was a turning point in his research. Since 1965, almost all of his papers related to fuzzy set theory and fuzzy logic. As he expected, his 1965 paper drew a mixed reaction, partly because the word "fuzzy" is generally used in a pejorative sense, but, more substantively, because the unsharpness of class boundaries was not considered in science and engineering. In large measure, comments on his paper were skeptical or hostile. An exception was Japan. In 1968, he began to receive letters from Japan expressing interest in the application of fuzzy set theory to pattern recognition. In the years that followed, in Japan fuzzy set theory and fuzzy logic became objects of extensive research and wide-ranging application, especially in the realm of consumer products. A very visible application was the subway system in the city of Sendai—a fuzzy-logic-based system designed and built by Hitachi and Kawasaki Heavy Industries. The system began to operate in 1987 and was considered to be a great success.

His term as Chair came to an end in 1968. He decided to switch from systems analysis to computer science. To this end, Lotfi spent a year at the IBM Research Laboratory, San Jose, California, and with project MAC at MIT. Upon his return to Berkeley in 1969 he began to teach courses in computer science, especially in the area of database systems and AI. To teach these courses he had to learn a lot of what was new to him. It was a challenge and it did Lotfi a lot of good. Freed from administrative duties, he could focus his efforts on the acquisition of knowledge about new theories and exploration of new ideas. One such idea was the concept of a linguistic variable. The concept occurred to him while he was watching a performance at the San Francisco Opera. Lotfi recognized at once that it was an idea that opened the door to important applications of fuzzy set theory. Today, the concept of a linguistic variable and the related concept of a fuzzy if–then rule are employed in almost all applications of fuzzy set theory and fuzzy logic. Lotfi described these concepts in an 1973 paper, "Outline of a new approach to the analysis of complex systems and decision processes," published in the IEEE Transactions on Systems, Man, and Cybernetics (SMC), and in his 1975 paper in Information Sciences, "The concept of a linguistic variable and its application to approximate reasoning." Both papers are among the highest-cited papers in SMC Transactions and Information Sciences, respectively.

The concept of a linguistic variable—a variable whose values are words rather than numbers—opened the door to a wide-ranging enlargement of the role of natural languages in science and engineering. It is a deep-seated tradition in science to accord much more respect for numbers than for words. As a counter-traditional concept, the concept of a linguistic variable raised questions regarding the validity of according more respect to numbers than to words. By raising the question, the concept of a linguistic variables sparked a great deal of opposition. Here is what Rudolf Kalman, a brilliant scientist and a student in Lotfi's courses at Columbia, had to say:

I would like to comment briefly on Professor Zadeh's presentation. His proposals could be severely, ferociously, or even brutally criticized from a technical point of view. This would be out of place here. But a blunt question remains: Is Professor Zadeh presenting important ideas or is he indulging in wishful thinking? No doubt Professor Zadeh's enthusiasm for fuzziness has been reinforced by the prevailing climate in the U.S.—one of unprecedented permissiveness. 'Fuzzification' is a kind of scientific permissiveness; it tends to result in socially appealing slogans unaccompanied by the discipline of hard scientific work and patient observation.

His 1973 paper found an immediate application to fuzzy control in a seminal 1974 work of Mamdani and Assilian. Today, fuzzy control is employed in a wide variety of applications ranging from cameras and household appliances to automobile transmission and ship stabilization.

Lotfi's first paper on fuzzy logic entitled "Fuzzy logic and approximate reasoning," was published in 1975 in Synthese. Lotfi's first paper on possibility theory was published in 1978 in Fuzzy Sets and Systems. This paper provided a basis for application of possibility theory to the semantics of natural languages. This is what he has done in a series of papers starting in 1978, leading to the concept of test-score semantics. He believed that eventually what will become widely recognized is that semantics of natural languages should be based on fuzzy logic.

An important direction in his work was the application of fuzzy logic in the construction of a computational system of perceptions, CTP. Basically, a natural language is a system for describing perceptions. In CTP, perceptions are described in a natural language. Then, fuzzy logic is employed to construct computational models of words and propositions, and employ these models as objects of computation and deduction.

For the past fifteen years, Lotfi's work has been focused on the construction of a system of computation which he calls Computing with Words (CW or CWW). CW is rooted in the concept of a linguistic variable. CW has two principal components. First, a precisiation module; and second, a computation module. The precisiation module translates natural language into what is referred to as a generalized constraint language, GCL. In the computation module, generalized constraints serve as objects of computation. He believed that in coming years, Computing with Words will gain recognition as an important formalism in the conception, design, and utilization of intelligent systems.

The year 2008 was memorable in Lotfi's life. He traveled to many conferences, covering 250,000 miles in the United. By next to the last conference, was BakuTel 2008, which was held in Baku. This was Lotfi's second visit to Baku, forty-three years after his first visit in 1965. Lotfi's participation in BakuTel was a memorable event. He had the honor of having a meeting with President Ilham Aliev—who impressed him as a truly outstanding personality and a leader with vision and initiative. Among memorable events were his visit to School Number 16 which brought back memories of his childhood. He for the end of his life carried fond memories of his 2008 visit to Baku.

His research was supported in part by the Office of Naval Research, Omron Grant, Tekes Grant, Azerbaijan Ministry of Communications and Information Technology Grant, Azerbaijan University of Azerbaijan Republic and the BISC Program of UC Berkeley.

The Azerbaijani government has launched the World Conferences of Soft Computing (WConSC) dedicated to Prof. Zadeh's scientific heritage since 2011. The 1st WConSC was held at San Francisco State University on May 24–27, 2011, and later 2nd WConSC in Baku, Azerbaijan on December 3–5, 2012, 3rd WConSC was held in San Antonio in Texas on November 16–18, 2013, 4th WConSC was held in Berkeley, USA on May 26–28, 2014, the 5th WConSC was held in Redmond, Washington State on August 13–15, 2015 jointly with the NAFIPS conference, 6th WConSC was held in Berkeley on May 26–29, 2016, 7th WConSC was held in Baku, Azerbaijan on May 28–31, 2018, and the last 8th WConSC was held in Baku, Azerbaijan on February 3–5, 2022.

The President of Azerbaijan awarded Lotfi A. Zadeh him a Friendship medal on the occasion of his 90th birthday anniversary in 2011. In 2016, on his 95th birthday anniversary, the Academy of Sciences of Azerbaijan awarded Zadeh Nizami Ganjavi Gold Medal which was presented at the UC Berkeley EECS department with the participation of faculty members.

The picture is from the 1st World Conference on Soft Computing, where the President of the Republic of Azerbaijan Ilham Aliyev awarded Prof. Lotfi A. Zadeh Friendship Medal on the occasion of his 90 Anniversary Birthday

I would like to share with you another aspect of his career—an impact of his work on us.

If you check the Google Scholar, a number of citations of prof. Zadeh's publications, as of January 21, 2022, is enormous.

All his work has 230,702 citations. His famous 1965 paper "Fuzzy Sets" alone has been cited 131,754 times. 13 of his papers are cited more than 2 500 times.

His work and individual engagement in multiple activities influenced us all.

Multiple areas of mathematics and statistics have experiences development and growth related to a new 'fuzzy look' at such topics as relations, intervals, numbers, just to name a few. Fuzzy versions of propositional logic, description logic and predicate calculus is very well known.

Also "softer" areas like economics, law, psychology, and sociology see valuable and interesting applications of fuzzy logic and fuzzy sets. Integration of fuzzy techniques with other state-of-the art techniques of artificial intelligence, especially neural networks and genetic algorithms, is very well established and continuously advancing. New innovative thinking about fuzzy-based modeling, control, and pattern recognition has become highly visible.

A simple yet solid evidence of the importance and popularity of fuzzy sets and fuzzy logic can be postulated by the fact that about 135 000 papers have a term "fuzzy" in their title.

This picture is from Lotfi's office at UC Berkeley, California, USA, 2008, where I got a Fulbright grant and spent 10 months for research under the supervisor of prof. Lotfi A. Zadeh

In one of the multiple interviews, Prof. Zadeh was asked a question about *a place fuzzy logic would hold in a future research.*

And his response was: I believe that, although much of modern science is based on bivalent logic, eventually most scientific theories will be based at least in part on fuzzy logic. It may, however, take a long time for this to become a reality. In coming years, I think that we will witness even more applications of fuzzy logic, a major application area is the Internet, particularly search and question answering.

This answer is one of many examples that prof. Zadeh is one of the most impressive thinkers of our times.

2 Lotfi Zadeh's Awards

In 1989, Prof. Lotfi A. Zadeh was awarded the Honda Prize by the Honda Foundation. It is awarded for "the efforts of an individual or group who contribute new ideas that may lead the next generation in the field of ecotechnology". In 1993, Dr. Zadeh received the Rufus Oldenburger Medal from the American Society of Mechanical Engineers "For seminal contributions in system theory, decision analysis, and theory of fuzzy sets and its applications to AI, linguistics, logic, expert systems and neural networks." He was also awarded the Grigore Moisil Prize for Fundamental Research, and the Premier Best Paper Award by the Second International Conference on Fuzzy Theory and Technology. In 1995, Dr. Zadeh was awarded the IEEE Medal of Honor "For pioneering development of fuzzy logic and its many diverse applications." In 1996, Dr. Zadeh was awarded the Okawa Prize "For outstanding contribution to information science through the development of fuzzy logic and its applications."

In 1997, Dr. Zadeh was awarded the B. Bolzano Medal by the Academy of Sciences of the Czech Republic "For outstanding achievements in fuzzy mathematics." He also received the J.P. Wohl Career Achievement Award of the IEEE Systems, Science and Cybernetics Society. He served as a Lee Kuan Yew Distinguished Visitor, lecturing at the National University of Singapore and the Nanyang Technological University in Singapore, and as the Gulbenkian Foundation Visiting Professor at the New University of Lisbon in Portugal. In 1998, Dr. Zadeh was awarded the Edward Feigenbaum Medal by the International Society for Intelligent Systems, and the Richard E. Bellman Control Heritage Award by the American Council on Automatic Control. In addition, he received the Information Science Award from the Association for Intelligent Machinery and the SOFT Scientific Contribution Memorial Award from the Society for Fuzzy Theory in Japan. In 1999, he was elected to membership in Berkeley Fellows and received the Certificate of Merit from IFSA (International Fuzzy Systems Association).

In 2000, he received the IEEE Millennium Medal; the IEEE Pioneer Award in Fuzzy Systems; the ASPIH 2000 Lifetime Distinguished Achievement Award; and the ACIDCA 2000 Award for the paper, "From Computing with Numbers to Computing with Words—From Manipulation of Measurements to Manipulation of Perceptions." In addition, he received the Chaos Award from the Center of Hyperincursion and Anticipation in Ordered Systems for his outstanding scientific work on foundations of fuzzy logic, soft computing, computing with words and the computational theory of perceptions. In 2001, Dr. Zadeh received the ACM 2000 Allen Newell Award for seminal contributions to AI through his development of fuzzy logic. In addition, he received a Special Award from the Committee for Automation and Robotics of the Polish Academy of Sciences for his significant contributions to systems and information science, development of fuzzy sets theory, fuzzy logic control, possibility theory, soft computing, computing with words, and computational theory of perceptions.

In 2003, Dr. Zadeh was elected a Foreign Member of the Finnish Academy of Sciences, and received the Norbert Wiener Award of the IEEE Society of Systems,

Man and Cybernetics "For pioneering contributions to the development of system theory, fuzzy logic and soft computing." In 2004, Dr. Zadeh was awarded Civitate Honoris Causa by Budapest Tech (BT) Polytechnical Institution, Budapest, Hungary. Also in 2004, he was awarded the V. Kaufmann Prize by the International Association for Fuzzy-Set Management and Economy (SIGEF). In 2005, Dr. Zadeh was elected a Foreign Member of the Polish Academy of Sciences, Korea Academy of Science and Technology, and Bulgarian Academy of Sciences. He was also awarded the Nicolaus Copernicus Medal of the Polish Academy of Sciences and the J. Keith Brimacombe IPMM Award.

In 2006, he was elected a Foreign Member of the National Academy of Sciences of Azerbaijan and was awarded the Pioneer Award for Outstanding Contributions to Soft Computing, Georgia State University, Atlanta, Georgia, and the Silicon Valley Engineering Hall of Fame.

In 2007, he was awarded the Egleston Medal, Columbia University, New York and became a member of the International Academy of Systems Studies (IASS). In 2009, he was awarded the Franklin Institute Medal, Philadelphia. In 2011, he was awarded the Medal of the Foundation by the Trust of the Foundation for the Advancement of Soft Computing, Spain, the High State Award 'Friendship Order', from the President of the Republic of Azerbaijan and the Transdisciplinary Award and Medal of the Society for Design and Process Sciences, Korea.

Dr. Zadeh is a recipient of twenty-four honorary doctorates from: Paul-Sabatier University, Toulouse, France; State University of New York, Binghamton, New York; University of Dortmund, Dortmund, Germany; University of Oviedo, Oviedo, Spain; University of Granada, Granada, Spain; Lakehead University, Canada; University of Louisville, Kentucky; State Oil Academy of Azerbaijan; Baku State University, Azerbaijan; the Silesian Technical University, Gliwice, Poland; the University of Toronto, Toronto, Canada; the University of Ostrava, the Czech Republic; the University of Central Florida, Orlando, Florida; the University of Hamburg, Hamburg, Germany; the University of Paris VI, Paris, France; Jahannes Kepler University, Linz, Austria; University of Waterloo, Canada; the University of Aurel Vlaicu, Arad, Romania; Lappeenranta University of Technology, Lappeenranta, Finland; Muroran Institute of Technology, Muroran, Japan; Hong Kong Baptist University, Hong Kong, China; Indian Statistical Institute, Kolkata, India; University of Saskatchewan, Saskatoon, Canada; the Polytechnic University of Madrid, Madrid, Spain, and Ryerson University, Toronto, Ontario, Canada.

3   Lotfi A. Zadeh's Scientific Heritage

Dr. Zadeh has single-authored over two hundred papers and serves on the editorial boards of over seventy journals. He was a member of the Advisory Committee, Center for Education and Research in Fuzzy Systems and Artificial Intelligence, Iasi, Romania; Senior Advisory Board, International Institute for General Systems Studies; the Board of Governors, International Neural Networks Society;

he was the Honorary President of the Biomedical Fuzzy Systems Association of Japan and the Spanish Association for Fuzzy Logic and Technologies. In addition, he was a member of the Advisory Board of the National Institute of Informatics, Tokyo; a member of the Governing Board, Knowledge Systems Institute, Skokie, IL; and an honorary member of the Academic Council of NAISO-IAAC.

3.1 Principal Contributions Prior to 1965

Prior to the publication of his first paper on fuzzy sets in 1965—a paper which was written while he was serving as Chair of the Department of Electrical Engineering at UC Berkeley—he had achieved both national and international recognition. He have published over sixty single-authored papers on a wide variety of problems centering on systems analysis, information systems, optimization, and control. His principal contributions were: development of a frequency-domain-based theory of time-varying networks; a generalization of Wiener's theory of prediction (with J.R. Ragazzini); initiation of system theory; a theory of nonlinear filtering; and the Z-transform method for the analysis of sampled-data systems (with J.R. Ragazzini). The Z-transform method which was introduced in 1952, is still in wide use today in digital signal processing and control. His most important contribution before 1965, was the development of the state space theory of linear systems, published as a co-authored book with C.A. Desoer in 1963. The state-space-based theory of linear systems was a path-breaking work. One of its important contributions was introduction into the theory of linear systems of concepts and techniques drawn from automata theory.

3.2 A New Direction—Development of Fuzzy Set Theory and Fuzzy Logic

The publication of his first paper on fuzzy sets in 1965 marked the beginning of a new phase of his scientific career. From 1965 on, almost all of his publications have been focused on the development of fuzzy set theory, fuzzy logic, and their applications. It should be noted that most of his papers were published prior to 1977, and all papers published since then, are single-authored. His publications list contains 247 papers and books. His publications are associated with 256099 Google Scholar citations.

Zadeh's first paper entitled "Fuzzy sets" got a mixed reaction. His strongest supporter was the late Professor Richard Bellman, an eminent mathematician and a leading contributor to systems analysis and control. For the most part, he encountered skepticism, derision, and sometimes outright hostility. There were two principal reasons: the word "fuzzy" is usually used in a pejorative sense; and, more importantly, his abandonment of the classical, Aristotelian, bivalent logic was a radical departure from deep-seated scientific traditions. What changed the situation was the enthusiastic acceptance of his ideas in Japan.

Starting in the early seventies, Japanese universities and industrial research laboratories began to play an active role in the development of fuzzy logic and its applications. Much has happened since that period. What is worthy of note is that as of 2014, his 1965 paper on fuzzy sets drew 53,172 Google Scholar citations. It was the highest cited paper in the literature of Computer Science (Web of Science); it was the seventh-highest cited paper in the literature of Science (Web of Science). The first five highest cited papers in Science were in biomedicine and the sixth highest ranking paper was in Chemistry.

3.3 The Concept of a Linguistic Variable. Decision-Making in a Fuzzy Environment

During the past forty-nine years, he played an active and visible role in the development of fuzzy logic and its applications. His 1973 paper entitled "Outline of a new approach to the analysis of complex systems and decision processes," was a path-breaking work in which the concept of a linguistic variable was introduced, and a calculus of fuzzy if–then rules was developed. Today, almost all applications of fuzzy logic employ the concept of a linguistic variable, and there is a huge literature centering on fuzzy-rule-based calculi. A related paper entitled, "The concept of a linguistic variable and its application to approximate reasoning," published in 1975, is the highest cited paper in Information Sciences. Another important paper was his 1970 paper entitled, "Decision-making in a fuzzy environment," co-authored with R.E. Bellman. This paper is widely viewed as a seminal contribution to the application of fuzzy logic to decision analysis.

3.4 Development of Possibility Theory

Zadeh's 1978 paper, "Fuzzy sets as a basis for a theory of possibility," laid the foundation for what he called "possibility theory." Greeted with skepticism at first, possibility theory has become a widely used tool for dealing with uncertainty, with the understanding that possibility theory and probability theory are complementary rather than competitive. Zadeh's 1978 paper on possibility theory is the highest cited paper in Fuzzy Sets and Systems.

3.5 Development of a Theory of Approximate Reasoning

Zadeh's 1979 paper entitled, "A theory of approximate reasoning," initiated a new the direction in the development of fuzzy logic as the logic of approximate reasoning. The basic ideas introduced in this paper underlie most of the techniques which are in use today for purposes of inference and deduction from information which is approximate rather than exact.

3.6 Soft Computing

In 1991, Zadeh introduced the concept of Soft Computing—a consortium of methodologies that collectively provide a foundation for the conception, design, and utilization of intelligent systems. One of the principal components of Soft Computing is fuzzy logic. Today, the concept of Soft Computing is growing rapidly in visibility and importance. In 2005, a European Center for Soft Computing was established in Spain. Many of his papers written in the eighties and early nineties were concerned, for the most part, with applications of fuzzy logic to knowledge representation and commonsense reasoning.

3.7 Computing with Words (CWW)

In 1996, a major idea occurred to him—an idea which underlies most of his latest research activities. This idea was described in two seminal papers entitled, "Fuzzy logic = Computing with words" and "From computing with numbers to computing with words—from manipulation of measurements to manipulation of perceptions." Zadeh's 1999 paper initiated a new direction in computation which he called, Computing with Words (CWW). CW opened the door to computation with the information described in natural language—a system of computation that is of intrinsic importance because much of human knowledge is described in a natural language. Computation with the information described in natural language cannot be dealt with through the use of the machinery of natural language processing (NLP). The problem is semantic imprecision of natural languages. More specifically, a natural language is basically a system for describing perceptions. Perceptions are intrinsically imprecise, reflecting the bounded ability of sensory organs, and ultimately the brain, to resolve detail and store information. Semantic imprecision of natural languages is a concomitant of imprecision of perceptions. Zadeh's book entitled, "Computing with Words—Principal Concepts and Ideas," was published by Springer in 2012.

3.8 Development of a Computational Theory of Perceptions (CTP)

Measurements of one kind or another have a position of centrality in science. In large measure, science is based on measurements but what is striking is that humans have a remarkable capability to perform a wide variety of mental and physical tasks without any measurements and any computations. Driving a car in heavy city traffic is an example. In a paper published in 1999, "From computing with numbers to computing with words—from manipulation of measurements to manipulation of perceptions," and in other papers which followed, Zadeh described an unconventional approach—computational theory of perceptions (CTP)—to mechanized reasoning and computation with perceptions rather than measurements. The key idea in CTP is that of describing perceptions in natural language employing the machinery of computing with words to act on perception-based information. This simple idea has a potential for wide-ranging applications in robotics, control, and related fields. A particularly important area is robotics. Another important application area is what is referred to as "perceptual computing."

3.9 Development of a Theory of Precisiation of Meaning

Raw natural language does not lend itself to computation. A prerequisite to computation is precisiation of meaning. What Zadeh considered to be one of his major contributions are the development of a theory of precisiation of meaning, starting with his 1978 paper, "PRUF—a meaning-representation language for natural languages" and continuing to present. Zadeh's theory of precisiation of meaning was a radical departure from traditional approaches to the semantics of natural languages, especially possible-world and truth-conditional semantics. The centerpiece of his theory was the concept of a restriction (generalized constraint)—a concept which was introduced in his 1975 paper, "Calculus of fuzzy restrictions," and extended in his 1986 paper, "Outline of a computational approach to meaning and knowledge representation based on the concept of a generalized assignment statement." The key idea involves representing the meaning of a proposition, drawn from a natural language as a restriction. A restriction is an expression of the form X is r R, where X is the restricted (constrained) variable, R is the restricting (constraining) relation and r is an indexical variable which defines the way in which R restricts X. Generally, X, R and r are implicit in p. At this juncture, restriction-based semantics of natural languages is the only system of precisiation of meaning which makes it possible to solve problems which are described in a natural language. The importance of his theory of precisiation of meaning has not as yet been widely recognized because it breaks away from traditional theories of natural language. He believed that eventually his theory will gain acceptance and wide use.

3.10 Development of a Generalized Theory of Uncertainty (GTU)

In a seminal paper published in 2002, "Toward a perception-based theory of proba-bilistic reasoning with imprecise probabilities" Zadeh initiated a significant gener-alization of probability theory. The ideas introduced in his 2002 paper were further developed in Zadeh's 2005 paper, "Toward a generalized theory of uncertainty (GTU)—an outline" and in his 2006 paper, "Generalized theory of uncertainty (GTU)—principal concepts and ideas." The Generalized Theory of Uncertainty (GTU) which is described in these papers adds to standard probability theory an essential capability which standard probability does not have—the capability to compute with probabilities, events, quantifiers, and relations which are described in a natural language. As we move further into the age of machine intelligence and automated everyday reasoning, this capability is certain to play an increasingly important role in decision analysis, planning, risk assessment, and economics. A key idea in GTU is that of equating information to restriction. The principal modes of restriction are possibilistic, probabilistic, and veristic.

3.11 Development of Extended Fuzzy Logic

In a short but important paper published in 2009, Zadeh outlined an extension of fuzzy logic which opens the door to the mechanization of reasoning with unprecisiated concepts. A model of extended fuzzy logic is f-geometry. In Euclidian f-geometry, figures are drawn by hand with a spray pen. There is no ruler and no compass. There are f-lines, f-triangles, and f-circles. There are f-definitions, f-axioms, and f-proofs. At this stage, extended fuzzy logic is in its early stages of development, but it has a potential for important applications in the future.

3.12 Introduction of the Concept of a Z-Number

In a 2011 paper entitled, "A note on Z-numbers," a new concept—the concept of a Z-number was introduced. Basically a Z-number is an ordered pair of two fuzzy numbers. The first fuzzy number is a restriction on the values which a real-valued variable can take. The second fuzzy number is a restriction on the probability of the first fuzzy number. Typically, the two fuzzy numbers are described in a natural language. The concept of a Z-number is intended to associate a measure of reliability with the value of a variable. The concept of a Z-number has the potential for important applications in economics, planning, risk assessment, and decision analysis. A new direction which is being explored is aimed at enhancing Web IQ (WIQ) through addition of deduction capability to search engines. Existing search engines have this capability to a very limited degree. The principal obstacle is the nature of world knowledge. In large measure, world knowledge is perception-based, e.g., "it is hard to find parking near the campus before late afternoon." Such knowledge cannot be dealt with through the use of methods based on classical, bivalent logic. In the approach that is being explored, world knowledge is dealt with through the use of PNL, in association with an epistemic lexicon and a modular, multiagent deduction database.

3.13 A New Approach to Truth and Meaning

The concepts of truth and meaning have a position of centrality in logic and theories of natural language. In 2013, in a short but important paper entitled, "Toward a restriction-centered theory of truth and meaning (RCT)" Zadeh described a new approach to truth and meaning based on the concept of restriction. In this paper, a proposition, p, drawn from a natural language is associated not just with one truth value—as in traditional theories, but with two truth values—internal truth value and external truth value. In the representation of meaning, the concept of explanatory database, ED, plays a pivotal role. RCT is the only system that offers the capability to represent the meaning of fuzzy propositions, that is, propositions that contain words that are labels of fuzzy sets, e.g., tall, fast, most, etc. Propositions drawn from a natural language are predominantly fuzzy propositions. Existing approaches to the semantics of natural languages, principally possible-world semantics and truth-conditional semantics, do not have this capability.

3.14 Similarity-Based Definitions of Possibility and Probability

The concept of probability has been around for more than two centuries. Probability theory is one of the most important and widely used theories in science. Probability theory is a deep and rigorous theory. But in real-world settings, there are many simple questions which relate to probability theory to which answers are hard to come up with. The problem is rooted in the fact that probability theory is based on the classical, Aristotelian, bivalent logic. Bivalent logic is intolerant of imprecision and partiality of truth. In the new approach, fuzzy logic is employed to construct a similarity-based definition of probability that lend itself to the use of machine learning techniques. The similarity-based definition of probability opens the door to a wide range of applications in which very large databases are involved, particularly in the realms of medical diagnostics and recognition technology. From its inception, fuzzy logic has been an object of controversy, skepticism and sometimes outright hostility. Eventually, the wide-ranging applications of fuzzy logic within science and technology have acquired visibility and acceptance. He believed that in the coming years, fuzzy logic will continue to grow in importance and visibility. With the passage of time, it's likely that the impact of fuzzy logic will be felt increasingly in many fields of science and technology. Strangely as it may seem, fuzzy logic may have a profound impact on both pure and applied mathematics. There is a reason, the concept of a set is one of the most fundamental concepts in mathematics. Progression from the concept of a set to the concept of a fuzzy set may eventually lead to a generalization of many theories and formalisms within mathematics which are based on the classical, Aristotelian, bivalent logic.

4 It was his Desire

We are saddened that today he is not around us. He was lucid and active almost until the last minute. Lotfi A. Zadeh died at 96, on September 6, 2017, in Berkeley. It was his desire to be buried in Azerbaijan, in his home country.

On September 29, the Azerbaijan National Academy of Sciences hosted Lotfi A. Zadeh's funeral. Lotfi A. Zadeh was laid to rest in the 1st Alley of Honor in Baku, Azerbaijan. President of Republic of the Azerbaijan Ilham Aliyev participated in this ceremony. With the support of the Azerbaijani government, Lotfi Zadeh's monument opened on the first anniversary of his death.

Lotfi A. Zadeh's monument in Baku, Azerbaijan

On September 29, 2018, we visited his monument on the anniversary of his death.

On September 8, 2017 College of Engineering and EECS Department of the UC Berkeley organized and provided a memorial service for Lotfi A. Zadeh where Dean prof Shankar, Chair prof James Demmel, Zadeh's son Norman Zada and myself, as well as other members of the faculty and Lotfi's great friends shared their memories about Lotfi.

5   Conclusion

Lotfi A. Zadeh İs No Longer With Us But His Legency And Spirit Are Still Leading Us.

From Fuzzy Optimization to Possibilistic-Probabilistic Optimization with Our Teacher Professor Lotfi Zadeh

A. V. Yazenin, Yu. E. Egorova, and I. S. Soldatenko

Abstract The article is dedicated to the 100th anniversary of Professor Lotfi Zadeh. It presents the history of the origin and development of one of many scientific areas associated with his name, currently classified as a possibilistic-probabilistic optimization. The article marks the main milestones in the development of the field and mentions scientific articles of Lotfi Zadeh containing results that influenced the evolution of this field from fuzzy to modern possibilistic-probabilistic optimization. It contains classification of models and methods of possibilistic-probabilistic optimization for various decision-making principles under conditions of possibilistic and possibilistic-probabilistic uncertainty, strongest and weakest triangular norms describing the interaction of fuzzy parameters. The problem of selecting an investment portfolio in conditions of hybrid uncertainty is considered as an application of the method.

Keywords Possibilistic-probabilistic optimization · Probability measure · Possibility measure · Expected possibility · Necessity measure · Fuzzy random variable · Indirect method · Direct method · Equivalent deterministic analog · Equivalent stochastic analog · Strongest t-norm · Weakest t-norm · Stochastic quasi-gradient method · Minimal risk portfolio · Quasi-efficient portfolio

Professor Lotfi Zadeh is the light of high ideas of science, of enlightenment and reason, freedom and justice.

This work is supported by RFBR under the project N20-01-00669.

A. V. Yazenin (✉) · Yu. E. Egorova · I. S. Soldatenko
Tver State University, 33 Zhelyabova Str., Tver, Russia
e-mail: yazenin.av@tversu.ru

1   Introduction

They say that "great things are seen from afar". However, this statement doesn't apply to the inventive scientific heritage of Prof. Lotfi Zadeh. Even now, we will say confidently that it's not simply large, but huge. Its greatness and scale, influence on the development of contemporary science, particularly the theory of artificial intelligence and methods of decision-making under conditions of uncertainty are noticeable from the middle of the last century to the current time. Prof. Lotfi Zadeh was the torchbearer who lit the road for his followers. His ideas of modeling knowledge with elements of uncertainty led to the formation of new scientific directions and the creation of information technologies based on soft computing. This article presents the author's concept of the development of one of them. At the origins of its creation were such titans as Lotfi Zadeh and Richard Bellman [2].

The authors of the cited paper [2] developed a symmetric approach for decision-making under fuzzy uncertainty. Mathematical tools for modeling this kind of uncertainty were introduced in the works of Lotfi Zadeh [32, 36–40]. All of this enabled us to advance and establish fuzzy optimization as a new scientific field. Classical in this context are the works of [43, 44] and of other authors.

The next step on the way to creating possibilistic-probabilistic optimization is the work [22, 41], which laid the foundations of the modern theory of possibilities. A possibilistic interpretation of the fuzzy subset was first presented in [41]. However, a mathematically justified transition to interpretation based on the possibility measure was made by Nahmias [22], although he called the corresponding set function—an evaluation function. He also developed the axiomatized calculus of possibilities using a Kolmogorov-like framework [12].

This calculus of possibilities was fully utilized in [27] and earlier works of the author for the first time in the field of linear programming with fuzzy parameters, although in relevant publications [15, 16] the fuzzy calculus based on the Lotfi Zadeh generalization principle was still applied. Their problems, though, were also classified as problems of possibilistic programming.

A mathematical model of a random experiment with a fuzzy outcome—a fuzzy random variable [13, 14, 23, 25]—was built in part thanks to Lotfi Zadeh's work on the idea of the probability of a fuzzy event [42]. As a result, we were able to create models and optimization techniques under conditions of hybrid uncertainty of possibilistic-probabilistic type [16, 28, 30]. The modern mathematical apparatus for representing uncertain knowledge of the aforementioned type, focused on its use in optimization problems, is systematized and developed in [32].

The structure of the article is as follows. The problem's necessary mathematical tools are presented in the second section. The third section presents models and methods of possibilistic optimization. The main methods are focused on a class of upper semi-continuous possibility distributions that characterize fuzzy parameters of the problem. Their interaction is described by both the strongest and weakest t-norms. It is demonstrated that the Bellman-Zadeh approach [2] can be immersed in a possibility context and presented as a special case of one of the models of possibilistic optimization considered here.

The fourth section is devoted to possibilistic-probabilistic optimization models and techniques. Principles for making decisions are defined, and optimization models for possibilistic-probabilistic hybrid uncertainty are developed. The methods presented in this section are comprised of two steps: construction of deterministic/stochastic equivalent problems and solving them with linear optimization/stochastic quasi-gradient methods.

The fifth section examines the application of possibilistic-probabilistic optimization models to the problem of portfolio optimization.

Conclusion contains the discussion about results that have been obtained in the fields of possibilistic and possibilistic-probabilistic optimization.

2 Definitions and Methods

Here we give several definitions and principles from the possibility theory following [7, 10, 21–24, 32]. Let $(\Gamma, P(\Gamma), \tau)$ and (Ω, B, P) be possibility and probability spaces, respectively, where Ω is a sample space with possible outcomes $\omega \in \Omega$, Γ is a pattern space with elements $\gamma \in \Gamma$, B is an σ-algebra of events, $P(\Gamma)$ is the discrete topology on Γ, $\tau = (\pi, \nu)$, π and ν are measures of possibility and necessity, whereas P is a probability measure and $\mathbb{E}^1$ is the real line.

Definition 1 Fuzzy variable X is a real function $X : \Gamma \to \mathbb{E}^1$, defined by its possibility distribution

$$\mu_X(t) = \pi\{\gamma \in \Gamma : X(\gamma) = t\}.$$

Definition 2 Fuzzy random variable Y is a real function $Y : \Omega \times \Gamma \to \mathbb{E}^1$, that is σ-measurable for each fixed γ and

$$\mu_Y(\omega, t) = \pi\{\gamma \in \Gamma : Y(\omega, \gamma) = t\}$$

is called its distribution function.

It follows from Definition 1 that a fuzzy random variable's distribution function is a random function since it depends on a random parameter.

Definition 3 Let $Y(\omega, \gamma)$ be a fuzzy random variable. We will denote expected value of $Y(\omega, \gamma)$ as E[Y]. It is a fuzzy variable with possibility distribution function

$$\mu_{\mathrm{E}[Y]}(t) = \pi\{\gamma \in \Gamma : \mathrm{E}[Y(\omega, \gamma)] = t\},$$

where E is the mathematical expectation operator

$$E[Y(\omega, \gamma)] = \int\limits_{\Omega} Y(\omega, \gamma) P(d\omega).$$

In this case, the distribution function of the expected value of a fuzzy random variable is deterministic since it is no longer dependent on a random parameter.

There exist at least two approaches to the definition of variance and covariance of fuzzy random variables. The variance and covariance are fuzzy within the *first* approach [11] and non-fuzzy within the *second* one [7].

In the first approach, the covariance is defined by the standard formula

$$cov(X, Y) = E[(X - E[X])(Y - E[Y])].$$

In the second approach, the covariance is defined as follows

$$cov(X, Y) = \frac{1}{2} \int\limits_{0}^{1} (cov(X_{\omega}^{-}(r), Y_{\omega}^{-}(r)) + cov(X_{\omega}^{+}(r), Y_{\omega}^{+}(r))) dr,$$

where $X_{\omega}^{-}(r), Y_{\omega}^{-}(r), X_{\omega}^{+}(r), Y_{\omega}^{+}(r)$ are left and right boundaries of α-level sets of fuzzy variables X_{ω} and Y_{ω} respectively.

Definition 4 The variance of a fuzzy random variable $Y(\omega, \gamma)$ is

$$D[Y] = cov(Y, Y).$$

One can see that second order moments in these approaches have different definitions in principle. In the first approach we identify possibility distribution and in the second one we make numeric calculations.

There are different ways to represent a fuzzy random variable. In our works we usually use a shift-scale representation [11]:

$$Y(\omega, \gamma) = a(\omega) + \sigma(\omega) Z(\gamma),$$

where $a(\omega), \sigma(\omega)$ are random variables defined on the probability space (Ω, B, P) and $Z(\gamma)$ is a fuzzy variable defined on the possibility space $(\Gamma, P(\Gamma), \tau)$. Components $a(\omega)$ and $\sigma(\omega)$ are called shift and scale respectively.

In order to have a better intuitive understanding of this type of representation of a fuzzy random variable imagine a scenario where a financial expert is requested to predict the return on a certain financial asset. The return itself and its estimation by the expert are both uncertain quantities. We assume that the uncertainty implied by market conditions is probabilistic in nature. On the other hand, uncertainty of the expert's estimation is best described by a possibility distribution. This model seems quite feasible if we assume that the degree of fuzziness of the expert depends mainly on the scale of variation of the estimated variable and not on its true value [32].

To aggregate fuzzy information, we employ triangular norms (t-norms). They generalize min operations that are used in actions on fuzzy variables [24].

In the present work we consider two extreme t-norms:

$$T_M(x, y) = \min\{x, y\}, \qquad T_W(x, y) = \begin{cases} \min\{x, y\}, & \text{if } \max\{x, y\} = 1, \\ 0, & \text{otherwise.} \end{cases}$$

T_M and T_W are the strongest and the weakest t-norm respectively, because for any t-norm T and $\forall x, y \in [0, 1]$ the following inequality holds [24]:

$$T_W(x, y) \leq T(x, y) \leq T_M(x, y).$$

The ability of t-norms to regulate the growth of uncertainty (or "fuzziness") is one of its key features. This effect can appear, for example, during arithmetic operations on fuzzy numbers: if two fuzzy numbers given by parametric distributions are added together with the help of the strongest t-norm T_M, corresponding coefficients of fuzziness are also added, which increases uncertainty. We can slow down the growth of fuzziness by using t-norms other than T_M. The aforementioned extreme triangular norms give us boundaries for control of fuzziness in our models.

Next we present the notion of mutual T-relatedness of fuzzy sets and fuzzy variables [10]. It serves as a tool for building joint possibility distribution functions.

Definition 5 Fuzzy sets $A_1, \ldots, A_n \in P(\Gamma)$ are mutually T-related, if for any $\{i_1, \ldots, i_k\} \subset \{1, \ldots, n\}, k = \overline{1, n}$, we have

$$\pi(A_{i_1} \cap \cdots \cap A_{i_k}) = T(\pi(A_{i_1}), \cdots, \pi(A_{i_k})),$$

where

$$T(\pi(A_{i_1}), \cdots, \pi(A_{i_k})) = T(T(\cdots T(T(\pi(A_{i_1}), \pi(A_{i_2})), \pi(A_{i_3}))), \pi(A_{i_k})).$$

Definition 6 Fuzzy variables $Z_1(\gamma), \ldots, Z_n(\gamma)$ are mutually T-related if for any $\{i_1, \ldots, i_k\} \subset \{1, \ldots, n\}, k = \overline{1, n}$ and for $t_{ij} \in \mathbb{E}^1$, we have

$$\mu_{Z_{i_1}, \ldots, Z_{i_k}}(t_{i_1}, \ldots, t_{i_k}) = \pi\{\gamma \in \Gamma : Z_{i_1}(\gamma) = t_{i_1}, \ldots, Z_{i_k}(\gamma) = t_{i_k}\}$$
$$= \pi\{Z_{i_1}^{-1}[t_{i_1}] \cap \cdots \cap Z_{i_k}^{-1}[t_{i_k}]\}$$
$$= T\{\pi(Z_{i_1}^{-1}[t_{i_1}]), \ldots, \pi(Z_{i_k}^{-1}[t_{i_k}])\}.$$

3   Possibilistic Optimization

3.1   *Possibilistic Optimization Models*

The basic principles of possibilistic optimization are based on ideas of fuzzy optimization which were discussed in [2]. We will show that when fuzzy sets are interpreted as fuzzy variables, the approaches to decision making presented in [2] can be naturally formulated as one of the following possibilistic models [27, 29, 31, 36].

Models of criterion

Optimization of the possibility/necessity of satisfying criterion

$$\tau_0\{f_0(x, \gamma)\mathfrak{R}_0\, 0\} \to \operatorname*{extr}_x \tag{1}$$

Level optimization

$$k \to \operatorname*{extr}_x,$$

$$\tau_0\{f_0(x, \gamma)\mathfrak{R}_0\, k\} \geq \alpha_0 \tag{2}$$

Models of constraints

Line-wise constraints on possibility/necessity

$$\begin{cases} \tau_i\{f_i(x, \gamma)\mathfrak{R}_i\, 0\} \geq \alpha_i, i = \overline{1, m}, \\ x \in X \end{cases} \tag{3}$$

Constraint on overall possibility/necessity

$$\begin{cases} \tau\{f_i(x, \gamma)\mathfrak{R}_i\, 0, i = \overline{1, m}\} \geq \alpha, \\ x \in X \end{cases} \tag{4}$$

In these models $X \subseteq \mathbb{E}^n_+ = \{x \in \mathbb{E}^n : x \geq 0\}$; and $f_i(x, \gamma)$ are possibilistic functions.

$$f_i : X \times \Gamma \to \mathbb{E}^1, i = \overline{0, m};$$

τ_i is the possibility measure π or necessity measure ν; $\mathfrak{R}_0$ and $\mathfrak{R}_i$ are binary relations $\{\leq, \geq, =\}$; $\alpha_i \in (0, 1]$ and k is some scalar variable.

Remark 1 The model of maximization of the possibility/necessity of satisfying criterion

$$\pi\{f_0(x, \gamma) = 0\} \to \max_x$$

is a classical fuzzy optimization problem [2] that is formulated in the context of possibility theory.

3.2 Equivalent Deterministic Models

In this section we will construct equivalent deterministic analogues to the models of criterion and constraints presented in the previous section. We will generalize methods studied in [27, 29, 31, 36].

Let functions $f_i(x, \gamma), i = \overline{0, m}$ can be represented by fuzzy variables with quasi-concave and upper semi-continuous possibility distributions $\mu_{f_i}(x, t)$

$$\mu_{f_i}(x, \lambda t_1 + (1 - \lambda)t_2) \geq min\{\mu_{f_i}(x, t_1), \mu_{f_i}(x, t_2)\}$$

for $t_1, t_2 \in \mathbb{E}^1, \lambda \in [0, 1]$.

Theorem 1 *If distributions of functions $f_i(x, \gamma), i = \overline{0, m}$ are quasi-concave and upper semi-continuous for each x, then the problem (1), (3) is equivalent to*

$$\lambda_0 \mu_{f_0}(x, 0) \to \underset{x}{extr}, \tag{5}$$

$$\begin{cases} \lambda_i f_{i,\beta_i}^-(x) \leq 0, \ i = \overline{1, k}, \\ \lambda_j f_{j,\beta_j}^+(x) \geq 0, \ j = \overline{1, l}, \\ \qquad x \in X, \end{cases} \tag{6}$$

where depending on $\mathfrak{R}_i$ and τ_i: $\lambda_0 = +1, \lambda_i = \pm 1, \beta_i = \alpha_i$ or $1 - \alpha_i$; $f_{i,\beta_i}^-(x)$ and $f_{i,\beta_i}^+(x)$ are left and right boundaries of β_i-level sets respectively; $k, l \in \{1, 2, \ldots, m\}$.

Theorem 2 *If distributions of functions $f_i(x, \gamma), i = \overline{0, m}$ are quasi-concave and upper semi-continuous for each x, then the problem (2), (3) is equivalent to the following*

$$\lambda_0 f_{0,\beta_0}^{\pm}(x) \to \underset{x}{min}, \tag{7}$$

$$\begin{cases} \lambda_i f_{i,\beta_i}^-(x) \leq 0, \ i = \overline{1, k}, \\ \lambda_j f_{j,\beta_j}^+(x) \geq 0, \ j = \overline{1, l}, \\ \qquad x \in X, \end{cases} \tag{8}$$

where depending on $\mathfrak{R}_i$ *and* τ_i: $\lambda_0 = \pm 1, \lambda_i = \pm 1, \beta_i = \alpha_i$ *or* $1 - \alpha_i$; $f_{i,\beta_i}^-(x)$
and $f_{i,\beta_i}^+(x)$ *are left and right boundaries of* β_i-*level sets respectively;* $f_{0,\beta_0}^-(x)$ *and*
$f_{0,\beta_0}^+(x)$ *are left and right boundaries of* β_0-*level set;* $k, l \in \{1, 2, \ldots, m\}$.

Sometimes we can obtain left and right boundaries of α-level sets of functions
$f_i(x)$. Let the functions $f_i(x, \gamma)$ are linear and have the following form

$$f_i(x, \gamma) = \sum_{j=1}^{n} b_{ij}(\gamma) x_j - b_i(\gamma), \tag{9}$$

where fuzzy variables $b_{ij}(\gamma)$ and $b_i(\gamma)$ are T-related for some arbitrary t-norm T and
are of LR-type [3]

$$b_{ij}(\gamma) = (\underline{m}_{ij}, \overline{m}_{ij}, \underline{d}_{ij}, \overline{d}_{ij},)_{LR}, \quad b_i(\gamma) = (\underline{m}_i, \overline{m}_i, \underline{d}_i, \overline{d}_i)_{LR}.$$

Then [3] the function $f_i(x, \gamma)$ will have the following possibility distribution

$$f_i(x, \gamma) = (\underline{m}_i(x), \overline{m}_i(x), \underline{d}_i(x), \overline{d}_i(x))_{LR},$$

where

$$\underline{m}_i(x) = \sum_{j=1}^{n} m_{ij} x_j - \overline{m}_i, \quad \overline{m}_i(x) = \sum_{j=1}^{n} \overline{m}_{ij} x_j - \underline{m}_i, \tag{10}$$

$$\underline{d}_i(x) = \bigoplus_{j=1}^{n} \left\{ \underline{d}_{ij} x_j, \overline{d}_i \right\}, \quad \overline{d}_i(x) = \bigoplus_{j=1}^{n} \left\{ \overline{d}_{ij} x_j, \underline{d}_i \right\}. \tag{11}$$

Here $\oplus$ denotes an aggregation operation for fuzziness coefficients [18, 19] that
depends on t-norm T.

According to [3], left and right boundaries of an α-level set will have the form (if
L(t) and R(t) are strictly monotonic)

$$f_{i,\alpha}^-(x) = \underline{m}_i(x) - L^{-1}(\alpha) \underline{d}_i(x), \quad f_{i,\alpha}^+(x) = \overline{m}_i(x) + R^{-1}(\alpha) \overline{d}_i(x).$$

3.3 The Case of the Strongest t-Norm

In this section we will discuss the equivalent deterministic analogues for possibilistic
optimization problems in the case when its fuzzy variables are T_M-related.

Let in the model (1), (3) functions $f_i(x, \gamma)$, $i = \overline{0, m}$ have the form of (9), where
$b_{i1}(\gamma), \ldots, b_{in}(\gamma)$ are characterized by quasi-concave and upper semi-continuous
distributions with finite supports. Then the following theorem can be proven.

Theorem 3 *Let* extr = "max", τ_i = "π", $\mathfrak{R}_i$ *be the equality relation "=" and the possibilistic variables* $b_{i1}(\gamma)$, ..., $b_{in}(\gamma)$ *be mutually* T_M*-related. Then, the problem (1), (3) is equivalent to*

$$x_0 \rightarrow \max_x, \tag{12}$$

$$\begin{cases} x_0 \le \mu_{b_j}(t_j), \ j = \overline{1, n} \\ x_0 \le \mu_{b_0}(t_0), \\ \sum_{j=1}^{n} t_j x_j = t_0, \\ \sum_{j=1}^{n} b_{ij}^- x_j \le b_i^+, \ i = \overline{1, m}, \\ \sum_{j=1}^{n} b_{ij}^+ x_j \ge b_i^-, \ i = \overline{1, m}, \\ (x_0, x, t) \in [0, 1] \times X \times \mathbb{E}^{n+1}, \end{cases} \tag{13}$$

where b_{ij}^-, b_{ij}^+ *and* b_i^-, b_i^+ *are boundaries of* α*-level sets of fuzzy variables* $b_{ij}(\gamma)$ *and* $b_i(\gamma)$ *respectively;* $t = (t_0, t_1, ..., t_n)$.

The problem (12), (13) can be reduced to a separable one. In order to do this, it is sufficient to eliminate the products of variables in the constraint $\sum_{j=1}^{n} t_j x_j = t_0$ in (13). The solution of separable programming problem can be obtained with the use of the methods developed in [21]. These methods are implemented in several commercial solver systems.

When functions $f_i(x)$ have the form of (9) with fuzzy variables of LR-type, the coefficients of fuzziness $\underline{d}_i(x)$ and $\overline{d}_i(x)$ in (10) will be [3]

$$\underline{d}_i(x) = \sum_{j=1}^{n} \underline{d}_{ij} x_j + \overline{d}_i, \quad \overline{d}_i(x) = \sum_{j=1}^{n} \overline{d}_{ij} x_j + \underline{d}_i.$$

Thus, functions in (6), (7) and (8) will be linear. On the other hand, criterion (5) can be transformed to a linear one by adding an additional variable [31].

This implies that the problems (5), (6) and (7), (8) are linear and can be solved by methods of linear programming, such as simplex method [21].

3.4 The Case of the Weakest t-Norm

In this section we will discuss equivalent deterministic analogues and approaches to solve them for the problems (5), (6) and (7), (8) when fuzzy variables are mutually T_W-related.

When functions $f_i(x)$ have the form of (9) with fuzzy variables of LR-type, the coefficients of fuzziness $\underline{d}_i(x)$ and $\overline{d}_i(x)$ in (10) will be [26]

$$\underline{d}_i(x) = \max_{j=1}^n \left\{ \underline{d}_{ij} x_j, \overline{d}_i \right\}, \quad \overline{d}_i(x) = \max_{j=1}^n \left\{ \overline{d}_{ij} x_j, \underline{d}_i \right\}.$$

These functions are convex. Nonetheless, since $\lambda_i = \pm 1$, some constraint functions in (6) and (8) may be concave. This leads us to non-convex optimization problems. In [26] a genetic algorithm is specified for solving the problem (7), (8) in the case of the weakest t-norm.

However, under certain assumptions we can reduce the problem (7), (8) to a convex one [1]. Let τ_i be necessity measure and $\mathfrak{R}_i = $ "$\geq$" for $i = \overline{1, m}$. Then the problem can be transformed to the equivalent deterministic one of the form:

$$m_0(x) - L^{-1}(1 - \alpha_i)d_0(x) \to \max_x,$$

$$\begin{cases} m_i(x) + R^{-1}(1 - \alpha_0)d_i(x) \leq 0, i = \overline{1, m}, \\ x \in X. \end{cases}$$

This problem is convex and can be solved, for example, by subgradient method [20] or the method of generalized linear programming [9]. In [1] we specify these methods and present a comparative analysis of their computational complexity.

4 Possibilistic-Probabilistic Optimization

4.1 Possibilistic-Probabilistic Optimization Models

Two types of uncertainty in possibilistic-probabilistic optimization models (we also call them fuzzy random for short) make it difficult to formulate principles on which solution methods for such problems can be based. However, there are two quite natural approaches that we can use. These are *expected possibility* [28, 30] principle and *probability constraints*.

The expected possibility principle allows us to remove randomness by calculating the expectations of possibilistic-probabilistic values. In the case of probability constraints, randomness is eliminated by applying a probability level of "fulfillment" of fuzzy random functions.

We can now define the following models of criterion and constraints for fuzzy random optimization problems.

Models of criterion

Optimization of the expected possibility/necessity of satisfying criterion

$$\tau_0 \left\{ E\left[f_0(x, \omega, \gamma) \right] \mathfrak{R}_0 \, 0 \right\} \to \underset{x}{\text{extr.}} \tag{14}$$

Level optimization with expected possibility

$$k \to \underset{x}{\text{extr}},$$

$$\tau_0\left\{E\left[f_0(x, \omega, \gamma)\right]\mathfrak{R}_0\, k\right\} \geq \alpha_0. \tag{15}$$

Level optimization with probability constraint

$$k \to \underset{x}{\text{extr}},$$

$$\tau_0\{P\{f_0(x, \omega, \gamma)\mathfrak{R}_0\, k\} \geq p_0\} \geq \alpha_0. \tag{16}$$

Models of constraints

Line-wise constraints on expected possibility/necessity

$$\begin{cases} \tau_i\left\{E\left[f_i(x, \omega, \gamma)\right]\mathfrak{R}_i\, 0\right\} \geq \alpha_i, i = \overline{1, m}, \\ x \in X. \end{cases} \tag{17}$$

Line-wise constraints on possibility/necessity and probability

$$\begin{cases} \tau_i\{P\{f_i(x, \omega, \gamma)\mathfrak{R}_i\, 0\} \geq p_i\} \geq \alpha_i, i = \overline{1, m}, \\ x \in X. \end{cases} \tag{18}$$

Here $X \subseteq \mathbb{E}_+^n = \{x \in \mathbb{E}^n : x \geq 0\}$; $f_i(x, \gamma)$ are possibilistic-probabilistic functions $f_i : X \times \Omega \times \Gamma \to \mathbb{E}^1, i = \overline{0, m}$; τ_i is the measure of possibility π or necessity ν; $\mathfrak{R}_0$ and $\mathfrak{R}_i$ are binary relations$\{\leq, \geq, =\}$; $\alpha_i \in (0, 1]$ and k is a scalar variable.

4.2 Equivalent Deterministic and Stochastic Models

Now we will construct equivalents to the presented in the previous section models of criterion and constraints.

Let the functions $f_i(x, \omega, \gamma)$ be linear and of the following form

$$f_i(x, \omega, \gamma) = \sum_{j=1}^{n} b_{ij}(\omega, \gamma)x_j - b_i(\omega, \gamma), \tag{19}$$

where fuzzy random variables $b_{ij}(\omega, \gamma)$ and $b_i(\omega, \gamma)$ have the shift-scale representation

$$b_{ij}(\omega, \gamma) = a_{ij}(\omega) + \sigma_{ij}(\omega)Z_{ij}(\gamma), \quad b_i(\omega, \gamma) = a_i(\omega) + \sigma_i(\omega)Z_i(\gamma).$$

Here $a_{ij}(\omega)$, $\sigma_{ij}(\omega)$ and $a_i(\omega)$, $\sigma_i(\omega)$ are independent random variables; fuzzy variables $Z_{ij}(\gamma) = \left(\underline{m}_{ij}, \overline{m}_{ij}, \underline{d}_{ij}, \overline{d}_{ij}\right)_{LR}$ and $Z_i(\gamma) = \left(\underline{m}_i, \overline{m}_i, \underline{d}_i, \overline{d}_i\right)_{LR}$ are mutually T-related [10, 26].

Then the distribution of expected value of the function $f_i(x, \omega, \gamma)$ will have the form [32]

$$E[f_i(x, \omega, \gamma)] = \left(E[\underline{m}_i(x, \omega)], E[\overline{m}_i(x, \omega)], E[\underline{d}_i(x, \omega)], E[\overline{d}_i(x, \omega)]\right)_{LR}$$
$$= (\underline{m}_{E[f_i]}(x), \overline{m}_{E[f_i]}(x), \underline{d}_{E[f_i]}(x), \overline{d}_{E[f_i]}(x))_{LR},$$

where

$$\underline{m}_{E[f_i]}(x) = \sum_{j=1}^n \left(\overline{a}_{ij} + \overline{\sigma}_{ij}\underline{m}_{ij}\right)x_j - (\overline{a}_i + \overline{\sigma}_i\overline{m}_i), \tag{20}$$

$$\overline{m}_{E[f_i]}(x) = \sum_{j=1}^n \left(\overline{a}_{ij} + \overline{\sigma}_{ij}\overline{m}_{ij}\right)x_j - (\overline{a}_i + \overline{\sigma}_i\underline{m}_i), \tag{21}$$

$$\underline{d}_{E[f_i]}(x) = E\left[\bigoplus_{j=1}^n \left\{\sigma_{ij}(\omega)\underline{d}_{ij}x_j, \sigma_i(\omega)\overline{d}_i\right\}\right], \tag{22}$$

$$\overline{d}_{E[f_i]}(x) = E\left[\bigoplus_{j=1}^n \left\{\sigma_{ij}(\omega)\overline{d}_{ij}x_j, \sigma_i(\omega)\underline{d}_i\right\}\right]. \tag{23}$$

Here $\oplus$ denotes an aggregation operation for fuzziness coefficients; $\overline{a}_{ij}$, $\overline{\sigma}_{ij}$, $\overline{a}_i$, $\overline{\sigma}_i$ are expected values of random variables $a_{ij}(\omega)$, $\sigma_{ij}(\omega)$, $a_i(\omega)$, $\sigma_i(\omega)$ respectively. We can prove the following theorems.

Theorem 4 *Under assumptions made above the problem* (14), (17) *has the following equivalent* [4]:

$$\frac{\lambda_0 m_{E[f_0]}(x)}{d_{E[f_0]}(x)} \to min, \tag{24}$$

$$\begin{cases} \lambda_i m_{E[f_i]}(x) + \beta_i d_{E[f_i]}(x) \le 0, i = \overline{1, k}, \\ \lambda m_{E[f_i]}(x) \le 0, \\ x \in X, \end{cases} \tag{25}$$

where depending on $\mathfrak{R}_i$ and τ_i for $i = \overline{0, k}$

$$\lambda = \pm 1, \lambda_i = \pm 1, \beta_i = S^{-1}(\alpha_i) \text{ or } S^{-1}(1 - \alpha_i),$$
$$m_i(x) = \underline{m}_{E[f_i]}(x) \text{ or } \overline{m}_{E[f_i]}(x), d_i(x) = \underline{d}_{E[f_i]}(x) \text{ or } \overline{d}_{E[f_i]}(x);$$

$S^{-1}(t)$ is either $\pm L^{-1}(t)$ or $\pm R^{-1}(t)$ and $k = \{m, m+1, \ldots, 2m\}$.

Theorem 5 *Under assumptions made above the problem (15), (17) has the following equivalent [4]:*

$$\lambda_0 m_{E[f_0]}(x) + \beta_0 d_{E[f_0]}(x) \to min, \tag{26}$$

$$\begin{cases} \lambda_i m_{E[f_i]}(x) + \beta_i d_{E[f_i]}(x) \le 0, i = \overline{1, k}, \\ x \in X, \end{cases} \tag{27}$$

where depending on $\mathfrak{R}_i$ and τ_i for $i = \overline{0, k}$

$$\lambda_i = \pm 1, \beta_i = S^{-1}(\alpha_i) \text{ or } S^{-1}(1 - \alpha_i),$$
$$m_i(x) = \underline{m}_{E[f_i]}(x) \text{ or } \overline{m}_{E[f_i]}(x), d_i(x) = \underline{d}_{E[f_i]}(x) \text{ or } \overline{d}_{E[f_i]}(x);$$

$S^{-1}(t)$ is either $\pm L^{-1}(t)$ or $\pm R^{-1}(t)$ and $k = \{m, m+1, \ldots, 2m\}$.

As for the models (16) and (18), their deterministic and stochastic analogues heavily depend on the triangular norm.

4.3 The Case of the Strongest t-Norm

Let us study the problems formulated in the previous section in case of the strongest triangular norm T_M.

When possibilistic variables $h_i(\gamma)$ are mutually T_M-related, the coefficients of fuzziness $\underline{d}_{E[f_i]}(x)$ and $\overline{d}_{E[f_i]}(x)$ in (22) have the form [32]

$$\underline{d}_{E[f_i]}(x) = \sum_{j=1}^{n} \overline{\sigma}_{ij} \underline{d}_{ij} x_j + \overline{\sigma}_i \underline{d}_i, \quad \overline{d}_{E[f_i]}(x) - \sum_{j=1}^{n} \overline{\sigma}_{ij} \overline{d}_{ij} x_j + \overline{\sigma}_i \underline{d}_i.$$

Therefore, functions in (25), (26) and (27) are linear. On the other hand, the criterion (24) can be transformed to a linear one with the help of one additional variable. This implies that problems (24), (25) and (26), (27) are linear and can be solved by methods of linear programming, such as, for example, simplex method.

Now we construct equivalent models for (16), (18). Let $\tau_i =$ "π" and functions $f_i(x, \omega, \gamma)$ have the form of (19) for $i = \overline{0, m}$, where random variables $a_{ij}(\omega), \sigma_{ij}(\omega)$ and $a_i(\omega), \sigma_i(\omega)$ are independent and normally distributed.

Theorem 6 *Let $p_i > 0.5$, $i \in \overline{0, m}$. Then the problem (16), (18) has the following deterministic analogue* [35]

$$\sum_{j=1}^{n} \left(\overline{a}_{0j} + \overline{\sigma}_{0j} Z_{0j}^{-}\right) x_j - \beta_0 \sqrt{\sum_{j=1}^{n} \sum_{k=1}^{n} C_{0jk}\left(Z_{0j}^{\pm}, Z_{0k}^{\pm}\right) x_j x_k} \to \min_x,$$

$$\begin{cases} \sum_{j=1}^{n} (\overline{a}_{ij} + \overline{\sigma}_{ij} Z_{0j}^{-}) x_j - \beta_i \sqrt{d_i(x, Z_{i1}^{\pm}, \ldots, Z_{in}^{\pm}, Z_i^{\pm})} \le Z_i^0 + \overline{\sigma}_i Z_i^{+} \\ \\ x \in X, \end{cases}$$

where $b_{ij}^{\pm}, b_i^{\pm}$ are right and left boundaries of α_i-level sets of possibilistic values $Z_{ij}(\gamma), Z_i(\gamma)$,

$$d_i\left(x, Z_{i1}^{\pm}, \ldots, Z_{in}^{\pm}, Z_i^{\pm}\right) = \sum_{j=1}^{n} \sum_{k=1}^{n} C_{ijk}\left(Z_{ij}^{\pm}, Z_{ik}^{\pm}\right) x_j x_k +$$

$$+2 \sum_{j=1}^{n} (-C_{a_{ij} a_i} - C_{a_{ij} \sigma_i} Z_i^{-} - C_{\sigma_{ij} a_i} Z_{ij}^{-} - C_{\sigma_{ij} \sigma_i} (Z_i Z_{ij})^{-}) x_j +$$

$$+ d_{a_i} + d_{\sigma_i} (Z_i^2)^{-} + 2 C_{\sigma_i b_i} (Z_i)^{-},$$

$$C_{ijk}(Z_{ij}^{\pm}, Z_{ik}^{\pm}) = C_{a_{ij} a_{ik}} + C_{a_{ij} \sigma_{ik}} Z_{ik}^{-} + C_{\sigma_{ij} a_{ik}} Z_{ij}^{-} + C_{\sigma_{ij} \sigma_{ik}} (Z_{ij} Z_{ik})^{-}.$$

Here we denote covariance of random variables $A(\omega)$ and $B(\omega)$ as C_{AB}. The boundaries of α-level sets of fuzzy variables Z_{ij}, Z_i and their products $(Z_i \cdot Z_{ij})$ can be found with the help of results from [3, 8, 32].

The criterion function in this problem is convex, and when $p_0 < 0.5$ it moves to the class of non-convex multi-extremal problems. However, the condition $p_0 > 0.5$ is in line with the requirements of actual problems.

Theorem 7 *Let in the model (16), (18) $p_i = 0.5$ for $i = \overline{0, m}$, $\tau = "\pi"$. Then the equivalent deterministic analogue will have the following form:*

$$\sum_{j=1}^{n} \left(\overline{a}_{0j} + \overline{\sigma}_{0j} Z_{0j}^{-}\right) x_j \to \min_x,$$

$$\begin{cases} \sum_{j=1}^{n} \left(\overline{a}_{ij} + \overline{\sigma}_{ij} Z_{ij}^{-}\right) x_j \le \overline{a}_i + \overline{\sigma}_i Z_i^{+}, i = \overline{1, m}, \\ \\ x \in X. \end{cases}$$

Remark 2 When $\tau = 'v'$ the equivalent deterministic analogue for the model (16), (18) has the form:

$$\sum_{j=1}^{n} \left(\overline{a}_{0j} + \overline{\sigma}_{0j} Z_{0j}^{-}\right) x_j \to \min_x,$$

$$\begin{cases} \sum_{j=1}^{n} (\overline{a}_{ij} + \overline{\sigma}_{ij} Z_{ij}^{+}) x_j \leq \overline{a}_i + \overline{\sigma}_i Z_i^{-}, i = \overline{1, m}, \\ x \in X, \end{cases}$$

where Z_{ij}^{+} and Z_i^{-} are boundaries of $(1 - \alpha_i)$-level sets.

In [35] we show that the problem (16), (18) when $p_i = 0.5, i = \overline{1, m}$ is equivalent to the problem (15), (17).

4.4 The Case of the Weakest t-Norm

In this section we will study approaches to solve problems of possibilistic-probabilistic optimization when fuzzy variables of models are mutually T_w-related.

In this case fuzziness coefficients $\underline{d}_{E[f_i]}(x)$ and $\overline{d}_{E[f_i]}(x)$ in (22) will have the form [32]

$$\underline{d}_{E[f_i]}(x) = E\left[\max_{j=1}^{n} \left\{ \sigma_{ij}(\omega) \underline{d}_{ij} x_j, \sigma_i(\omega) \overline{d}_i \right\} \right]$$

$$\overline{d}_{E[f_i]}(x) = E\left[\max_{j=1}^{n} \left\{ \sigma_{ij}(\omega) \overline{d}_{ij} x_j, \sigma_i(\omega) \underline{d}_i \right\} \right].$$

Even though these functions are convex, the problems (24), (25) and (26), (27) will be in general non-convex, because there can be negative β_i that make some constraint functions concave.

With some additional assumptions, the problem (24), (25) can be transformed to a convex one. Let τ_0 be possibility measure π and $\mathfrak{R}_0 \in \{=, \leq, \geq\}$, while τ_i be necessity measure and $\mathfrak{R}_i \in \{\leq, \geq\}$ for $i - \overline{1, m}$. Let us denote

$$t = \frac{1}{d_{E f_0}(x)}, \quad y_j = \frac{x_j}{d_{E f_0}(x)}, \quad j = \overline{1, n}.$$

Now we can transform the initial problem to

$$\lambda_0 m_{E f_0}(y, t) \to \min_{y,t},$$

$$\begin{cases} \lambda_i m_{E[f_i]}(y, t) + \beta_i d_{E[f_i]}(y, t) \leq 0, i = \overline{1, m}, \\ \lambda m_{E[f_0]}(y, t) \leq 0, \\ d_{E[f_0]}(y, t) \leq 1, \\ t > 0, \frac{y}{t} \in X, \end{cases}$$

where

$$m_{E[f_i]}(y, t) = \sum_{j=1}^{n}(\overline{a}_{ij} + \overline{\sigma}_{ij}m_{ij})y_j - (\overline{a}_i + \overline{\sigma}_{ij}m_i)t,$$

$$d_{E[f_i]}(y, t) = E[\max\{\sigma_{i1}(\omega)d_{i1}y_1, \ldots, \sigma_{in}(\omega)d_{in}y_n, \sigma_i(\omega)d_i t\}].$$

This model is convex because all β_i in the case of necessity measure are positive.

We now move to the construction of equivalent stochastic analogues for (16), (18). Let $\tau_i =$ "π" and functions $f_i(x, \omega, \gamma)$ have the form of (19) for $i = \overline{0, m}$, where random values $a_{ij}(\omega)$, $\sigma_{ij}(\omega)$, $a_i(\omega)$, $\sigma_i(\omega)$ are independent and normally distributed.

Theorem 8 *Let in the model* (16), (18) *fuzzy random variables* $b_{ij}(\omega, \gamma)$, $b_i(\omega, \gamma)$ *have shift-scale representation, where fuzzy parameters* $Z_{ij}(\gamma)$ *and* $Z_i(\gamma)$ *are* T_W-*related and have symmetrical distribution of LR-type; the shift and scale parameters are random gaussian variables. Then the problem* (16), (18) *has the following equivalent stochastic analogue*

$$M_0^-(x) - \beta_0\sqrt{D_0^-(x)} \to \min_{x}, \tag{28}$$

$$\begin{cases} M_i^-(x) - \beta_i\sqrt{D_i^-(x)} \le 0, i = \overline{1, m}, \\ x \in X, \end{cases} \tag{29}$$

where for $i = \overline{0, m}$

$$M_i^-(x) = m_{E[f_i]}(x) - L^{-1}(\alpha_i)d_{E[f_i]}(x),$$

$$D_i^-(x) = \min\{E[m_{D_i}(x, \omega) - L^{-1}(\alpha_i)d_{D_i}(x, \omega)]^2,$$
$$E[m_{D_i}(x, \omega) + L^{-1}(\alpha_i)d_{D_i}(x, \omega)]^2\},$$

β_i *is solution of the equation* $F(t) = 1 - p_i$, *where* $F(t)$ *is the standard normal distribution function.*

Optimization problems that we discussed in this section can be solved by deterministic approaches of mathematical programming. However, it requires the calculation of precise values of the expectations and variances of random functions.

In case of the weakest triangular norm expectation and variance of a fuzzy random function are of such a complex nature that it is almost impossible to represent them in a closed analytical form. Moreover, this representation will be nonlinear non-convex, and that leads us to non-convex optimization area.

An alternative approach would be to use the stochastic quasi-gradient (SQG) methods [6, 20]. The basic principle behind these approaches is to substitute statistical estimates for the exact values of functions and their derivatives. SQG generalize the well-known stochastic approximation methods for unconstrained optimization of the expected value of random functions.

We described the stochastic quasi-gradient method for the problems (24), (25) and (26), (27) in [4, 33].

4.5 Numerical Example

Let us consider the following possibilistic-probabilistic optimization problem

$$k \rightarrow \min_{x}, \tag{30}$$

$$\begin{cases} v\{E[b_{01}(\omega, \gamma)x_1 + b_{02}(\omega, \gamma)x_2] \leq k\} \geq 0.6 \\ v\{E[b_{11}(\omega, \gamma)x_1 + b_{12}(\omega, \gamma)x_2 - b_1(\omega, \gamma)] \leq 0\} \geq 0.8, \\ v\{E[b_{21}(\omega, \gamma)x_1 + b_{22}(\omega, \gamma)x_2 - b_2(\omega, \gamma)] \leq 0\} \geq 0.5, \\ x_1, x_2 \in [5, 15], \end{cases} \tag{31}$$

where $b_{ij}(\omega, \gamma)$ and $b_i(\omega, \gamma)$ have a shift-scale form, and $a_{ij}(\omega), a_i(\omega)$, $\sigma_{ij}(\omega), \sigma_i(\omega)$ are drawn from standard uniform distribution and $Z_{ij}(\gamma), Z_i(\gamma)$ are mutually T_w-related fuzzy intervals with shape functions $L(t) = R(t) = max\{0, 1 - t\}, t \geq 0$

$Z_{01} = (2, 2, 3, 1)_{LR}$, $Z_{11} = (1, 1, 2, 3)_{LR}$, $Z_{21} = (1, 1, 2, 3)_{LR}$, $Z_1 = (5, 7, 2, 3)_{LR}$, $Z_{02} = (0, 3, 5, 2)_{LR}$, $Z_{12} = (2, 3, 2, 4)_{LR}$, $Z_{22} = (3, 4, 1, 2)_{LR}$, $Z_2 = (7, 8, 1, 3)_{LR}$.

We can transform the problem (30), (31) with the help of approaches discussed earlier to

$$2x_1 + 3x_2 + 0.6 \, E[max\{\sigma_{01}(\omega)x_1, 2\sigma_{02}(\omega)x_2\}] \rightarrow \min_{x}, \tag{32}$$

$$\begin{cases} x_1 + 3x_2 + 0.8 \, E[max\{3\sigma_{11}(\omega)x_1, 4\sigma_{12}(\omega)x_2\}] \leq 3.4, \\ 4x_1 + 4x_2 + 0.5 \, E[max\{2\sigma_{21}(\omega)x_1, 2\sigma_{22}(\omega)x_2\}] \leq 6.5, \\ x_1, x_2 \in [5, 15]. \end{cases} \tag{33}$$

The criterion (32) is a continuous strictly convex function and the constraints (33) define a convex area [33]. Thus, we can use the stochastic quasi-gradient method in order to solve that optimization problem.

Let the initial approximation be $x_0 = (11.7, 10.3)$, $u_0 = (1.0, 1.0)$ and the step multiplier be $\rho_s = \frac{1}{2s-1}$. We obtain the following solution $x^* = (5.0051, 5.002)$ with optimal value $f_0(x^*) = 28.28$. We present values during some iterations in Table 1.

Table 1 Some intermediate results of SQG algorithm

Iteration	200	400	600	800	1000
x^s	(7.68, 6.65)	(6.42, 5.83)	(5.97, 5.41)	(5.41, 5.16)	(5.01, 5.00)
$f_0(x^s)$	39.74	33.30	28.99	29.47	28.17

5 Portfolio of Minimum Risk

5.1 Models and Methods

One of the most important applications of possibilistic-probabilistic optimization is portfolio analysis. Modern portfolio theory was first introduced in Markowitz's publication [17]. The expected value and the risk were the two criteria Markowitz used to determine portfolio selection in his probabilistic market model.

In some situations financial assets can be characterized by two uncertainty factors—stochastic and fuzzy. Financial information is usually represented with the help of three parameters: minimum, maximum, and average weighted asset prices. Therefore, fuzzy random variable is a natural approach for modelling such combined ambiguities.

We now construct model of minimum risk portfolio under hybrid uncertainty of possibilistic-probabilistic type. Its return can be represented as a linear fuzzy random function of portfolio's capital shares x_i

$$R_P(x, \omega, \gamma) = \sum_{i=1}^{n} R_i(\omega, \gamma)x_i, \qquad (34)$$

where $R_i(\omega, \gamma)$ model the profitability of individual financial assets and are fuzzy random variables with a shift-scale form [11]:

$$R_i(\omega, \gamma) = a_i(\omega) + \sigma_i(\omega)Z_i(\gamma).$$

Under these assumptions the possibility distribution for portfolio return (34) has the form

$$R_P(x, \omega, \gamma) = (\underline{m}_{R_P}(x, \omega), \overline{m}_{R_P}(x, \omega), \underline{d}_{R_P}(x, \omega), \overline{d}_{R_P}(x, \omega))_{LR}$$

where

$$\underline{m}_{R_P}(x, \omega) = \sum_{i=1}^{n}(a_i(\omega) + \sigma_i(\omega)\underline{m}_i)x_i, \quad \underline{d}_{R_P}(x, \omega) = \bigoplus_{i=1}^{n}\{\sigma_i(\omega)\underline{d}_i x_i\},$$

$$\overline{m}_{R_P}(x, \omega) = \sum_{i=1}^{n}(a_i(\omega) + \sigma_i(\omega)\overline{m}_i)x_i, \quad \overline{d}_{R_P}(x, \omega) = \bigoplus_{i=1}^{n}\{\sigma_i(\omega)\overline{d}_i x_i\}.$$

We can create a risk function for the portfolio and move its expected return into the system of restrictions in accordance with the traditional Markowitz approach [17] and with the assistance of [7]. Since the expected return in the case of possibilistic-probabilistic model is fuzzy we therefore should somehow eliminate the uncertainty of possibilistic type. It can be achieved by putting restrictions on the possibility/necessity of meeting an investor's constraints on an acceptable level of expected portfolio return. We can therefore represent the model of feasible portfolios by Markowitz in the following form

$$D[R_P(x, \omega, \gamma)] \to \min_x, \tag{35}$$

$$\begin{cases} \tau\{E[R_P(x, \omega, \gamma)] \,\Re\, m_d\} \ge \alpha, \\ \sum_{i=1}^{n} x_i = 1, \\ x \in X. \end{cases} \tag{36}$$

Here $D[R_P(x, \omega, \gamma)]$ is variance and $E[R_P(x, \omega, \gamma)]$ is expected return of a portfolio; τ is possibility measure π or necessity measure ν; $\Re \in \{\ge, =\}$; $\alpha \in (0, 1]$ and m_d is an acceptable by investor level of the expected return.

The model of minimum risk portfolio (35), (36) under hybrid uncertainty has been thoroughly studied and various approaches to solve it were proposed. For the strongest triangular norm methods for construction of its deterministic equivalent were presented in [34]. For the weakest triangular norm methods for construction of its stochastic equivalent were discussed in [5].

5.2 Numerical Example

Here we will show numerical example for $n = 2$. Let shift and scale coefficients $a_i(\omega)$ and $\sigma_i(\omega)$ be uniformly distributed on $[0, 1]$, all possibilistic variables have symmetrical triangular forms with $R(t) = L(t) = \max\{0, 1 - t\}, t \ge 0$.

Let $Z_1 = (0.3, 0.3, 3.5, 3.5)_{LR}$ and $Z_2 = (2.8, 2.8, 1.5, 1.5)_{LR}, \alpha = 0.65$, the relation $\Re$ is "greater than or equal" $(\ge)$ and fuzzy variables Z_1 and Z_2 are mutually T_W-related. Then we can construct equivalent stochastic analogue of (35), (36) in possibility context

$$\frac{1}{12}\left(1.09x_1^2 + 8.84x_2^2\right) + d_{D[R_P]}(x) \to \min_x, \tag{37}$$

$$\begin{cases} -0.65x_1 - 1.9x_2 - 0.35d_{E[R_P]}(x) \ge -m_d, \\ x_1 + x_2 = 1, x \in X, \end{cases} \tag{38}$$

and in necessity context

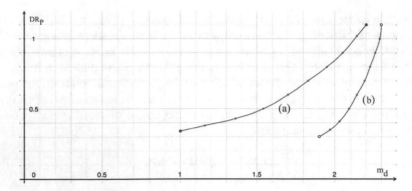

Fig. 1 Quasi-efficient portfolio frontiers in necessity context (**a**) and possibility context (**b**)

$$\frac{1}{12}\left(1.09x_1^2 + 8.84x_2^2\right) + d_{D[R_P]}(x) \to \min_x, \tag{39}$$

$$\begin{cases} -0.65x_1 - 1.9x_2 + 0.65d_{E[R_P]}(x) \geq -m_d, \\ x_1 + x_2 = 1, x \in X, \end{cases} \tag{40}$$

where

$$d_{E[R_P]}(x) = E[d_{R_P}(x, \omega)], \quad d_{D[R_P]}(x) = D[d_{R_P}(x, \omega)]$$

and

$$d_{R_P}(x, \omega) = \max\{3.5\sigma_1(\omega)x_1, 1.5\sigma_2(\omega)x_2\}.$$

Figure 1 shows quasi-efficient frontiers for the problems (37), (38) and (39), (40). Curves (a) and (b) represent portfolio frontier in necessity context (39), (40) and in possibility context (37), (38) respectively.

6 Conclusion

In this paper we made an outline of the current state of the theory of possibilistic and possibilistic-probabilistic optimization. We showed how ideas of professor Lotfi Zadeh have influenced this field.

We can state that methods of possibilistic optimization are well developed. As for the possibilistic-probabilistic optimization, there are some open problems waiting to be studied.

References

1. Antropov, A.: On solution methods for one task of possibility-necessity optimization. Nechetkie Sistemy i Myagkie Vychisleniya **5**(2), 5–24 (2010). (in Russian)
2. Bellman, R.E., Zadeh, L.A.: Decision making in fuzzy environment. Manag. Sci. **17**, 141–162 (1970)
3. Dubois, D., Prade, H.: Possibility Theory. Plenum Press, New York (1988)
4. Egorova, Y.E., Yazenin, A.V.: The problem of possibilistic-probabilistic optimization. J. Comput. Syst. Sci. Int. **56**(4), 642–667 (2017). (in Russian)
5. Egorova, Y.E., Yazenin, A.V.: A method for minimum risk portfolio optimization under hybrid uncertainty. J. Phys. Conf. Ser. **973**, 012033 (2018). (in Russian)
6. Ermolev, Y.M.: Methods of Stochastic Programming. Nauka, Moscow (1976). (in Russian)
7. Feng, Y., Hu, L., Shu, H.: The variance and covariance of fuzzy random variables and their applications. Fuzzy Sets Syst. **120**, 478–497 (2001)
8. Fuller, R., Keresztfalvi, T.: On generalization of Nguyen's theorem. Fuzzy Sets Syst. **41**(3), 371–374 (1991)
9. Goldstein, E.G., Judin, D.V.: New Trends in Linear Programming. Sovetskoe Radio, Moscow (1996). (in Russian)
10. Hong, D.H.: Parameter estimations of mutually t-related fuzzy variables. Fuzzy Sets Syst. **123**, 63–71 (2001)
11. Khokhlov, M.Y., Yazenin, A.V.: The calculation of numerical characteristics of fuzzy random data. Vestnik TvGU. Seriya: Prikladnaya Matematika **2**, 39–43 (2003). (in Russian)
12. Kolmogorov, A.N.: Foundations of the Theory of Probability. Chelsea Publishing Company, New York (1956)
13. Kwakernaak, H.: Fuzzy random variables i. Inf. Sci. **15**, 1–29 (1978)
14. Kwakernaak, H.: Fuzzy random variables ii. Inf. Sci. **17**, 253–278 (1979)
15. Luhandjula, M.K.: On possibilistic linear programming. Fuzzy Sets Syst. **18**, 15–30 (1986)
16. Luhandjula, M.K.: Optimization under hybrid uncertainty. Fuzzy Sets Syst. **146**, 187–203 (2004)
17. Markowitz, H.M.: Portfolio selection. J. Financ. **7**, 77–91 (1952)
18. Mesiar, R.: A note to the t-sum of l-r fuzzy numbers. Fuzzy Sets Syst. **2**, 259–261 (1996)
19. Mesiar, R.: Triangular norm-based additions of fuzzy intervals. Fuzzy Sets Syst. **91**, 231–237 (1997)
20. Minoux, M.: Mathematical Programming: Theory and Algorithms. Wiley, Hoboken (1986)
21. Murtagh, B.A.: Advanced linear programming: Computation and Practice. McGraw-Hill, New York (1981)
22. Nahmias, S.: Fuzzy variables. Fuzzy Sets Syst. **1**, 97–110 (1978)
23. Nahmias, S.: Fuzzy variables in a random environment. In: Gupta, M.M., Ragade, R.K., Yager, R.R. (eds.) Advances in Fuzzy Sets Theory and Applications, pp. 165–180. NHCP, Amsterdam (1979)
24. Nguen, T., Walker, E.A.: A First Course in Fuzzy Logic. CRC Press, Boca Raton (1997)
25. Puri, M.L., Ralescu, D.A.: Fuzzy random variables. J. Math. Anal. Appl. **114**, 409–422 (1986)
26. Soldatenko, I., Yazenin, A.: Possibilistic optimization problems with mutually t-related parameters: Comparative study. J. Comput. Syst. Sci. Int. **47**(5), 752–763 (2008)
27. Yazenin, A.V.: Fuzzy and stochastic programming. Fuzzy Sets Syst. **22**, 171–180 (1987)
28. Yazenin, A.V.: Linear programming with random fuzzy data. Soviet J. Comput. Syst. Sci. **30**, 86–93 (1992)
29. Yazenin, A.V.: On the problem of possibilistic optimization. Fuzzy Sets Syst. **81**, 133–140 (1996)
30. Yazenin, A.V.: On a method of solving a problem of linear programming with random fuzzy data. J. Comput. Syst. Sci. Int. **36**(5), 737–741 (1997)
31. Yazenin, A.V.: On the problem of maximization of the possibility to attain a fuzzy goal. J. Comput. Syst. Sci. Int. **38**(4), 621–624 (1999)

32. Yazenin, A.V.: Basic Concepts of Possibility Theory: A Mathematical Apparatus for Decision Making Under Hybrid Uncertainty Condition. Fizmatlit Publ., Moscow (2016). (in Russian)
33. Yazenin, A.V., Egorova, Y.E.: On solution methods for tasks of possibilistic-probabilistic optimization. Vestnik TvGU. Seriya: Prikladnaya Matematika **4**, 85–103 (2013). (in Russian)
34. Yazenin, A.V., Soldatenko, I.S.: A portfolio of minimum risk in a hybrid uncertainty of a possibilistic-probabilistic type: comparative study. In: Kacprzyk, J., Szmidt, E., Zadrozny, S., Atanassov, K.T., Krawczak, M. (eds.) IWIFSGN/EUSFLAT-2017. AISC, vol. 643, pp. 551–563. Springer, Cham (2018)
35. Yazenin, A.V., Soldatenko, I.S.: On the problem of possibilistic-probabilistic optimization with constraints on possibility/probability. In: Fullér, R., et al. (eds.) WILF 2018, LNAI 11291, pp. 43–54. Springer, Switzerland (2019)
36. Yazenin, A.V., Wagenknecht, M.: Possibilistic Optimization, vol. 6/96. Brandenburgische Technische Universitat, Cottbus: Aktuelle Reihe (1996)
37. Zadeh, L.A.: Fuzzy sets. Inf. Control **8**(3), 338–353 (1965)
38. Zadeh, L.A.: The concept of a linguistic variable and its application to approximate reasoning-i. Inf. Sci. **8**(3), 199–249 (1975)
39. Zadeh, L.A.: The concept of a linguistic variable and its application to approximate reasoning-ii. Inf. Sci. **8**(4), 301–357 (1975)
40. Zadeh, L.A.: The concept of a linguistic variable and its application to approximate reasoning-iii. Inf. Sci. **9**(1), 43–80 (1975)
41. Zadeh, L.A.: Fuzzy sets as a basis for a theory of possibility. Fuzzy Sets Syst. **1**, 3–28 (1978)
42. Zadeh, L.A.: Probability measures of Fuzzy events. J. Math. Anal Appl. **23**(2), 421–427 (1968)
43. Zimmermann, H.J.: Description and optimization of fuzzy systems. Int. J. of Gen. Systems **16**, 209–215 (1976)
44. Zimmermann, H.J.: Fuzzy programming and linear programming with several objective functions. Fuzzy Sets Syst. **1**, 45–55 (1978)

Ideas of Lotfi Zadeh in Explainable Artificial Intelligence

Alexey Averkin ⓘ

Abstract Various aspects of fuzzy systems applications for explainable artificial intelligence are described. It is shown how the pioneering works of L. A. Zadeh can be used for the current explainable artificial intelligence challenges.

Keywords Fuzzy logic · Explainable artificial intelligence · Computing with words · Interpretable fuzzy systems

1 Introduction

This year marks the 100th anniversary of the birth of the great scientist of our time, the founder of several major scientific trends in applied mathematics, automatic control theory, computer science and artificial intelligence, Professor Lotfi Zadeh. He belonged to the cohort of very few pioneering scientists who generate new, original scientific ideas and form the basic scientific paradigms that change our world. Professor L. Zadeh was the founder of the theory of fuzzy sets and linguistic variables, the "father" of fuzzy logic and approximate reasoning, the author of the theory of possibility and general theory of uncertainty, the creator of Z-numbers theory and generalized restrictions, the ancestor of granular and soft computing. His ideas and theories not only opened a new epoch in the development of scientific thought, free from the limitations of narrow scientific directions and contributing to their synergy. They made a significant contribution to the development of new information and cognitive technologies, led to the creation of effective industrial technologies, such as fuzzy computers and processors, fuzzy regulators, fuzzy clustering and recognition systems, and many others. Professor L. Zadeh has been deservedly included in

A. Averkin (✉)
Federal Research Centre of Informatics and Computer Science of RAS, Vavilova str., 42, Moscow, Russia
e-mail: averkin2003@inbox.ru

Educational and Scientific Laboratory of Artificial Intelligence, Neuro-technologies and Business Analytics, Plekhanov Russian University of Economics, Stremyanny lane, 36, Moscow, Russia

© The Author(s), under exclusive license to Springer Nature Switzerland AG 2023   45
Sh. N. Shahbazova et al. (eds.), *Recent Developments and the New Directions of Research, Foundations, and Applications*, Studies in Fuzziness and Soft Computing 422,
https://doi.org/10.1007/978-3-031-20153-0_3

the IEEE Computer Society's gallery of fame scientists who have made pioneering contributions to the field of artificial intelligence and intelligent systems.

Soviet scientists were among the first to support the new direction. Speaking at the ICSCCW-2001 conference in June 2001, L. Zadeh stressed that his first paper on fuzzy sets took place in 1965 at a conference on cybernetics held in the USSR aboard the liner «Admiral Nakhimov» [1].

2 L. A. Zadeh and Explainable AI

The role of L. Zadeh in AI is also hard to overestimate, and I would especially like to focus on the concept of soft computing, originally combining hybrid models based on fuzzy sets, neural networks, and soft computing. The emergent properties of these models were one of the foundations of the current hype in artificial intelligence and machine learning. The study of fuzzy logic culminated in the late twentieth century and has since begun to slow down a bit. This slowdown may be due in part to the temporary absence of fuzzy math results in machine learning. Current research will pave the way for fuzzy logic researchers to develop AI applications and solve complex problems that are also of interest to the machine learning community. Experience and expertise in fuzzy logic is well suited to model ambiguities in big data, model uncertainty in knowledge representation, and provide transfer learning with noninductive inference.

L. Zadeh has made fundamental and invaluable contributions to the field of fuzzy logic and artificial intelligence. These contributions were pioneering ideas with great potential, which are now being actively developed by other researchers. Explainable artificial intelligence now represents a key area of research in artificial intelligence and an unusually promising one in which many of Zadeh's contributions could become crucial. Research in the of area of explainable artificial intelligence can be divided into three stages: in the first stage (starting from 1970) expert systems were developed; in the second stage (mid-1980s), the transition was made from expert systems to knowledge-based systems; and in the third phase (since 2010), deep architectures of artificial neural networks, which required new global research on the construction of explainable systems, have been studied. On each stage the ideas of L. Zadeh played the important roles [2].

For example, there are two main problems for explainable artificial intelligence. The first challenge is to create dialogic agents capable of providing people with semantic reasoning, convincing and trustworthy interactive explanations for any user. The second problem requires measuring the effectiveness and naturalness of automatically generated explanations. Computing with words, fuzzy measures and Z-numbers introduced by Zadeh will contribute to the success of both problems and the results needed for the Fifth Industrial Revolution [3, 4].

Since 2011, L. Zadeh had been involved in Z Advanced Computing, Inc. (ZAC) and was one of the ZAC's inventors. ZAC is the pioneer Cognitive Explainable Artificial Intelligence (Cognitive XAI) technologies, e.g., for the detailed complex

3D Image/Object Recognition from any view angle. Z Advanced Computing, Inc. (ZAC), the pioneer Cognitive Explainable Artificial Intelligence (Cognitive XAI) software startup, has been recognized as the Top 5 Leading Global Companies in the Fourth and Fifth Industrial Revolutions. ZAC has had major AI and Machine Learning (ML) breakthrough demos in the recent projects for the US Air Force (USAF) and for Bosch/BSH (the largest appliance maker in Europe). ZAC has achieved detailed complex 3D Image Recognition using only a few training samples and using only an average laptop with low power CPU, for both training and recognition. This is simpler realization than the other algorithms in industry, such as deep convolutional neural networks [5].

3 Various Contributions of Fuzzy Systems to Explainable AI

Nowadays, XAI is a prominent and fruitful research field where many of Zadeh's contributions can become crucial if they are carefully considered and thoroughly developed. It is worth noting that about 30% of publications in Scopus related to XAI, dated back to 2017 or earlier, came from authors well recognized in the fuzzy logic field. This is mainly due to the commitment of the fuzzy community to produce interpretable fuzzy systems, since interpretability is deeply rooted in the fundamentals of fuzzy logic [4, 6].

This talk will examine fuzzy models to improve the effectiveness of XAI systems in explaining their decisions and actions to the user, through fuzzy models. and to establish a concrete and fundamental connection between two important fields in artificial intelligence i.e., symbolic systems and connectionist systems, more specifically, between deep learning and fuzzy logic. Several authors show how deep learning could benefit from the comparative research by re-examining many heuristics in the field of fuzzy logic.

Very effective is also, the use of fuzzy layers in deep learning networks. In [7] L. Fan established a fundamental connection between two important fields in artificial intelligence i.e. deep learning and fuzzy logic. He shows, how deep learning could benefit from the comparative research by re-examining many trail-and-error heuristics in the lens of fuzzy logic, and consequently, distilling the essential ingredients with rigorous foundations. The author proposed deep generalized hamming network as such not only lends itself to rigorous analysis and interpretation within the fuzzy logic theory but also demonstrates fast learning speed, well-controlled behavior and state-of-the-art performances on a variety of learning tasks.

Very interesting form is the extraction of rules using neuro-fuzzy models. Systems based on fuzzy rules, developed using fuzzy logic, have become a field of active research in the last few years. These algorithms have proven their strengths in tasks such as managing complex systems, creating fuzzy controls. The relationship between production rules and neural networks of both worlds has been thoroughly studied and shown to be equivalent. This means that we can translate the knowledge embedded in the neural network.

The ideas of L. A. Zadeh in XAI will help fuzzy logic researchers to develop AI applications and solve complex problems that are also of interest to the machine learning community. The expertise in fuzzy logic is well suited to modeling ambiguities in big data, modeling uncertainty in knowledge representation, and providing transfer learning with noninductive inference.

As part of this ideology, the Russian Association of Artificial Intelligence is currently actively developing fuzzy situational management of complex systems based on their composite hybrid modelling, which uses the capabilities of analytical, neural network and fuzzy approaches to construct composite hybrid models for the realization of the explainable artificial intelligence concepts.

Acknowledgements The paper is partially supported be the grants RFBR 20-07-00770 "Facing Fundamental Problems of Constructing "Understanding" Cognitive Agents, Multi-Agent Systems and Artificial Societies on the Basis of Synergetic Artificial Intelligence Approaches, Information Granulation Techniques, Dynamic Bipolar Scales and Dialogical Worlds" and RSCF 22-71-10112 "Hybrid Decision Support Models Based on Augmented Artificial Intelligence, Cognitive Modeling and Fuzzy Logic in Problems of Personalized Medicine".

References

1. Pospelov, D.A., Stefanuk, V.L., Averkin, A.N., et al.: Remembering Lotfi Zadeh. Open Semant. Technol. Design. Intell. Syst. **8**, 27–39 (2018)
2. Averkin, A.N., Yarushev, S.A.: Review of research in the field of developing methods to extract rules from artificial neural networks. J. Comput. Syst. Sci. Int. **60**(6), 966–980 (2021)
3. IT Software & Hardware Development Company Homepage. https://kotaielectronics.com/ind ustries-that-need-explainable-ai-principles-to-survive. Accessed 21 Feb 2022
4. Alonso, J.M.: From Zadeh's computing with words towards explainable artificial intelligence. In: WILF2018-12th International Workshop on Fuzzy Logic and Applications, pp. 244–248. Springer, Heidelberg (2019)
5. Z Advanced Computing Homepage. https://www.zadvancedcomputing.com. Accessed 09 Jan 2022
6. Bouchon-Meunier, B., Lesot, M.-J., Marsala, C.: Lotfi A. Zadeh, the visionary in Explainable Artificial Intelligence. TWMS J. Pure Appl. Math **12**(1), 5–13 (2021)
7. Fan, L.: Revisit fuzzy neural network: demystifying batch normalization and ReLU with generalized hamming network. In: Conference on Neural Information Processing Systems (NIPS2017), pp. 1923–1932 (2017)

Different Concepts, Similar Computational Complexity: Nguyen's Results About Fuzzy and Interval Computations 35 Years Later

Hung T. Nguyen and Vladik Kreinovich

Abstract When we know for sure which values are possible and which are not, we have crisp uncertainty—of which interval uncertainty is a usual case. In practice, we are often not 100% sure about our knowledge, i.e., we have fuzzy uncertainty—i.e., we have fuzzy knowledge, of which crisp is a particular case. Usually, general problems are more difficult to solve that most of their particular cases. It was therefore expected that processing fuzzy data is, in general, more computationally difficult than processing interval data—and indeed, Zadeh's extension principle—a natural formula for fuzzy computations—looks very complicated. Unexpectedly, Zadeh-motivated 1978 paper by Hung T. Nguyen showed that fuzzy computations can be reduced to a few interval ones—and in this sense, fuzzy and interval computations have, in effect, the same computational complexity. In this paper, we remind the readers about the motivations for (and proof of) this result, and show how and why in the last 35 years, this result was generalized in various directions.

1 Crisp, Interval, and Fuzzy Uncertainty: A Brief Reminder

Emergence of modern science. In the ancient times, there was no clear separation between speculations, prejudices, feeling, and confirmed scientific facts. For example, Johannes Kepler made great discoveries in astronomy—in particular, he discovered that planets follow elliptical orbits—but he also described horoscopes predicting people's fate. Actually, his salaried position required him to deal both with

H. T. Nguyen (✉)
Department of Mathematical Sciences, New Mexico State University Las Cruces, New Mexico 88003, USA
e-mail: hunguyen@nmsu.edu

Faculty of Economics, Chiang Mai University, Chiang Mai, Thailand

V. Kreinovich
Department of Computer Science, University of Texas at El Paso, El Paso, Texas 79968, USA
e-mail: vladik@utep.edu

astronomy and astrology—which at that time, were not clearly separated. Chemists were analyzing chemical reactions—and at the time, tried different magical incantations that would help them turn matter into gold—and were paid for doing both chemistry and alchemy.

The situation gradually changed when the need for reliable practical applications necessitates a clear separation between science—that studies well-established facts and relations—and semi-poetic imprecise speculations and feeling. The great Newton was interpreting the Bible, the grest Goethe came up with his theory of vision—but these activities were clearly outside what was then considered as science.

Traditional scientific approach to uncertainty. Of course, usually, no matter what object we study, we do not have a complete knowledge of this object. In science, knowledge about objects is usually described in terms of numbers. For example, in mechanics, each object is characterized by its mass, by the coordinates of its spatial location, by the components of its velocity, and—if this object is rotating—by a unit vector describing its rotation axis and by the angular velocity. The fact that we do not have a complete knowledge means that we do not know the exact values of each of these quantities x; at best, we know the bounds $\underline{x}$ and $\overline{x}$ on the actual value. In such situations, all we know about the actual (unknown) value x is that this value belongs to the interval $[\underline{x}, \overline{x}]$.

In some cases, we know that some values inside this interval are not possible; in this case, the set of all possible values of x has a more general form than an interval. But even in this case, for some real numbers, we are, at this stage, 100% sure that the value of the physical quantity x cannot be equal to this number, while for other numbers, we are 100% sure that the value of the quantity x can be equal to this number. We may have an additional gut feeling that some numbers from this interval are more possible than others, but such gut feeling was not taken into account in the traditional scientific paradigm.

Comment. Sometimes, in addition to the interval (or, more generally, set) of possible values of x, we also have some information about the frequency (probability) with different numbers from this interval appear in different situations. But again, what traditional science considered was only guaranteed knowledge about these probabilities.

Need to go beyond traditional scientific paradigm. In the early 1960s, Lotfi Zadeh, one of the world's leading specialists in automatic control, a co-author of the then most widely used book on automatic control, noticed that in many practical situations, automatic controllers—that take into account all scientific information about the object of control—perform much worse than human controllers. He realized that human expert controllers use additional knowledge, knowledge which is—in contrast to what traditional science considered—not precise. The rules describing this knowledge were usually formulated in terms of imprecise ("fuzzy") words from natural language, such as "small", "approximately", etc., words that the traditional scientific approach ignored.

So, instead of ignoring these words, Zadeh proposed to incorporate the corresponding scientific knowledge into the automatic controllers. For this purpose, he came up with a methodology that he called *fuzzy*; see, e.g., [1, 12, 17, 21, 22, 26].

Fuzzy methodology: a brief description. For precise ("crisp") properties like "smaller than 10", every number either satisfies this property or does not. Such properties can be described by describing the set of all the values that satisfy this property, or, equivalently, by a function $\mu(x)$ that assigns, to each possible value x, the value 1 if x has this property and 0 if not. In mathematics, such a function is known as a *characteristic function*.

In contrast, for fuzzy words like "small", for some values x, experts are not 100% sure whether x is small or not, they are sure to some degree. We want to process such information in a computer. Computers were not designed to process words from natural languages, they were designed to process numbers. So, to be able to use computers to process expert information, we need to be able to describe this degree of confidence by a number. A natural way to do it is to ask the expert to mark his/her degree on a scale from 0 to 1, where 1 means absolute confidence, and 0 means no confidence at all. Alternatively, we can use a scale from 0 to 10 or from 0 to 5, and then divide the result, correspondingly, by 10 or by 5. This is how students evaluate their instructors, this is how we evaluate the quality of different services. Thus, to describe a property, we need to describe a function that assigns, to each value x, a degree $\mu(x) \in [0, 1]$ to which each value x satisfies the given property (e.g., is small). The corresponding function is known as a *membership function* or, alternatively, as a *fuzzy set*.

An additional complication comes from the fact that many rules describe what happens when several properties are satisfied. For example, a rule may describe what to do if the temperature t in a chemical reactor is slightly below the desired one *and* the pressure p is slightly higher than desired. We can, in principle, ask the expert to mark the degrees $\mu_{\text{temp}}(t)$ corresponding to different values t and the degrees $\mu_{\text{press}}(p)$ corresponding to different values p, but what we real need is, for all possible pairs (t, p), to estimate the degree of the above "and"-statement. It may be still possible to ask the expert about all such pairs, but what if there are 5 inputs? Ten inputs? Even if we consider only 10 different values for each quantity, this would still make 10^5 or even 10^{10} combinations—and we cannot ask that many questions to an expert.

Since we cannot always directly elicit the expert's degree of certainty in an "and"-statement—of the type $A \& B$—we need to be able to estimate this degree based on whatever information we have—i.e., based on the degrees of certainty a and b of the statements A and B. The algorithm for this estimation is known as an *"and"-operation* (or, for historical reasons, a *t-norm*); we will denote it by $f_{\&}(a, b)$. The most widely used "and"-operations are $\min(a, b)$ and $a \cdot b$.

Similarly, we need an algorithm to estimate the degree of certainty of $A \vee B$; such an algorithm is known as an *"or"-operation*, or a *t-conorm*; we will denote it by $f_{\vee}(a, b)$. The most widely used "or"-operations are $\max(a, b)$ and $a + b - a \cdot b$. To describe the degree of certainty of a negation $\neg A$, we need an algorithm which is known as a *negation operation* $f_\neg(a)$. The mostly widely used negation operation is $f_\neg(a) = 1 - a$. Because of the important role of these logical operations, fuzzy methodology is often called *fuzzy logic methodology*.

2 Data Processing Under Interval and Fuzzy Uncertainty: Reminder

Need for data processing. In many real-life situations, we are interesting in quantities which cannot be measured directly—e.g., in future values of some quantities. Since we cannot measure these quantities y directly, we need to estimate y based on available information—i.e., based on the known values $\tilde{x}_1, \ldots, \tilde{x}_n$ of related quantities $x_1, \ldots, x_n$. We will denote the estimating algorithm by $y = f(x_1, \ldots, x_n)$.

Need to take uncertainty into account. The values $\tilde{x}_i$ come either from measurements or from expert estimates. In both cases, the available value $\tilde{x}_i$ is somewhat different from the actual (unknown) value x_i of the corresponding quantity. As a consequence, even if the relation $y = f(x_1, \ldots, x_n)$ is exact, the resulting estimate $\tilde{y} = f(\tilde{x}_1, \ldots, \tilde{x}_n)$ for y is, in general, different from the actual value y of the quantity of interest. To make decisions, we need to know how accurate is this estimate.

Case of interval uncertainty. In the interval case, for each quantity x_i, the only thing we know is the interval $\mathbf{x}_i = [\underline{x}_i, \overline{x}_i]$ that contains x_i. So, the set $\mathbf{y}$ of all possible values of y is the set of all possible values $f(x_1, \ldots, x_n)$ when each x_i is in the corresponding interval:

$$\mathbf{y} = \{f(x_1, \ldots, x_n) : x_i \in \mathbf{x}_i \text{ for all } i\}.$$

In mathematical terms, the right-hand side of this equality is known as the *range* of the function $f(x_1, \ldots, x_n)$; it is usually denoted by $f(\mathbf{x}_1, \ldots, \mathbf{x}_n)$.

For many important classes of problems, there are feasible algorithms which either compute this range or at least compute a reasonable approximation for this range; see, e.g., [11, 16, 19]. The problem of computing this range is known as the problem of *interval computation*.

Comment. While there exist efficient interval computation algorithms for many classes of problems, it should be mentioned that, in general, the problem of computing this range is NP-hard; see, e.g., [15]. This means, in effect, that unless P = NP (which most computer scientists believe to be not true), no feasible algorithm is possible that would *always* compute the exact range.

Case of fuzzy uncertainty. What if for each i and for each x_i, we only know the degree $\mu_i(x_i)$ to which this value x_i is possible? In this case, the value y is possible if and only if for some tuple $(x_1, \ldots, x_n)$ for which $y = f(x_1, \ldots, x_n)$, the value x_1 is possible, *and* the value x_2 is possible, *and* ...

We know the degree $\mu_i(x_i)$ to which x_i is possible. So, to get the degree $\mu(y)$ to which y is possible, we need to apply an "and"-operation for "and" and an "or"-operation for "for some" (which is nothing else but "or"). Thus, we get

$$\mu(y) = f_\vee\{f_\&(\mu_1(x_1), \ldots, \mu_n(x_n)) : y = f(x_1, \ldots, x_n)\}.$$

Here, the "or"-operation is applied to infinitely many values. For most "or"-operations, e.g., for $f_\vee(a, b) = a + b - a \cdot b$, if we apply this operation to infinitely many positive terms, we get 1. The only exception is when we use $f_\vee(a, b) = \max(a, b)$. In this case, the above expression takes the following form:

$$\mu(y) = \sup_{(x_1,\ldots,x_n):y=f(x_1,\ldots,x_n)} f_\&(\mu_1(x_1), \ldots, \mu_n(x_n)). \tag{1}$$

In particular, for the most commonly used "and"-operation $f_\&(a, b) = \min(a, b)$, we get:

$$\mu(y) = \sup_{(x_1,\ldots,x_n):y=f(x_1,\ldots,x_n)} \min(\mu_1(x_1), \ldots, \mu_n(x_n)). \tag{2}$$

The formula (2) was first proposed by Zadeh and is therefore called *Zadeh's extension principle.*

3 Nguyen's Theorem: Brief History, Formulation, and the Main Idea Behind the Proof

What was expected. The formulas (1) and (2) looks much more complex than the corresponding interval formulas—which are, of course, a particular case of the fuzzy formulas when all fuzzy sets are crisp, i.e., when for each i and each x_i, we have $\mu_i(x_i) = 1$ or $\mu_i(x_i) = 0$. This is a known phenomenon—that general computational problems are usually more complex than their particular cases. For example, solving linear and quadratic equations is straightforward, we have explicit formulas for these solutions, but solving general polynomial equations is complicated. Solving systems of linear equations is feasible, but already solving systems of quadratic equations is NP-hard.

It was therefore expected that fuzzy computation—i.e., fuzzy data processing—is much more complex than interval computation. Lotfi Zadeh himself understood the complexity of this problem, and realized that to make fuzzy methodology practically useful, it is important to develop efficient algorithms for at least some cases of fuzzy computing.

Nguyen's theorem: unexpected result. In 1975, Professor Zadeh invited Hung T. Nguyen, a promising recent PhD in Mathematics and Statistics from University of Paris, to spend two years at the University of California-Berkeley—to have a mathematician's look at fuzzy theory. For this purpose, he asked Hung T. Nguyen to read his papers and related papers of others—including a paper by a Japanese researcher visiting Berkeley on the computational aspects of fuzzy computing.

Hung T. Nguyen started working on this topic and came up with a general result about fuzzy computation, published in [20]. To explain this result, we need to recall the notion of an α-cut. The notion was known in fuzzy methodology because in many situations, we need to make a decision—e.g., whether to perform a certain action or not. When the satellite deviates a little bit from the desired trajectory, we need to decide whether we should use the precious fuel to correct its trajectory or not yet. If a chemical process starts deviating a little bit from the desired parameter, we need to decide whether to apply an appropriate control—e.g., shut down the reactor.

If we know for sure that a sufficiently large deviation took place, then yes, we should perform the corresponding action. But what if we can only conclude that this deviation occurred with some degree of confidence d? In this case, we need to select some threshold value $\alpha \in (0, 1]$, and perform the action if our degree d is larger than or equal to α: $d \geq \alpha$. Correspondingly, when this degree depends on the value of some quantity x ($d = \mu(x)$), then we perform the action if and only if $\mu(x) \geq \alpha$.

For each fuzzy set $\mu(x)$, the corresponding set $\{x : \mu(x) \geq \alpha\}$ is known as the α-cut of this fuzzy set. What Nguyen proved was that under reasonable conditions, the α-cut $\mathbf{y}(\alpha)$ of y is equal to the range of the function $f(x_1, \ldots, x_n)$ when each x_i is in the α-cut $\mathbf{x}_i(\alpha)$ of x_i:

$$\mathbf{y}(\alpha) = f(\mathbf{x}_1(\alpha), \ldots, \mathbf{x}_n(\alpha)). \tag{3}$$

The value α corresponds to an expert's degree of confidence in a statement. An expert cannot estimate his/her degree with accuracy higher than 0.1. Thus, it is sufficient to consider only values α differing by 0.1: 0, 0.1, 0.2, ..., 1.0. So, fuzzy computation can reduced to a few cases of interval computation.

This theorem is the main tool behind fuzzy computing. This theorem shows that there is no need to come up with new algorithms for fuzzy computing—it is sufficient to use well-developed interval algorithms, and this is exactly what most practitioners are doing.

Every year, there are sessions on interval computations at fuzzy conferences—and sessions on possible fuzzy applications at interval conferences.

Such situations happen. The fact that a more general case turned out to be no more computationally complex than a particular case was unexpected, but such situations happened before. For example, to come up with equations of General Relativity that describe gravity—i.e., forces caused by masses (= energy)—Einstein came up with a completely new idea of curved space-time; see, e.g., [5, 18, 25]. The general feeling was that without this new physical idea, we cannot come up with a reasonable explanation for these complex nonlinear partial differential equations. However, later, it turned out that the same equations appear if we consider a simple tensor field in flat (not-curved) space time, by making a natural-for-gravity assumption that the source of this field includes both the energy-momentum of other fields and the energy-momentum of the gravity field itself; see, e.g., [8–10, 13, 18].

In physics. there have been many examples of this type—when a seemingly completely revolutionary theory turned out to be derivable from the previous physics. For example, even the notion of the black hole—which was originally perceived as specific for general relativity—follows already from Newtonian mechanics. Indeed, in Newtonian mechanics, for each celestial body, there is an escape velocity—so that any object travelling slower than that will fall back to the body. If this escape velocity exceeds the speed of light—the largest possible speed—then nothing can leave this body, including light. Even the thresholds for when the body with a given mass and radius becomes a black hole are very similar in General Relativity and in Newton's mechanics; see, e.g., [5, 18, 25].

From this viewpoint, it is not very surprising that the general fuzzy computing was reduced to a simpler interval computing case. Not surprising but still not trivial, since each such reduction requires mathematical and physical ingenuity. It took almost 40 years to show that General Relativity can be derived from field theory, it took several centuries after Newton to conclude that black holes can exist in Newtonain physics, and it took more than 10 years to realize that fuzzy computation can be reduced to interval computations.

So how is this theorem proved: main idea. According to the formula (2), the value $\mu(x)$ is the maximum of several values. When is the maximum of several numbers larger than or equal to α? When one of these numbers if larger than or equal to α:

$$\mu(y) \geq \alpha \Leftrightarrow$$

$$\exists x_1, \ldots, x_n \, (y = f(x_1, \ldots, x_n) = y \,\&\, \min(\mu_1(x_1), \ldots, \mu_n(x_n)) \geq \alpha). \quad (4)$$

Thus,

$$y \in \mathbf{y}(\alpha) \Leftrightarrow \exists x_1, \ldots, x_n \, (y = f(x_1, \ldots, x_n) = y \,\&\, \min(\mu_1(x_1), \ldots, \mu_n(x_n)) \geq \alpha). \quad (5)$$

When is the smallest of n numbers larger than or equal to α? When all of them are larger than or equal to α:

$$\min(\mu_1(x_1), \ldots, \mu_n(x_n)) \geq \alpha \Leftrightarrow \mu_1(x_1) \geq \alpha \,\&\, \ldots \,\&\, \mu_n(x_n) \geq \alpha. \quad (6)$$

By definition of the α-cut, this means that

$$\min(\mu_1(x_1), \ldots, \mu_n(x_n)) \geq \alpha \Leftrightarrow x_1 \in \mathbf{x}_1(\alpha) \,\&\, \ldots \,\&\, x_n \in \mathbf{x}_n(\alpha). \quad (7)$$

Thus, the value y belongs to the α-cut $\mathbf{y}(\alpha)$ if and only if there exist values $x_1, \ldots, x_n$ for which $y = f(x_1, \ldots, x_n)$ and each x_i belongs to the corresponding α-cut $\mathbf{x}_i(\alpha)$. In other words, the set $\mathbf{y}(\alpha)$ is indeed equal to the range $f(\mathbf{x}_1(\alpha), \ldots, \mathbf{x}_n(\alpha))$. This is exactly what the theorem says.

Important warning. What we described is an idea, but not the full proof. It would be a full proof if we had the maximum of finitely many terms—but in our case, we have infinitely many terms, and it is known that in this case, the supremum may be larger than or equal to α without any of the maximized numbers being larger than or equal to α. For example, the supremum of the values $1 - 2^{-n}$ corresponding to $n = 0, 1, 2 \ldots$ is equal to $\alpha = 1$, while all the values $1 - 2^{-n}$ are smaller than 1.

Thus, to have a real proof, we need to guarantee that the supremum is attained for some tuple $(x_1, \ldots, x_n)$. This can be guaranteed, e.g., if all the membership functions $\mu_i(x_i)$ are continuous and all the α-cuts are compact—which for continuous functions of real numbers is equivalent to requiring that all the α-cuts are bounded, a requirement which is true for most practical membership functions.

4 Extensions of Nguyen's Theorem: A Brief Overview

Extensions beyond real numbers. The above proof does not depend on the fact that x_i are real numbers:

- they could be vectors, tuples;
- they could be, more generally, elements of a general metric (or even general topological) space.

Such extensions have indeed been published; see, e.g., [2–4].

Extensions to interval-valued, type-2, and more general fuzzy sets. In the above description, we implicitly assumed that an expert can always describe his/her degree of confidence by an exact number. In reality, however, just like people are not 100% confident about their estimates, they are also not 100% confident about their degrees of confidence. A natural idea is to allow an interval of possible degrees (this leads to interval-values fuzzy sets) or even to fuzzy sets describing each degree (this leads to general type-2 fuzzy sets). In both cases, we face a natural computational question: how to propagate such type-2 uncertainty through a data processing algorithm?

It turned out that such computations can also be reduced to interval computations; see, e.g., [14].

Extensions to other t-norms. In the above text, we considered the case when we use $f_\&(a, b) = \min(a, b)$. What if we use a different t-norm? Several extensions of Nguyen's theorem to different t-norms have been proposed; see, e.g., [7].

It turns out that for other t-norms, we can also have an efficient data processing algorithm [23]—although this time the reduction is not to interval algorithms but to algorithms from convex optimization; see, e.g., [24].

Other extensions. An interesting and promising extension was proposed in [6], where the authors represented a fuzzy number—a fuzzy generalization of an interval—as an interval $[\ell(\alpha), r(\alpha)]$ formed by two what they called *gradual numbers* $\ell(\alpha)$ and $r(\alpha)$: mappings from $(0, 1]$ to the real line. The left gradual number $\ell(\alpha)$ is formed by lower endpoints of the α-cut intervals, while the right gradual number $r(\alpha)$ is formed by its right end-points, so that $[\ell(\alpha), r(\alpha)] = \mathbf{x}(\alpha)$.

Each of these gradual numbers may not have a clear meaning, but this subdivision seems to simplify computations—just like in physics, while it is not possible to actually separate, e.g., a proton into three quarks, many computations are simplified if we represent a proton this way [5, 25].

We hope that other fruitful extensions will occur.

Acknowledgements This work was supported in part by the National Science Foundation grants 1623190 (A Model of Change for Preparing a New Generation for Professional Practice in Computer Science), and HRD-1834620 and HRD-2034030 (CAHSI Includes), and by the AT&T Fellowship in Information Technology.

It was also supported by the program of the development of the Scientific-Educational Mathematical Center of Volga Federal District No. 075-02-2020-1478, and by a grant from the Hungarian National Research, Development and Innovation Office (NRDI).

References

1. Belohlavek, R., Dauben, J.W., Klir, G.J.: Fuzzy Logic and Mathematics: A Historical Perspective. Oxford University Press, New York (2017)
2. Carlsson, C., Fuller, R.: On additions of interactive fuzzy numbers. Acta Polytechica Hugarica **2**, 59–73 (2005)
3. de Barros, L.C., Bassanezi, R.C., Tonelli, P.A.: On the continuity of the Zadeh's extension. In: Proceedings of the 7th World Congress of the International Fuzzy Systems Association IFSA'1997, Prague, June 25–29, 1997, pp. 22–26 (1997)
4. de Mota Ferreira, J.C., Bassanezi, R.C., Brandão, A.J.V.: On the Nguyen Theorem for Topological Spaces (2018). https://www.researchgate.net/publication/328538986
5. Feynman, R., Leighton, R., Sands, M.: The Feynman Lectures on Physics. Addison Wesley, Boston, Massachusetts (2005)
6. Fortin, J., Dubois, D., Fargier, H.: Gradual numbers and their application to fuzzy interval analysis. IEEE Trans. Fuzzy Syst. **16**(2), 388–402 (2008)
7. Fuller, R., Keresztfalvi, T.: On generalization of Nguyen's theorem. Fuzzy Sets Syst. **41**, 371–374 (1991)
8. Gupta, S.N.: Quantization of Einstein's gravitational field: general treatment. Proc. Phys. Soc. **A65**, 608–619 (1952)
9. Gupta, S.N.: Gravitation and electromagnetism. Phys. Rev. **96**, 1683–1685 (1954)
10. Gupta, S.N.: Einstein's and other theories of gravitation. Rev. Modern Phys. **29**, 334–336 (1957)
11. Jaulin, L., Kiefer, M., Didrit, O., Walter, E.: Applied Interval Analysis, with Examples in Parameter and State Estimation, Robust Control, and Robotics. Springer, London (2001)
12. Klir, G., Yuan, B.: Fuzzy Sets and Fuzzy Logic. Prentice Hall, Upper Saddle River, New Jersey (1995)
13. Kreinovich, V.: Gupta's derivation of Einstein equations. Soviet Phys. Doklady **20**(5), 341–342 (1975)
14. Kreinovich, V.: From processing interval-valued fuzzy data to general type-2: towards fast algorithms. In: Proceedings of the IEEE Symposium on Advances in Type-2 Fuzzy Logic Systems T2FUZZ'2011, Part of the IEEE Symposium Series on Computational Intelligence, Paris, France, April 11–15, 2011, pp. ix–xii (2011)
15. Kreinovich, V., Lakeyev, A., Rohn, J., Kahl, P.: Computational Complexity and Feasibility of Data Processing and Interval Computations. Kluwer, Dordrecht (1998)
16. Mayer, G.: Interval Analysis and Automatic Result Verification. de Gruyter, Berlin (2017)
17. Mendel, J.M.: Uncertain Rule-Based Fuzzy Systems: Introduction and New Directions. Springer, Cham, Switzerland (2017)

18. Misner, C.W., Thorne, K.S., Wheeler, J.A.: Gravitation. W. H. Freeman and Company, San Francisco, California (1973)
19. Moore, R.E., Kearfott, R.B., Cloud, M.J.: Introduction to Interval Analysis. SIAM, Philadelphia (2009)
20. Nguyen, H.T.: A note on the extension principle for fuzzy sets. J. Math. Anal. Appl. **64**, 369–380 (1978)
21. Nguyen, H.T., Walker, C.L., Walker, E.A.: A First Course in Fuzzy Logic. Chapman and Hall/CRC, Boca Raton, Florida (2019)
22. Novák, V., Perfilieva, I., Močkoř, J.: Mathematical Principles of Fuzzy Logic. Kluwer, Boston, Dordrecht (1999)
23. Pownuk, A., Kreinovich, V., Sriboonchitta, S.: Fuzzy data processing beyond min t-norm. In: Berger-Vachon, C., Gil Lafuente, A.M., Kacprzyk, J., Kondratenko, Y., Merigo Lindahl, J.M., Morabito, C. (eds.) Complex Systems: Solutions and Challenges in Economics, Management, and Engineering, pp. 237–250. Springer (2018)
24. Rockafeller, R.T.: Convex Analysis. Princeton University Press, Princeton, New Jersey (1997)
25. Thorne, K.S., Blandford, R.D.: Modern Classical Physics: Optics, Fluids, Plasmas, Elasticity, Relativity, and Statistical Physics. Princeton University Press, Princeton, New Jersey (2017)
26. Zadeh, L.A.: Fuzzy sets. Inf. Control **8**, 338–353 (1965)

Fuzzy Applications and Fuzzy Control

Solid Development Teams Search Approach

Alexey Zhelepov⦿, Vladislav Moiseev⦿, and Nadezhda Yarushkina⦿

Abstract Solid engineering teams are the most valuable units in IT. However, the search of these teams is a complex procedure. But it can be simplified and speeded up by the implementation of specific algorithms that will allow finding these teams by data analysis. The article describes such an approach based on open data of project repositories. Required benchmark results are also presented.

Keywords Solid development team · Project repository · Big data · Data analysis · Social graph · Social engineering

1 Introduction

According to the research of the IT market made by reputable organizations like Gartner, ICDC, the growth and potential of the IT-sphere exceeds the available human resources [1, 2]. Hence, many organizations have to compete for software developers in the personnel market [3]. 2020 made the situation even worse, the pandemic made almost all IT companies restructure their development processes [4].

But the pandemic brought positive opportunities as well. The perspective of global hiring became available for all companies. Although beforehand, most of the organizations were unable to do this due to their closed corporate culture or even habit. The remote employees were something unusual and most companies did not want to adjust their processes for this minor group [5].

A. Zhelepov (✉) · V. Moiseev · N. Yarushkina
Ulyanovsk State Technical University, 32, Severny Venetz Str., 432027 Ulyanovsk, Russia
e-mail: a.zhelepov@gmail.com

© The Author(s), under exclusive license to Springer Nature Switzerland AG 2023
Sh. N. Shahbazova et al. (eds.), *Recent Developments and the New Directions of Research, Foundations, and Applications*, Studies in Fuzziness and Soft Computing 422,
https://doi.org/10.1007/978-3-031-20153-0_5

On the other hand, the remote model is a good opportunity for an employee to grow as a professional working for top companies and not moving from his home region. Anyway, companies started following modern trends and improving themselves to be on top. Finally, the global remote work opportunity became mutually favorable for employers and employees [6].

2 Solid Development Teams Searching Problem

Hiring new software engineers is a complex task because they are required to be found first. Secondly, they are needed to be onboarded and merged into company processes and culture. Afterwards, separate developers still need some time to tune their work habits and turn into a solid team [7].

On the contrary, ready solid teams are more valuable human resources for IT companies. Such teams do not need to spend much time on the onboard procedures. They can be almost instantly pushed into the development of vital projects or MVPs [8].

However, these engineering units are quite complex to be searched. HR managers have to make a thorough analysis of the teams' previous developments to identify whether they are solid or not [9].

3 Data Source Based Solution

The main problem of solid teams search procedure is to find working relations between engineers and check their previous development artifacts.

Generally, while creating applications developers use version control systems such as Git to keep track of co-working changes and conflict resolution [10]. Therefore, the project repository becomes the main artifact of developers' cohesion within a team.

There are several global network systems that allow managing repositories and sharing developments such as GitHub, GitLab. These information systems help developers to work with each other not only on the main work but to solve common software engineering problems (e.g. via open source frameworks). These networks contain a vast number of differently purposed project repositories and information about its developers [11].

Fig. 1 A relationship model
of software engineers
working on the same projects

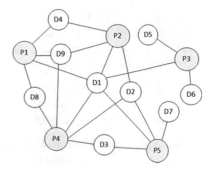

In terms of these networks, the relations between developers can be presented as a social graph as presented in Fig. 1.

The vertices describe developers (D) and projects (P). The edges show relations between developers and projects. Ordinarily, if the same developers relate to the common projects they can be stated as a team.

This article suggests an algorithm to find these relations and form potential solid teams. The efficiency of the algorithm was confirmed by various tests.

4 Look-Alike Searching Algorithm

The algorithm for searching solid development teams consists of two sub-algorithms: look-alike searching algorithm and forming teams algorithm. This section will cover the first sub-algorithm with all required descriptions. The second one will be presented in the next section. The algorithms were implemented and tested with GitHub data.

The look-alike principle is commonly used in targeted marketing [12] when the initial dataset of users is swelled by the users which are similar to the users from the initial set. The initial dataset is being prepared by a marketing manager who is an expert in the field of the subject area.

The same principle lies in a look-alike searching algorithm. However, the look-alike approach is different: the algorithm forms a graph of development relations starting from the initial user and finishing collecting data on the specific conditions.

The choice of the initial developer is up to experts who need to find an appropriate GitHub user profile. For example, the HR department needs to find the team of Java developers, they ought to seek an initial developer's profile who is an expert in Java, worked with many Java-based projects, etc. Summing up, the first input parameter to the algorithm is the GitHub profile of most suitable engineer (E) for the open position.

The main principle of the algorithm is a recursive search. It scans the initial developer's repositories, analyzes its participants, then begins scanning them and etc. Such an algorithm can work infinitely due to the vast volume of data in GitHub. Therefore, the next input parameters were included:

- Depth (D), which confines the recursive iterations of the algorithm;
- Time limit (TL), which simply stops the searching procedure after the moment when the time will have passed.

The goal of the algorithm is to collect social graph data in the form of a hash table. The example of this hash table is presented in Listing 1.

```
{
    "13805649": [ "jbaruch", "yoav", "quidryan", "fred-
dy33", "elig", "Jeka1978" ],
    "126888110": [ "victor-cr", "Jeka1978" ],
    ...
}
```

Listing 1 The example of the output dataset. The key is the ID of repository. The value is the set of repository developers

5 Forming Teams Algorithm

The goal of this sub-algorithm is to detect the potential teams among all extracted development relations in the previous step. The principle of the algorithm relies on data management operations.

Basically, the algorithm consists of two functions:

- *create_teams_table()*, the goal is to filter the data received as an output on the previous algorithm;
- *form_teams()*, the goal is to find all possible teams based on input data.

Listings 2 and 3 cover in detail both functions written in Python.

```
# Input data
# repos - a hash table, { repo_1: [c1, c2, .. cN], re-
po_2: [c3, c4, .. cM] ... }
# The data was received as an output of the previous sub-
algorithm.
def create_teams_table(repos):
    data = []
    # STEP 1. Goal: to unwrap the dictionary and prepare
data to store in table frame.
    for i, rg in enumerate(repos):
        for r in rg:
            # Not include repositories with the only con-
tributor.
            if (len(rg[r]) > 1):
                data.extend(list(map(lambda x: (r, x, 1),
rg[r])))
    # STEP 2. Goal: creation of dataframe that opens the
opportunity to apply the set operation.
    df = pd.DataFrame(data, columns=['r_id', 'c', 'val'])
    # STEP 3. Goal: grouping contributors records and
leave only those who participate in 2 or more reposito-
ries.
    grouped_c = df.groupby('c').aggregate(sum).query('val
> 1')
    # STEP 4. Goal: to get a temporary selection of those
contributors who take part in more than 2 repositories.
    pt = df.merge(grouped_c, left_on='c',
right_index=True)[['r_id', 'c']]
    # STEP 5. Goal: to get full table of relations be-
tween the contributors and repos.
    full = pd.merge(pt, pt, on='r_id', how='inner')
    # STEP 6. Goal: to drop all duplicates.
    full = full.drop_duplicates().query('c_x != c_y')
    # STEP 7. Goal: to find pairs of contributors who
worked together with more than 2 projects (CRC).
    flt = full
    flt['count'] = 1
    flt = full.set_index(['c_x', 'c_y']).sum(level=[0,1])
    flt = flt.query('count > N')
    # STEP 8. Goal: to form final selection with all pos-
sible pairs of developers who worked together.
    teams = pd.merge(full, flt, left_on = ['c_x', 'c_y'],
right_on= ['c_x', 'c_y'])
    teams = teams[['r_id', 'c_x', 'c_y', 'count_y']]
    teams = teams.rename(columns={'r_id': 'repo_id',
'c_x': 'c1', 'c_y': 'c2', 'count_y': 'count'})
    return teams
```

Listing 2 Create_teams_table() function description

```
# This function is purposed to find all possible teams.
# Input data: the result of function cre-
ate_teams_table().
# Output data (dictionary):
# Output example: {"33977002|30590969|24880409|56431832":
["fsouza","jvshahid","lilyball"], ... }
# Key is a bunch of common repositories.
# Value is a group of contributors of these repositories.
 def form_teams(tf):
     teams = {}
     frames = list(set(tf['count']))
     for f in frames:
         df = tf.query(f'count == {f}')
         i = 0
         while(i < len(df)):
             data = df[['repo_id', 'c1', 'c2',
'count']][i:f+i]
             key = tuple(data['repo_id'])
             if key not in teams:
                 teams.update({key: []})
             contributors = list(data[['c1',
'c2']].iloc[0])
             for c in contributors:
                 if c not in teams[key]:
                     teams[key].append(c)
             i += f
     return teams
```

Listing 3 Form_teams() function description

Eventually, the presented function constructs the data table with all available pairs of software developers who previously worked together with projects. The number of these projects is restricted by the global input parameter CRC (common repository count).

In sum, the presented function is based on the previous data table and returns all possible combinations of contributors and their co-projects. MTMC (minimum team members count) is another restrictive parameter that helps to exclude the teams whose members' quantity is less than the set value. It is recommended to set the value of 3 or more.

6 Tests

To prove the efficiency of the algorithm three groups of tests were provided. Each group of tests varies by the number of runs and TL values. For each group of tests, the same initial engineers were taken. To avoid the same run conditions for the look-alike algorithm, the order of the received repositories was shuffled in a random way

Table 1 Initial engineers per programming language

GitHub profile name	Specialization
Jeka1978	Java
Scitator	Python, C++
MortiMer	Kotlin

Table 2 Input parameters for the test groups for each of initial engineer

	Group 1	Group 2	Group 3
D	3	3	3
TL, min	10	60	180
CRC	3	3	3
MTMC	3	3	3
Number of runs per spec	20	20	20

for each iteration of the recursion. The overall number of provided tests is 300. The input parameters of the algorithm are presented in Tables 1 and 2.

The initial engineers were chosen according to the modern software development trends. All developers specialize in one of the languages: Java, Python, Kotlin. The engineers are leaders in the development society. The search results are presented in Table 3.

The average number of solid teams is an arithmetical mean of all results for the programming technology within the test group. The number of teams increases due to the longer time of the search procedure. The accuracy of the algorithm is another characteristic that shows the part of relevant teams by technology. The tables also contain the average value provided by tests. The accuracy grows as the time limit increases.

Figures 2 shows search statistics for randomly selected runs from test groups 1 and 2. The first histogram shows the distribution of the most used programming languages by the teams. The second chart shows the distribution of teams by the number of its members. The presented figures give a clear picture that most of the found teams are basically small by number of members.

Table 3 Test results (ANST—average number of solid teams, A—accuracy)

Technology	Group 1		Group 2		Group 3	
	ANST	A	ANST	A	ANST	A
Java	104	0.79	798	0.84	2149	0.93
Python	98	0.82	945	0.86	2451	0.95
Kotlin	64	0.81	598	0.88	1891	0.90

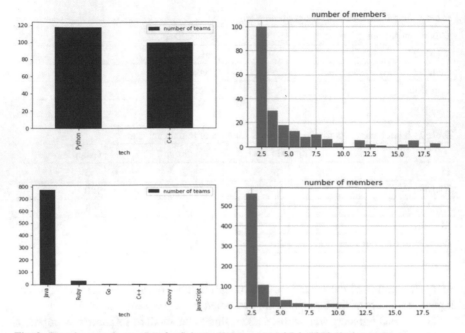

Fig. 2 Search example run stats for Scitator (1st group) and Jeka1978 (2nd group)

7 Conclusion

The developed algorithm can be applied in practice for searching development teams by required programming languages. However, this aspect requires an extension. Furthermore, it is planned to upgrade the analysis of search results by an additional module that will identify not only programming languages that are used in the team's repositories but frameworks. Finally, it will help to extend input parameters and make the search more flexible for HR managers.

Another future improvement is the development of a rating system for search results. Now the search output is an unordered data set. Hence, the recommendation module is required that will range teams by various metrics within their repository activity [13]. It is expected that this is going to save more time while making a decision on which team is the best for the job offer.

Probably, the ontology analysis will be applied via the injection of the semantic layer. This may be achieved by the description of the subject area by OWL2 [14]. The knowledge base will help to form a various set of axioms for the logical inference system [15, 16]. The semantics will help to find implicit data dependencies.

Also, the developed method can be applied not only for a global search but for the inner analysis of the development company: to find out which team of software engineers is the best option for the start of a new project. However, this approach will be useful when the enterprise has a big staff and detailed repository-based history of project developments.

References

1. Solovyova, N.: Russian IT-sphere expects extreme employee shortage, IT World. Available via DIALOG (2018). https://www.itworld.ru/it-news/it/140881.html. Accessed 12 Jan 2022
2. OSP: Russian IT services market. Annual report and forecast (2019). https://www.osp.ru/par tners/13054737/. Accessed 12 Jan 2022
3. McGill, J.P.: The software engineering shortage: a third choice. IEEE Trans. Softw. Eng. **1**, 42–49 (1984)
4. Elias, J.: Google warns pandemic could hinder its ability to 'maintain corporate culture', TECH (2021). https://vk.cc/c0erAj. Accessed 12 Jan 2022
5. Noll, J., Beecham, S., Richardson, I.: Global software development and collaboration: barriers and solutions. ACM inroads **1**(3), 66–78 (2011)
6. Newton, C.: Facebook says IT will permanently shift tens of thousands of jobs to remote work, The Verge (2020). https://vk.cc/c0erGb. Accessed 12 Jan 2022
7. Rodeghero, P., Zimmermann, T., Houck, B., Ford, D.: Please turn your cameras on: remote onboarding of software developers during a pandemic. In: 2021 IEEE/ACM 43rd International Conference on Software Engineering: Software Engineering in Practice (ICSE-SEIP), pp. 41–50 (2021)
8. Gil, E.: High Growth Handbook: Scaling Startups from 10 to 10,000 People, pp. 36–37. Stripe Press (2018)
9. Richmond, W.B., Seidmann, A.: Software development outsourcing contract: structure and business value. J. Manag. Inf. Syst. **10**(1), 57–72 (1993)
10. Kalliamvakou, E., Damian, D., Blincoe, K., Singer, L., German, D.M.: Open source-style collaborative development practices in commercial projects using GitHub. In: 2015 IEEE/ACM 37th IEEE International Conference on Software Engineering, vol. 1, pp. 574–585 (2015)
11. GitHub: Where the world builds software. https://github.com. Accessed 12 Jan 2022
12. Semeradov, T., Weinlich, P.: Computer estimation of customer similarity with Facebook lookalikes: advantages and disadvantages of hypertargeting. IEEE Access **7**, 153365–153377 (2019)
13. IT: The overview of personnel market and top 15 most wanted specialties (2021). https://uly anovsk.hh.ru/article/24562. Accessed 12 Jan 2022
14. Baader, F., Calvanese, D., McGuinness, D., Patel-Schneider, P., Nardi, D.: The Description Logic Handbook: Theory, Implementation and Applications. Cambridge University Press (2003)
15. Konev, B.: Ontologies and Knowledge Representation-Description Logics, [Publication in Russian]. https://bit.ly/3w6xDE3. Accessed 12 Jan 2022
16. Barwise, J., Feferman, S.: Model-Theoretic Logics, pp. 3–23. Springer, New York (1985). https://projecteuclid.org/euclid.pl/1235417266. Accessed 12 Jan 2022

Mathematical Modelling of Distributed Memory System

Vadim L. Stefanuk

Abstract Several mathematical models are considered in the paper to make a strict comparison of the storage models used both in artificial systems and probably in Nature. The very first artificial model was proposed by M. Tsetlin, who called it "The Pile of Books" in his publication in 1961, where he expected some valuable properties in comparison to the typical librarian book storage. Same model was re-discovered by Hendriks in 1972. In 1973 Burville and Kingman obtained mathematical results for corresponding Markov Chain. This Pile of Books model later found use in the Internet search approaches. The main attention in the present paper is paid to our "Distributed Memory" model, which contained additional parameters—the numbers of copies of the same elements stored in the memory. This model demonstrated some new useful applications for search in Internet as multiple storage sites in Internet is widely used. The most important result of the paper is that some features exhibited in the distributed model turned out to be in close similarity with the properties described in a book of biology by A. Levi and S. Sikevits who mentioned rather unusual organization used in biological cells for storage and extraction of enzymes.

Keywords Distributed memory · Pile of books · Biological cells · Enzimes

1 Introduction

The term Artificial Intelligence may be found in Alan Turing publications when he worked out the concept of a complicated calculating device that later was referred to as Computer. It consists from two essential parts: a *Finite Automaton* and an *Endless*

V. L. Stefanuk (✉)
Laboratory of Distributed Systems of Institute for Information Transmission Problems, Tsvetnoy Bulevar 19, 115407 Moscow, Russia
e-mail: stefanuk@iitp.ru

Department of Information Technologies, Peoples' Friendship University of Russia, Miklukho-Maklaya Str.6, 117198 Moscow, Russia

© The Author(s), under exclusive license to Springer Nature Switzerland AG 2023
Sh. N. Shahbazova et al. (eds.), *Recent Developments and the New Directions of Research, Foundations, and Applications*, Studies in Fuzziness and Soft Computing 422, https://doi.org/10.1007/978-3-031-20153-0_6

Memory Tape. A. Turing planned to use such a device to decipher cablegrams created by enemy with *Enigma* during the Second World War.

The Finite Automation provides logic manipulation with data available. The Endless Memory Tape was to store input data as well as some intermediate results of the manipulations with symbols and numbers. The calculation part of the computer was developed to a great extend being considered by engineers as the most essential part of the device, however the endless memory tape was treated mostly as a theoretical concept.

Nevertheless, in practice it was well understood that the bigger memory amount— the better. In the present paper most attention is paid to the memory, using mathematical models and attracting experiments to show some properties of the considered models.

Several mathematical models are considered with the purpose to make a strict comparison of the memory models used both in artificial systems and probably in alive creatures and in a variety of technical applications.

The very first model was proposed by M. Tsetlin, who called it "The Pile of Books" in his publication [1] in the year of 1961. M. Tsetlin predicted some valuable properties of his memory model. He left his model for further study to his students in his seminar held in Physical Department of Moscow State University.

The Tsetlin's model was rediscovered ten years later by Hendriks in [2]. Next year it was rediscovered again by Burville and Kingman in [3] who obtained mathematical result concerning the depth of the books in the pile.

The Pile of Books model later found applications in the Internet search tools and since then was attributed to Google.

The main attention in the present paper is paid to our "Distributed Memory" model, which contains additional parameters—the numbers of copies of the same memory elements.

Our model has demonstrated some new useful applications for the information search in Internet. Also some features of the model turned out to be similar to properties described in a book by Levi and Sikevits [4] concerning some "strange" organization in biological cells for enzymes storage and etc. Thus, the mathematical model of memory that came from strong Artificial Intelligence provides an explanation for the mentioned biological phenomena.

This concluding results may be considered as a qualitative proof of our approach to the optimal storage memory, demonstrating that our mathematical model is valuable one for both artificial and alive worlds. In the discussion section of our paper we will touch this problem again as we actually proved that the strong mathematical AI approach and the Natural evolution have came to the same final result using obviously quite different technology! One may say that the same theorem may be proved theoretically and experimentally.

2 Weak AI Versus Strong AI

Before going further, please note that Alan Turing from the very start planned to overcome human intelligence. Probably, he did not care very much on how human brain was constructed. The machine built by A. Turing was not aimed to *imitate* human intelligence, or *simulate* it.

We have proposed to refer to simulation of human intelligence as the *Weak AI*, i.e. WAI. The majority of applied research in AI belongs undoubtedly to weak AI.

However, the approach by Alan Turing presents a bright example of *Strong AI*.

The *Strong AI* differs from simulation of human intelligence, because its goal is to *overcome* natural intelligence and create new possibilities for Humanity. Hence, *Strong AI* is not necessary maybe traced in the human brain activity, or in its neuron schemas.

Methods and approaches of these two directions in AI are quite different. The approach of Weak AI is to observe the natural intelligence with the purpose to imitate it with computer. However, the approach of *Strong AI* is to discover some new intelligent structures.

Prof. McCarthy [5] in his paper on creative problem solving specially stressed that a solution may be considered to be creative only if it was never discovered before and it was not contained in the problem formulation or in the problems that somehow close to the original one.

In this paper we will try to present some memory models based on the *Strong AI* approach. The models [1–3] were not observed in some way in Alive Nature and we do not know whether our own memory operates following this models. Yet, in the main part of our paper it will be shown that our distribute memory does exist in alive beings, proving that our *creative idea* may be already known to our Creator.

However, this observation raises serious philosophical question as we came to the construction via rather complicated mathematical procedure of optimization. In contrary, the optimization may be obtained in Nature only via Process of Evolution, and we do not know reliable examples of complex theorem proving or designing complicated apparatus like TV or Mobile Communication with help of evolution. Moreover, many things must be invented. One way to resolve this problem is to create an intelligent brain in the process of evolution such a complicated machine as the intelligent brain.

The average human memory depth is a standard value for many mathematical models of biological situations, which has been created in the school of Prof. Michael Tsetlin in Moscow University. We have tried to make a comparison between these two models, namely pile of the book and distributed memory, using the formulas for the **average depth of memory** obtained for both models.

The average memory depth is a standard value for many mathematical models.

For the description of both models the concept of *Pattern Matching* is used, which is taken from Artificial Intelligence.

The Pattern Matching is the basic way of searching for an "associative" information. It was first used in the COMIT and COBOL in order to extract information in accordance with its content. It is used extensively in AI, where the information is sought usually among symbolic data.

It is assumed below that an item is extracted from the memory in this way or another and is matched against the pattern. If the match is successful the search is being completed.

Otherwise the current item is returned to the memory (or disposed of), the next item is being extracted, and the process continues till a datum, matching the pattern, is founded.

The pattern matching is rather complex procedure, hence it is desirable that the average number of intermediate matches to be minimal. (Or, the time spent for search should be made minimal.)

Assume that there are exist the search probabilities $p_1, ..., p_n$, where

$$\sum_{k-1}^{n} p_k = 1 \tag{1}$$

Below, all the items (such as books, triplets, words, programs) will be referred to as *words*.

We will try to optimize the pattern matching and propose a mathematical model of a random extraction. Next, we will consider a number of algorithms of adaptive search optimization.

However, first we will remind *the pile of the books model* proposed by Michael Tsetlin in 1963. Instead of storing books in a catalog M. L. Tsetlin proposed to return a book, that had been recently red, to the very top of the pile. Thus, the most needed books gradually will be collected at the top of the pile, while less popular books are moving to its bottom.

In 10 years Hendriks [2] described the same model.

Burville and Kingman [3] have obtained the expression for the average depth of the books in the pile:

$$\Lambda_b = 1 + 2 \sum_{i<j}^{n} \frac{p_i p_j}{p_i + p_j} \tag{2}$$

In the distributed random memory [2] similar effect of *optimality* is reached due to the storing of items (books, words, and etc.) in copies: the most requested items have to be in greater numbers as we will try to arrange below.

Such *storage of duplicates* is typical for many practical applications [5].

Our random memory has some interesting properties based on the result proved in [5] and shown in the next chapter.

3 Square Root Law

Let l_i be the average number of random visits to the memory in search for i-th word. Then the total average of visits to the memory is equal [2] to:

$$\Lambda = \sum_{i=1}^{n} p_i l_i \tag{3}$$

It is assumed that 'the next' distinct word will be searched for only after the successfully finding the previous one. Let N_i be the number of copies of the i-th word and $\sum_{i=1}^{n} N_i = N$ — be the total number of words.

When the random memory is visited, each of the N words is extracted with an equal probability and the unmatched words are returned to the memory. Thus, it is true that

$$l_i = \frac{N_i}{N} + 2\left(1 - \frac{N_i}{N}\right)\frac{N_i}{N} + 3\left(1 - \frac{N_i}{N}\right)^2 \frac{N_i}{N} + \cdots = \frac{N}{N_i} \equiv \frac{1}{\tilde{p}_i} \tag{4}$$

From this we may obtain the total average of the numbers of visits to our random memory in process of search.

The value of $\tilde{p}_i$ is the *fraction* of the i-th word in the memory, or the probability to find this word at a first visit to the memory.

From $\Lambda_r = \sum_{i-1}^{n} \frac{p_i}{\tilde{p}_i}$ it is seen that if the numbers of copies N_i are proportional to the a priori probabilities of their search, then the average number of visits to the memory will be equal to n.

It is possible to show that the following inequality is valid:

$$\Lambda_r \geq \left(\sum_{i=1}^{n} \sqrt{p_i}\right)^2 \tag{5}$$

The minimal average number of visits to the memory (in search for matches), is equal to

$$\Lambda_r^{opt} = \left(\sum_{i=1}^{n} \sqrt{p_i} \right)^2 \tag{6}$$

This optimal value is obtained when

$$\frac{N_i}{N} = \tilde{p}_i = \frac{\sqrt{p_i}}{\sum_{i=1}^{n} \sqrt{p_i}}, \; i = 1, ..., n \tag{7}$$

Thus, for an *optimal search* of a match in the randomly organized distributed memory *the number of copies of each word must be proportional to the square root of the probability of its search.*

4   An Algorithm to Generate and Remove Copies

The appropriated algorithm will be built in two steps.

4.1   Basic Part

After each successful match at the moment t we will add exactly σ such words. The numbers of copies of the most searchable words will increase with time. To save a storage space some of the copies might be deleted:

$$N_i(t + 1) = N_i(t) + \xi_i(t) - \eta_i(t), \; i = 1, ...n \tag{8}$$

The values $\xi(t)$ are independent random values σ, 0 with probabilities p_i, $(1 - p_i)$ correspondingly. The random values $\eta_i(t)$ depend only on $N_i(t)$, taking values m_i with probabilities $C_{N_i(t)}^{m_i} \alpha^{m_i} (1 - \alpha)^{N_i(t) - m_i}$.

Taking averages $\hat{N}_i$, we may see that the solution will be as follows

$$\hat{N}_i = \frac{\sigma}{\alpha} p_i, \; i = 1, ..., n, \tag{9}$$

Thus the average depth will be equal to n, which is not optimal. Yet, some geometry consideration might be used to correct the situation.

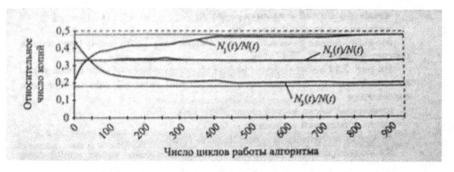

Fig. 1 Simulation for the first part of the natural algorithm

Computer simulation is shown in the next figure

The below figure is obtained for the case (see Fig. 1)

> added words $\sigma = 1000$ copies of words;
>
> the factor of deletions $\alpha = 0, 01$;
>
> initials numbers are 200, 150, 100;
>
> the search probabilities 0.1, 0.3, 0.6;
>
> square roots are 0.193, 0.334, 0.473;

.

4.2 Geometrical Considerations

Imagine that all similar words are being put on a piece of a plane and are "trying to keep close to each other", creating clouds of a roundish shape on this plain. And let the items are taken from the memory only from the borders of such clouds. Then the fractions of the words, among which the matches are being sought, *will be proportional roughly* to the square root of p_i from (9). And the *search becomes optimal*!

This algorithm has a rather good parallel in Alive Nature.

In the book by biologists Levi and Sikevits [4] it is said: "*Probably the lysosomes and peroxysomas serves in the cell for storage of a large quantity of enzymes. With this form of storage the large portion of these enzymes are taken out of circulation …*".

See also [7] concerning similar observation for lysosomes.

However, we do not have strict optimality as for modeling of alive cell it is important to use 3D geometry. If the reading of the items is performed only from the surface of the 3D bag, we obtain that the average number of visits to the memory in the pattern matching search will be $\Lambda_{rd} = \sum_{i=1}^{n} p^{1/3} \sum_{i-1}^{n} p^{2/3}$

This expression may be considered only as semi optimal. However, if $p_i \to 1$ then we have $\Lambda \to 1$. It means that the items constituting strong fraction will be obtained from the memory first.

There are some other algorithms to withdraw items from the distributed memory listed in [5].

5 Direct Optimization Algorithm

The way of the optimization depends on the information available. If $p_1, ..., p_n$ are known the above formula (7) may be used for an optimization.

If $p_1, ..., p_n$ are not known, but the current values $N_i(t)$ are known, one may add instead of σ the corrected value σ / N_i. Then the following equation maybe written down for the average values of $\overline{N}_i(t)$:

$$\overline{N}_i(t+1) = \overline{N}_i(t) + \frac{\sigma}{\overline{N}_i(t)} p_i - \alpha \overline{N}_i(t) \tag{10}$$

Obviously for the stationary point the "The law of square root" is fulfilled here.

And at last, if all these information is absent an effective algorithm, which we call the algorithm with an adequate replacement is the following one.

After each successful match one has to count the number of "wrong" words obtained right before the perfect match and all these words are replaced with the "correct" words and added to the storage.

An approximate calculations and computer simulation show that the *algorithm with adequate replacement* produces optimal values for copies of words despite the fact of absence of any a priori evidences or any knowledge of the current state of the memory.

We do not know whether such algorithms may be found in Nature, and will consider them temporally belonging to the *Strong Artificial Intelligence (see the part2 in this paper).*

The algorithm with adequate replacement

An approximate theory is shown in [5].

$N_i \approx \sqrt{\left(\frac{1}{4} + \frac{p_i}{\delta} N\right)} - \frac{1}{2}$, where $\delta \leq 1$.

Hence, for $N_i \to \infty$ one asymptotically has the following expression

$$\tilde{p}_i = \frac{N_i}{N} = \frac{\sqrt{p_i}}{\sum_{k=1}^{n} \sqrt{p_k}}, i = 1, ..., n$$

i.e. optimal values are iobtained for N_i despite the absence of any a priori information.

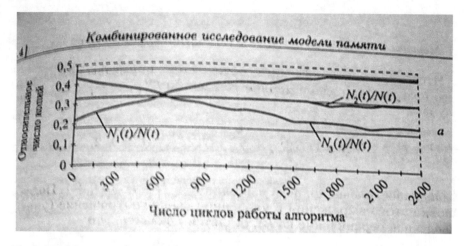

Fig. 2 Combined research for the memory model

In our approximate theory the moments of all order, except averages, have been neglected. Also, the probabilities of obtaining "zeroes" were neglected. Thus, computer simulation is necessary indeed to confirm the theory (see Fig. 2).

Initial numbers are 2000, 1500, 1000

Final numbers are 693, 1400, 2117

The search probabilities 0.1, 0.3, 0.4

Square roots are 0.193, 0.334, 0.473

Number of visits to the memory 3000

6 Conclusions

- Further calculations demonstrate that the second probabilistic moment has rather high value in our models.
- In the worst case the Pile of the Books is two times better than the "randomly organized memory".
- However in case of interconnected requests our distributed model gives better results.
- Yet, in practice those models have different areas of applications and there is a choice which model should be preferred [2].

Also, there is an impression that the volume of memory (number of molecules in alive cell) maybe increased without limit. However, the time spent for the search of an item may not be made very small. Hence early or later in any application there will be a need for *adaptive algorithms* of the type shown above, which reduce the time of total search to a minimum.

These models demonstrate that the adaptive memory, which is heavily used in humans and in many biological systems, is arranged reasonably well.

For people it helps to keep the knowledge in a way facilitating its fast extraction. *For biological systems* [4, 7] the models explain arrangement of various microbodies storage in an optimal manner to provide better conditions for an organism.

Note that our distributed random memory described above is rather complex from mathematical point of view, and it looks like that it would be impossible to design and implement the model, using the geometrical considerations of 4.2, in Alive Nature in the process of simple Evolution. Presently we do not have a scientific explanation to this observation.

This is our final and important comment!

References

1. Tsetlin, V.L.: Finite Automata and Modelling of Simplest Forms of Behaviour, Russian Mathematical Surveys, vol. 18, no. 4, pp. 3–28 (1963). [Цетлин М.Л., Конечные автоматы и моделирование простейших форм поведения: Успехи математических наук.]
2. Stefanuk, V.L.: A Model of Random Pattern Matching in a Data Base, MIP-R-110, University of Edinburgh, Machine Intelligence Research Unit, 15p (1974)
3. Burville, P.J., Kingman, J.F.C.: On a model for storage and search. J. Appl. Probab. (10), 697–701 (1973)
4. Levi, A., Sikevits, F.: Structure and Functions of a Cell [Struktura i funktsii kletki. S Il.. Publisher: MIR. Year: 1971. Moskva: P. 584.]
5. McCarthy, J.: Creative Problem Solving (1999), http://www-formal.stanford.edu/jmc/
6. Stefanuk, V.L.: Local Organization of Intelligent Systems. Models and Applications [Локальная организация интеллектуальных систем. Модели и приложения. Москва: Физико-математическая литература (2004), С. 328.]
7. Christian de Duve, A Guided tour of the living cell, New York, W.H. Freeman and Company, New York, pp. 444. (1985)
8. Hendricks, W.J.: The stationary distribution of an interesting Markov chain. J. Appl. Prob. **9**, 231–233 (1972)

Explainable Correlation of Categorical Data and Bar Charts

Imre J. Rudas and Ildar Z. Batyrshin ⓘ

Abstract We propose a method of calculating the correlation between frequency distributions defined on a finite set of categories. The method is based on the general approach to constructing an invertible correlation function on a set with involutive operation. This correlation function is constructed using a suitable similarity (or dissimilarity) function defined on such a set. For constructing a correlation function on the set of frequency distributions, we use the recently introduced involutive negation of probability distributions and a suitable dissimilarity function between them. Surprisingly, the obtained correlation function coincides with the Pearson correlation coefficient. The proposed approach is illustrated in an example of calculating the correlation between categories of a categorical variable corresponding to strings of a contingency table. The traditional graphical interpretation of the Pearson correlation by linear regression gives an explanation of obtained correlations.

Keywords Categorical data · Probability distribution · Correlation · Involutive negation · Dissimilarity function

1 Introduction

Categorical data are widely used in social and behavioral sciences, business analytics, machine learning and pattern recognition, linguistics, and other areas where the counts or frequency of appearance of data in different categories or classes are considered. An analysis of possible relationships between such frequency distributions is important for many applications [1–3]. One can compare the distributions of votes for several political parties in two states or regions, the distributions of gender preferences of social or professional activity, distributions of preferences for two types

I. J. Rudas
Obuda University, Bécsi ut 96/B, 1034 Budapest, Hungary

I. Z. Batyrshin (✉)
CIC-IPN, Instituto Politécnico Nacional, Av. Juan de Dios Bátiz S/N, Nueva Industrial Vallejo, 07738 Gustavo A. Madero, CDMX, Mexico

© The Author(s), under exclusive license to Springer Nature Switzerland AG 2023   81
Sh. N. Shahbazova et al. (eds.), *Recent Developments and the New Directions of Research, Foundations, and Applications*, Studies in Fuzziness and Soft Computing 422, https://doi.org/10.1007/978-3-031-20153-0_7

of customers, distributions of sales of car models in two countries, etc. Usually, such distributions are presented in contingency tables or bar charts. In this work, we propose the method of calculation of correlation between categories of one variable based on their frequency distributions on the set of categories of another variable. The proposed approach uses recently developed methods of constructing invertible correlation functions from similarity and dissimilarity functions (fuzzy relations) [4] defined on a set with involutive operation [5–7]. It was shown that many classical correlation and association coefficients [8, 9] are invertible correlations and can be constructed from similarity and dissimilarity functions by the proposed in [5–7] methods. For constructing the correlation function between distributions, we use the recently introduced involutive negation of probability distributions [10, 11] and a suitable dissimilarity function between them [4].

The paper has the following structure. In Sect. 2, we give a short introduction to invertible correlation functions. In Sect. 3, we introduce the involutive negation of finite probability distributions, a dissimilarity function between them, and construct a correlation function between probability distributions. Section 4 presents an example of calculating the correlation between categories of a categorical variable given in the contingency table. Section 5 contains conclusions.

2   Invertible Correlation Functions

Invertible correlation functions (association measures) were introduced and studied in [5–7]. Consider the basic definitions [6, 7, 12]. Let Ω be a set with two or more elements. A function $N : \Omega \rightarrow \Omega$ is a *reflection* or a *negation* on Ω if, for all x in Ω, it satisfies the *involutivity* property:

$$N(N(x)) = x,$$

and $N(x) \neq x$ for some x in Ω, i.e., N is not an identity function. An element x in Ω such that $N(x) = x$ is called a *fixed point* and the set of all fixed points of the negation N on Ω is denoted as $FP(\Omega)$.

Let N be a reflection on Ω and V be a non-empty subset of $\Omega \setminus FP(\Omega)$ closed under N, i.e., for any x in V, we have $N(x) \in V$. An *invertible correlation function* (association measure) on V is a function $A : V \times V \rightarrow [-1, 1]$ satisfying for all x, y in V the properties:

> A1. $A(x, y) = A(y, x)$,   (symmetry)
> A2. $A(x, y) = 1$,   (reflexivity)
> A3. $A(x, N(y)) = -A(x, y)$.   (inverse relationship)

Invertible correlation functions satisfy for all x, y in V the following properties:

$$A(x, N(x)) = -1, \quad \text{(opposite elements)}$$
$$A(N(x), N(y)) = A(x, y), \quad \text{(co} - \text{symmetry I)}$$
$$A(x, N(y)) = A(N(x), y) \quad \text{(co} - \text{symmetry II)}$$

Example 1. Consider Pearson's linear correlation coefficient [8, 9, 12]:

$$r(x, y) = \frac{\sum_{i=1}^{n}(x_i - \bar{x})(y_i - \bar{y})}{\sqrt{\sum_{i=1}^{n}(x_i - \bar{x})^2}\sqrt{\sum_{i=1}^{n}(y_i - \bar{y})^2}}, \tag{1}$$

where $x = (x_1, \ldots, x_n)$ and $y = (y_1, \ldots, y_n)$ are real-valued n-tuples, e.g., n measurements of real-valued variables x and y. It is clear that the Pearson's correlation satisfies the properties of symmetry and reflexivity on the set V of non-constant n-tuples $x = (x_1, \ldots, x_n)$, such that $x \neq (c, \ldots, c)$ for any real c. Define $N(y) = (-y_1, \ldots, -y_n)$, then $\overline{N(y)} = \frac{1}{n}\sum_{i=1}^{n}(-y_i) = -\frac{1}{n}\sum_{i=1}^{n} y_i = -\bar{y}$, and A3 is fulfilled:

$$r(x, N(y)) = \frac{\sum_{i=1}^{n}(x_i - \bar{x})\left(-y_i - \overline{N(y)}\right)}{\sqrt{\sum_{i=1}^{n}(x_i - \bar{x})^2}\sqrt{\sum_{i=1}^{n}\left(-y_i - \overline{N(y)}\right)^2}}$$

$$= \frac{\sum_{i=1}^{n}(x_i - \bar{x})(-y_i - (-\bar{y}))}{\sqrt{\sum_{i=1}^{n}(x_i - \bar{x})^2}\sqrt{\sum_{i=1}^{n}(-y_i - (-\bar{y}))^2}} =$$

$$\frac{-\sum_{i=1}^{n}(x_i - \bar{x})(y_i - \bar{y})}{\sqrt{\sum_{i=1}^{n}(x_i - \bar{x})^2}\sqrt{\sum_{i=1}^{n}(y_i - \bar{y})^2}} = -r(x, y).$$

Let N be a reflection on Ω and V be a non-empty subset of $\Omega \setminus FP(\Omega)$ closed under N. A function $D : V \times V \to [0, 1]$ is a *bipolar dissimilarity function* on V [4, 6, 7] if for all x, y in V, it is *symmetric*:

$$D(x, y) = D(y, x),$$

irreflexive:

$$D(x, x) = 0,$$

and *bipolar*:

$$D(x, y) + D(x, N(y)) = 1.$$

It was shown in [6, 7] that any bipolar dissimilarity function D on V defines an invertible correlation function (association measure) A by:

$$A(x, y) = 1 - 2D(x, y). \tag{2}$$

In [12], it was shown that the function:

$$D(x, y) = \frac{1}{4} \sum_{i=1}^{n} \left(\frac{x_i - \overline{x}}{\sqrt{\sum_{i=1}^{n} (x_i - \overline{x})^2}} - \frac{y_i - \overline{y}}{\sqrt{\sum_{i=1}^{n} (y_i - \overline{y})^2}} \right)^2, \tag{3}$$

is a bipolar dissimilarity function that defines by (2) Pearson's correlation coefficient.

3 Invertible Correlation on the Set of Probability Distributions

Consider the set Ω of probability distributions $P = \{p_1, \ldots, p_n\}$ defined on a finite set $C = \{c_1, \ldots, c_n\}$ of n categories. The following properties are fulfilled for all probability distributions $P = \{p_1, \ldots, p_n\}$:

$$0 \le p_i \le 1, \quad \sum_{i=1}^{n} p_i = 1. \tag{4}$$

For us, it is more convenient to write probability distributions as n-tuples: $P = (p_1, \ldots, p_n)$. In [10], it was introduced an involutive negation on the set of probability distributions as follows. Denote $\max(P) = \max\{p_1, \ldots, p_n\}$, $\min(P) = \min\{p_1, \ldots, p_n\}$ and $MP = \max(P) + \min(P)$. The involutive *negator* N of all elements p_i of P is defined as follows:

$$N(p_i) = \frac{\max(P) + \min(P) - p_i}{n(\max(P) + \min(P)) - 1} = \frac{MP - p_i}{nMP - 1}.$$

It defines an involutive negation $neg(P)$ of probability distributions by:

$$neg(P) = (N(p_1), \ldots, N(p_n)),$$

such that $neg(neg(P)) = P$ for all P in Ω. This negation has a *uniform distribution*:

$$P_U = \left(\frac{1}{n}, \ldots, \frac{1}{n} \right),$$

as a unique fixed point, such that

$$neg(P_U) = P_U.$$

Consider the set $V = \Omega \setminus \{P_U\}$. We can prove the following theorems.

Theorem 1. The function $D : V \times V \to [0, 1]$ defined for any probability distributions P and Q in V as follows:

$$D(P, Q) = \frac{1}{4} \sum_{i=1}^{n} \left[\frac{p_i - \frac{1}{n}}{\sqrt{\sum_{i=1}^{n} \left(p_i - \frac{1}{n}\right)^2}} - \frac{q_i - \frac{1}{n}}{\sqrt{\sum_{i=1}^{n} \left(q_i - \frac{1}{n}\right)^2}} \right]^2 =$$

$$\frac{1}{4} \sum_{i=1}^{n} \left[\frac{np_i - 1}{\sqrt{\sum_{i=1}^{n} (np_i - 1)^2}} - \frac{nq_i - 1}{\sqrt{\sum_{i=1}^{n} (nq_i - 1)^2}} \right]^2, \tag{5}$$

is a bipolar dissimilarity function, such that for all probability distributions P and Q in V we have:

$$D(P, Q) + D(P, neg(Q)) = 1.$$

Theorem 2. The dissimilarity function (5) defines by

$$A(P, D) = 1 - 2D(P, D)$$

the following correlation function on the set of probability distributions V:

$$A(x, y) = \frac{\sum_{i=1}^{n} \left(p_i - \frac{1}{n}\right)\left(q_i - \frac{1}{n}\right)}{\sqrt{\sum_{i=1}^{n} \left(p_i - \frac{1}{n}\right)^2} \sqrt{\sum_{i=1}^{n} \left(p_i - \frac{1}{n}\right)^2}}$$

$$= \frac{\sum_{i=1}^{n} (np_i - 1)(nq_i - 1)}{\sqrt{\sum_{i=1}^{n} (np_i - 1)^2} \sqrt{\sum_{i=1}^{n} (nq_i - 1)^2}}. \tag{6}$$

Surprisingly, we obtained Pearson's correlation coefficient, taking into account that $\frac{1}{n}$ is the mean value of any probability distribution $P = (p_1, \ldots, p_n)$!

4 Calculating Correlation of Categories in Contingency Tables

Example 2. Consider the contingency table given in Fig. 1 adapted from [1].

- Convert into relative frequencies (probabilities) the frequency values in strings 1, 2, and 3, dividing them by total values 424, 1660, and 642, respectively.
- Calculate correlations between obtained probability distributions using formula (6). We obtained the following correlation values between these three distributions corresponding to the categories 1: "Less than high school," 2: "High school or junior college," and 3: "Bachelor or graduate,": $A(1, 2) = 0.55$, $A(1, 3) = -0.90$, $A(2, 3) = -0.14$.
- These three categories we can consider as variables taking corresponding frequency values on the set of categories "Fundamental," "Moderate," and "Liberal." Figure 2 depicts linear regressions for pairs of variables 1 and 2 (left), 1 and 3 (center), and 2 and 3 (right).

	Religious Beliefs			
Highest Degree	Fundamentalist	Moderate	Liberal	Total
1_Less than high school	178	138	108	424
2_High school or junior college	570	648	442	1660
3_Bachelor or graduate	138	252	252	642
Total	886	1038	802	2726

Fig. 1 Contingency table adapted from Table 3.2 in [1]

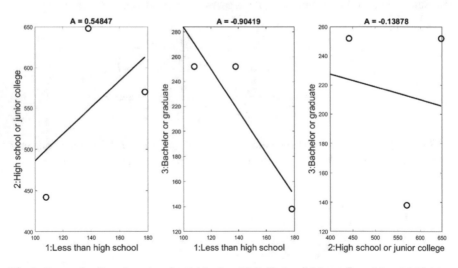

Fig. 2 Regression lines for pairs of variables 1 and 2 (left), 1 and 3 (center), and 2 and 3 (right)

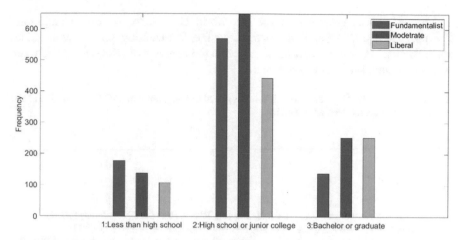

Fig. 3 Bar charts of three frequency distributions

- Graphically represent frequency distributions from strings 1, 2, and 3 (Fig. 1) by bar charts (Fig. 3).

These linear regressions give a good explanation of the relationships between the three categories. For example, the central image in Fig. 2 shows that categories 1: "Less than high school," and 3: "Bachelor or graduate," are almost opposite categories with a high negative correlation $A(1, 3) = -0.90$ and an inverse relationship. When the frequencies of the first category decrease in the direction "Fundamental" $\rightarrow$ "Moderate" $\rightarrow$ "Liberal," (see Figs. 1 and 3), the frequencies of the third category increase. Category 2: "High school or junior college" occupies an intermediate position between the other two categories. It is more similar to category 1: "Less than high school," with a positive correlation $A(1, 2) = 0.55$, than to category 3: "Bachelor or graduate," with a negative correlation $A(2, 3) = -0.14$.

5 Conclusion

The paper proposes a method of calculating the correlation between finite probability and frequency distributions. These correlations can be used for the analysis of relationships between categories in contingency tables. Since the proposed correlation coefficient coincides with Pearson's linear correlation, the obtained correlation can be easily visualized and explained by linear regressions between variables corresponding to considered categories. Because bar charts can represent the frequency distributions, the proposed approach also can be applied to calculating correlations between them. The proposed method of analysis of relationships between categorical data can be used as a regular tool in descriptive statistics, data mining, and data analytics.

This work was presented at the 8th World Conference on Soft Computing, February 3–5, 2022, Baku, Azerbaijan. In the forthcoming paper, we consider some extensions of the proposed approach to the analysis of relationships between categorical data and between bar charts.

Acknowledgements The work was partially supported by the grants SIP 20220857 and 20211874 of Instituto Politecnico Nacional, Mexico.

References

1. Agresti, A.: Categorical Data Analysis, 2nd edn. Wiley, Hoboken, New Jersey (2002)
2. Simonoff, J.S.: Analyzing Categorical Data, vol. 496. Springer, New York (2003)
3. Hancock, J.T., Khoshgoftaar, T.M.: Survey on categorical data for neural networks. J. Big Data **7**(1), 1–41 (2020)
4. Batyrshin, I.: Towards a general theory of similarity and association measures: similarity, dissimilarity, and correlation functions. J. Intell. Fuzzy Syst. **36**(4), 2977–3004 (2019)
5. Batyrshin, I.Z.: On definition and construction of association measures. J. Intell. Fuzzy Syst. **29**, 2319–2326 (2015)
6. Batyrshin, I.Z.: Data science: Similarity, dissimilarity and correlation functions. In: Artificial Intelligence, pp. 13–28. Springer, Cham (2019)
7. Batyrshin, I.Z.: Constructing correlation coefficients from similarity and dissimilarity functions. Acta Polytechnica Hungarica **16**(10), 191–204 (2019)
8. Chen, P.Y., Popovich, P.M.: Correlation: Parametric and Nonparametric Measures. Sage, Thousand Oaks, CA (2002)
9. Gibbons, J.D., Chakraborti, S.: Nonparametric Statistical Inference, 4th edn. Dekker, New York (2003)
10. Batyrshin, I.Z.: Contracting and involutive negations of probability distributions. Mathematics **9**(19), 2389 (2021)
11. Batyrshin, I.Z., Kubysheva, N.I., Bayrasheva, V.R., Kosheleva, O., Kreinovich, V.: Negations of probability distributions: a survey. Computación y Sistemas **25**(4), 775–781 (2021)
12. Batyrshin, I.Z., Tóth-Laufer, E.: Bipolar dissimilarity and similarity correlations of numbers. Mathematics **10**(5), 797 (2022)

ImageRobo: Controlling Robot by Some EEGs from Right Frontal Area on Recalling Image of Its Movements

Takahiro Yamanoi, Mika Otsuki, Hisashi Toyoshima, Yuzu Uchida, and Shin-ichi Ohnishi

Abstract First part is a brief history of preliminary researches on BCI. Then the paper introduces the recent research by the authors. The authors measured electroencephalograms (EEGs) from subjects on recalling ten types of movement images of a robot. They have measured a set of Single-Trial-EEGs from six subjects during watching the movement of the robot images. The canonical discriminant analysis was applied to these EEGs data specially sampled. Four channels of EEGs at the right frontal and temporal were used in the discrimination. Results of the canonical discriminant analysis were more than 90% for six subjects. This system is called ImageRobo. After that the ImageRobo was optimized for three channels of the Single-Trial EEGs and also sampling data were one third of the four channel systems. The discriminant ratios were slightly improved though the amount of data was reduced to one quarter.

Keywords Electroencephalogram · Image of robot movement · Image recalling · Single-Trial-EEGs · Robot control · Brain computer interface · Canonical discriminant analysis · Bluetooth

1 Dawn of BMI/BCI

On November 10th in 2015, the Japanese Ministry of Health, Labor and Welfare expert meeting approved the auxiliary device "Robot Suit HAL" [1] that can be

T. Yamanoi (✉) · M. Otsuki · H. Toyoshima
Research Faculty of Health Sciences, Department of Emergent Neurocognition, Hokkaido University, Kita 12, Nishi 5, Kita-ku, Sapporo 060-0812, Japan
e-mail: yamanoi@hgu.jp

T. Yamanoi
Hokkai Gakuen Univ, Sapporo, Japan

Y. Uchida · S. Ohnishi
Faculty of Engineering, Department of Electronics and Information Engineering, Hokkai-Gakuen University, Minami 26, Nishi 11, Chuo-ku, Sapporo 064-0926, Japan

worn on the human body to improve walking ability as a medical device for patients with intractable diseases with weakened muscles throughout the body. The Japanese government's support for the physically handicapped had begun officially at that moment. The HAL is named after the computer robot HAL9000 in the film "2001: A Space Odyssey" by Stanley Kubrick in the sense that it is one letter ahead of IBM in the order of the alphabet.

As the purpose of the robot suit HAL shows, the humans get older, the hardware of living organisms often breaks down because of fatigue or aging. The critical death judgment symbolizes as the brain death, it is the brain that ultimately maintains its function as long as humans are alive. In the Discours de la Méthode, Rene Descartes wrote "Je pense donc je suis (I am because I think)". Here, he described in the word "penser (think)" is really the activities of the human brain. The electroencephalogram (EEG) is currently used to determine the "alive or death" problem for us. Therefore, EEG may be the final or ultimate means of human-to-human communication, including the judgment of alive or death. Research in this area was preceded by the term brain-machine interface (BMI). BMI here is to take out fine signals of EEG by any measurement and to interface with humans and the machine to be well controlled. These days, most machines use computer, so now it's called the Brain Computer Interface (BCI) or widely the Human Computer Interface (HCI). A series of studies by John Donoghue's group in the United States was a pioneer in this field [2]. On the other hand, in Japan as well, we are conducting large-scale BCI research using EEG and fMRI, led by the group of Dr. Mitsuo Kawato of the International Advanced Telecommunications Research Institute International (ATR). Although it is not a direct signal from the brain, Dr. Emi Tamaki's PossessedHand [3], which was selected as one of the 50 inventions in the TIME world in 2011, cannot be overlooked. This is to move the finger by applying an electrical stimulus that is almost equivalent to the electrical signal that moves the finger from the brain with the band attached to the arm to the muscle in the upper arm.

As above mentioned, it can be said that BCI research began with the Brown Brain Science Program, which was founded by John Donoghue in 1999 and later became the Brown Brain Science Institute. In 2001, spin-offed from Brown University, the Cyberkinetics company was founded for piloting clinical trials. This company became later the Cyberkinetics Neurotechnology Systems and is now active as BrainGate.

Initially, the BrainGate Neural Interface System was developed at Brown University [4], which implants penholders-shaped microelectrodes in the brain, directly invasively extracts.

EEG signals, analyzes them, and sends to control signals. This is called the Neuro-Motor Prosthetic (NMP) and is paralyzed in humans by outputting motion-related signals from the brain to external devices around the damaged part of the nervous system. It was intended to act for or restore motor function. Furthermore, NMP was intended to enable useful work by converting human voluntary neuronal activity into control signals. The study using BrainGate was a pilot experiment and used NMP for humans with quadriplegia. The neuronal activity of the intended hand movement by the subject was extracted through the penholders-shaped microelectrodes with a diameter of less than 5 mm, which had 96 electrodes inserted into the motor cortex of the brain. The volunteer MN was able to form a cortical spike pattern after three years spinal cord injury. While he was talking, he was able to open the simulated email displayed on the TV screen prepared for operation, and could move the "nerve cursor". In addition, he was able to open and close artificial limbs to control a multi-link robot arm designed to perform basic actions with nerve output, however, that is an invasive method.

Among these series of research projects by John Donoghue et al. [5], the one that attracted a lot of attention from BCI researchers was probably the BrainGate 2 project [6] article published in Nature in May 2015. According to this article by Nature, a patient who suffered a stroke in 1997 and was paralyzed underwent BrainGate 2 implantation surgery in 2005. This patient operated a robot arm developed by DEKA Research and Development Corporation, picked up a sponge ball with a success rate of 46%, and succeeded in grasping a coffee bottle with a success rate of 66% and bringing it to the mouth. Furthermore, patients who became paralyzed in 2006 and underwent this electrode implantation surgery 6 years later were able to pick up sponge balls with a success rate of 62% via the robot arm described above. However, this movement speed was slow, unlike the movements performed by ordinary adults. DEKA, which developed this robot arm, is also the company that developed Segway, which controls its movement by moving the center of gravity without operating the accelerator or brake.

By a highly invasive method such as directly measuring the cortical potential through electrodes into the skull by surgery, it is possible to obtain highly accurate EEG because it is not affected by EMG, etc. On the other hand, there are side effects such as weakening of brain signals due to the damage of surgery. And according to a study by Abdulkader et al., the surgery may not be possible depending on the symptoms [7].

In addition, a slightly less invasive method of passing electrodes through the skull without penetrating the cortex is also being studied. This method is considered to be a more effective method because it is not easily affected by myoelectricity and has few side effects that weaken brain signals. By use of this technique, Gulati et al. conducted a clinical trial of BCI using EEG measured from the cortex degenerated by stroke [8].

An EEG device, as a representative of noninvasive technology for the brain, has a wide application in both clinical and research fields [9]. For example, Rocsa et al. used noninvasive EEG measurements to control a drone [10]. And the brain death judgement is performed by EEGs, the EEG is the ultimate and the last communication tool among human.

For examples, it is possible to confirm brain activity as a medical device as well as some kind of BCI tools. When the brain is physically damaged due to a stroke or trauma some parts of the motor function may be lost. Several research groups have been investigating methods by use of BCI in order to restore motor function in patients with motor function impairment [11–13]. In these studies, by confirming whether the patient can correctly image the movement by BCI, they are considering to help the recovery of the patient's motor function, such as setting training to image the movement of the patient. They are also studying the use of BCI to assist patients with impaired consciousness [14]. It can be provided as a communication assistance tool, such as allowing the patient to communicate basic requirements such as bed position and adjustment of air conditioning equipment [15–17].

In recent years, there have been many cases where brain mapping is performed during brain surgery. The method is to avoid damaging to the motor area and the language area as much as possible by surgery and to remove the only minimum necessary brain tissue. Recently, BCI researchers and doctors have collaborated to seek improvements in brain mapping, contributing to the identification of various brain function-related sites based on the results of BCI research [18].

2   Recent BCI Research by Present Authors

According to researches on the human brain, the primary processing of a visual stimulus is treated in visual area 1 (V1) and visual area 2 (V2) in the occipital lobe. The processing, then, gradually moves on to the temporal association areas and the parietal association area [19]. Successive higher order processing is strongly lateralized in the brain. For example, the language processing areas in the brain (e. g. the Wernicke's area and the Broca's area) are located in the left hemisphere for the 99% of right-handed people and for the 70% of left-handed people [20, 21]. In general, the brain activity for language is also processed in the left angular gyrus (AnG), the left fusiform gyrus (FuG), the left inferior frontal gyrus (IFG), and the prefrontal cortex (PFC) [22].

In the previous study, some of the present authors had experimented for input stimuli with a directional meaning comprised of Chinese characters (Kanji), which are generally used in writing Japanese, or arrow-symbols. They had measured EEGs on silent reading task and those EEGs were averaged and summed by each 19channel for each image, and the event related potentials (ERPs) were obtained. Then, by use of the equivalent current dipole source localization (ECDL) techniques for those ERPs [23], they had concluded that for input stimuli of arrow-symbols, equivalent current dipoles (ECDs) were estimated to the right middle temporal gyrus (MTG), which is said to be working memory for the spatial perception: to the right IFG or to the right middle frontal gyrus (MFG). Further, in case of Kanji for visual stimuli, ECDs were also estimated to the PFC and the precentral gyrus [24, 25].

However, in the case of silent reading to these stimuli with directional meanings, spatiotemporal activities were estimated in the same areas around the same latencies independently of the stimulus (Kanji or arrow). That is, ECDs were estimated to the Broca's area, known as the language area for controlling speech. After the processing on the right frontal lobe, the spatiotemporal activities go to the working memory. It was found that latencies of main peak were almost the same, while the polarities of potentials were opposite in the frontal lobe in the higher order processing [24].

The middle frontal lobe is related to the central executive system, including the working memory. Functions of the central executive system include selecting information from the outer world, holding it temporarily in memory, ordering subsequent actions, evaluating these orders, making decisions, and also erasing temporarily stored information. Indeed, this art of the frontal cortex is the higher order functions in the central nervous system in the brain.

On the basis of these facts, some of the present authors compared EEGs from electrode channels, and found that the channel 4 (F_4), 6 (C_4) and 12 (F_8) were effective in discriminating EEGs during silently reading four types of arrows and Kanji characters. Each discrimination ratio was more than 80% [26]. The EEGs data were analyzed by the canonical discriminant analysis with the jack knife (cross validation) statistical technique in that case, speaking generally, their discriminant ratios are decreased. To cope with this phenomenon, we had improved discriminant ratios by adding another EEG channel (channel 2: Fp_2) [27, 28]. With this modification, discriminant ratios in spite of the jack knife method had been increased.

On the other hand, the super-aging society is highly advanced in Japan and other developed countries, as the term the brain death judgement indicates, the EEG are crucial communication tool among the human. And also, by use of EEGs with non-invasive technique is the safest toll to us. The present authors are planning to control a wheel chair by EEGs in this super-aging society, not only for the aged people but also for the handicapped people such as the muscle dystrophy etc. The present paper is a fundamental step for that. The authors have attempted to control the robot Palmi (DDM.com) by using EEGs. They made use of four electrodes attached on the right frontal area channel 2 (Fp_2), 4 (F_4), 6 (C_4) and 12 (F_8), so this fact is very convenient in making the attachment because the four electrodes on the channels lie very close, then the system is improved by three channels 4 (F_4), 6 (C_4) and 12 (F_8) and EEGs data are also reduced.

3 Measurement of EEGs on Recalling of Robot Movement Images

Subjects were six 22-year-old university students with normal visual acuity and right-handed. They wore an electrode cap with 19 active electrodes and watched visual stimuli that were presented on a 21-inch LCD monitor placed 30 cm in front of them. Subjects kept their heads steady by placing their chins on a chin rest fixed to an experimental table. Electrode positions were set on subject according to the international 10–20 system and two other electrodes were fixed on the upper and lower eyelids of the dominant eye for eye blink monitoring. Impedances were adjusted to less than 50kΩ. Reference electrodes were put on both earlobes and the ground electrode was attached to the base of the nose. Cannels of EEGs signals to be analyzed in the present paper are at most four, however, EEGs were taken by 19channels according to the 10–20 system for ECDL analysis in afterward for other researches.

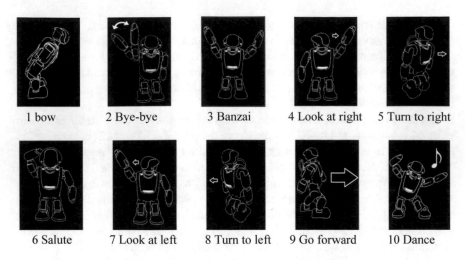

Fig. 1 Ten types images of Palmi robot movement presented in the experiment (Copyright FUJISOFT)

EEGs were recorded on a multi-purpose portable bio-amplifier recording device (Polymate, TEAC). The frequency band was set between 1.0 Hz and 2000 Hz. The outputs were transmitted to a recording computer and sampled at a rate of 1 kHz.

During the experiments, subjects were presented ten movement images of communication robot Palmi (Fig. 1). These ten images are fundamental movements of Palmi. In addition, we included images showing the opposite direction, e. g. No. 5 and No. 8 in Fig. 1 ('Turn to the right' and 'Turn to the left'), on the basis of the results from previous works [24, 25]. Each trial consisted of four periods, A: the first 3000 ms fixation period, B: the 2000 ms encoding period with a robot's movement image, C: the second fixation (delay) period for 3000 ms, and D: the final 2000 ms recalling period. During the recall period D, subjects were asked to recall the robot's image presented just before in period B. These images were randomly presented in order to prevent prejudges of the subjects, and measurements were repeated about ten times for each movement. EEGs were recorded in the encoding period B and in recalling period D (Fig. 2). In the present study, we used EEGs in the recalling period D especially, because they are considered to be less affected by the visual evoked potentials (VEP), and no need to present images to any subjects in a practical usage.

4   Discrimination of Single-Trial-EEGs

From the results of the precedent our research [24], the silent reading pathway with directional symbols was supposed to reach the right frontal area at the latency after 400 ms. Therefore, in the experiments, we also sampled EEGs from 400 to 900 ms at 25 ms intervals according to the precedent research [27]. We call it the data 1. We

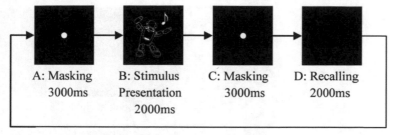

A: Masking	B: Stimulus	C: Masking	D: Recalling
3000ms	Presentation	3000ms	2000ms
	2000ms		

Repeated about 100 times

Fig. 2 Experimental procedure

also sampled the data from 399 to 899 ms (data 2) and from 398 to 898 ms (data 3). Each set of samples yields 21 data points from each channel for each sampling period. By these three sets of data, the number of sampling EEG data are tripled, we call these the tripled EEG data. We have been using this sampling to discriminate EEGs.

According to optimal number of the electrodes to be used in the discrimination at that time under those presented images, we have been using four channels 2 (Fp_2), 4 (F_4), 6 (C_4) and 12 (F_8) according to the International 10–20 system (Fig. 3 upper), since these points of channels lie on above mentioned the right frontal area. Initially in the previous study on the discrimination of the Single-Trial-EEGs, they made use of three channels Fp_2, F_4 and C_4 according to four types of directional symbols.

Although EEGs are time series data, we regarded them as vector-valued in the 84-dimensional space (4 channels × 21 time-points) (Fig. 3 lower). In this research, we used EEGs measured in the same time period, however, the sampling interval was changed from 25 to 50 ms.

5 Results of Discriminant Analysis for Tripled EEG Data

For each type of image recalling, the experiments were conducted about three hundred times, except for one subject about six hundred times. These data were resampled three times: in three sample-timings as described in Fig. 3, we call them tripled EEG data.

We have one set of criterion variable, i.e., ten types of images of Palmi's movement, and have 84 explanatory variates. Since explanatory variates consist of four channels with 21 sampling data, the learning data are 360 with 84 varieties. We applied so called jack knife statistics where we picked up one sample to discriminate and used the other samples as learning data; the method was repeated. For ten types of images recalling, the experiments were conducted several times (Fig. 4). In a similar manner, the data were resampled three times in three different sample time periods to triple the data.

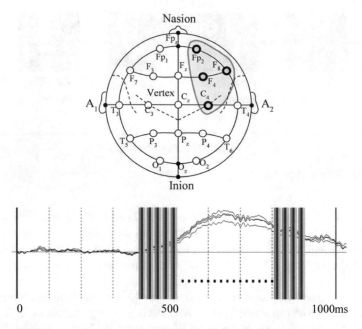

Fig. 3 Selected channels of EEGs and their sampling points. Upper: Selected 4 channels of International 10–20 system, Lower: Colored bold lines denote sampling point of the EEGs

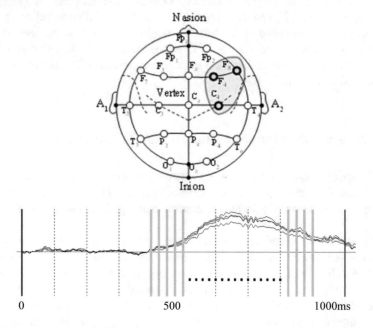

Fig. 4 Selected channels of EEGs and their sampling points. Upper: Selected 3 channels of International 10–20 system, Lower: Colored bold lines denote sampling point of the EEGs on data 1

Subject	Result
A	Success: 270 Trial: 300 Ratio: 90.00%
B	Success: 270 Trial: 300 Ratio: 90.00%
C	Success: 271 Trial: 300 Ratio: 90.33%
D	Success: 572 Trial: 600 Ratio: 95.33%
E	Success: 272 Trial: 300 Ratio: 90.67%
F	Success: 242 Trial: 270 Ratio: 89.63%

Table 1 Results of discrimination for subjects

We repeated the experiments with six students for a couple of days. Then, we tried to discriminate the Single-Trial-EEG samples for ten types of movement-images, where each canonical discriminant coefficients were determined for each subject and each series of the experiments. As a result, the discriminant ratios were more than 90% in all cases. Results are shown in Table 1 even in the worst case the discrimination rate was 89.63%.

6 Results of Discriminant Analysis for Simple EEG Data of Data 1

According to tripled EEG Data, we obtained satisfactory results, however, an optimization of the system is needed in order to make this BCI system into a product. On this point of view, we have tried to reduce the data. At the beginning of research on BCI, some of the present authors had started by three electrodes and by simple sampling as data 1 [27]. So, we go back the original data sampling mode and tried to discriminate the Single-Trial-EEGs data: data 1. The results are shown in Table 2. Although the number of the Single-Trial-EEGs data to discriminate is decreased to a quarter of the tripled mode, result of dicriminant rates are slightly imploved.

Table 2 Results of Discrimination for Subjects

Subject	Result
A	Success: 91 Trial: 100 Ratio: 91.00%
B	Success: 91 Trial: 100 Ratio: 91.00%
C	Success: 90 Trial: 100 Ratio: 90.00%
D	Success: 191 Trial: 200 Ratio: 95.50%
E	Success: 90 Trial: 100 Ratio: 90.00%
F	Success: 82 Trial: 90 Ratio: 91.11%

7 Concluding Remarks

We tried to control a robot by using Single-Trial-EEG data taken from the six subjects when they are recalling images of the movements of a robot. We call this system ImageRobo.

The canonical discriminant analysis with the jack knife method for ten objective variates achieved was that each discriminant rate was more than 90% for the four subjects for both of two ways of sampling, i. e. four electrodes and three electrodes. In order to increase the efficiency and accuracy of control, we have parallelly revised the method of data collection and processing, by comparing with two types of sampling of Single-Trial-EGs method, triple and single sampling. And it is confirmed that the single sampling EEGs by three channels is efficient to control ten types of images of robot movement.

As a result, we could control the robot Palmi wirelessly from PC in which EEGs are stored previously. Although the present work depends on off line EEGs data, theoretically we could improve our system for taking EEGs directly from subjects. The communication robot Palmi has Wi-Fi network communication, we could use this and send the 10 discriminated result to Palmi and control its movement.

Acknowledgements This research was partially supported by a grant from the Ministry of Education, Culture, Sports, Science and Technology for the national project of the High-tech Research Center of Hokkai-Gakuen University ended in March 2013. The experiments were approved by the ethical review board of Hokkaido University. The authors express their gratitude to former under graduate students of Yamanoi Laboratory in Hokkai-Gakuen University, especially Mr. Shunki Ito, Mr. Kaito Sato, and Mr. Tatsuki Sato, for their assistance in the EEG experiments.

References

1. Kawamoto, H., Kamibayashi, K., Nakata, Y., Yamawaki, K., Ariyasu, R., Sankai, Y., Sakane, M., Eguchi, K., Ochiai, N.: Pilot study of locomotion improvement using hybrid assistive limb in chronic stroke patients. BMC Neurol., 13–141 (2013)
2. Serruya, M.D., Hatsopoulos, N.G., Paninski, L., Fellows, M.R., Donoghue, J.P.: Instant neural control of a movement signal. Nature 416(6877), 141–2 (2002)
3. Tamaki, E., Miyaki, T., Rekimoto, J.: PossessedHand: techniques for controlling human hands using electrical muscles stimuli. Conference: Proceedings of the International Conference on Human Factors in Computing Systems, CHI 2011, Vancouver, BC, Canada, May 7–12 (2011)
4. Donoghue, J.P.: From mind to movement: Neurotechnologies to reconnect the brain to the world. 16th International Symposium on ALS/MND, p. 64 (2005)
5. Hochberg, L.R., et al.: Neuronal ensemble control of prosthetic devices by a human with tetraplegia. Nature 442, 164–171 (2006)
6. Hochberg, L.R., Bacher, D., Jarosiewicz, B., Masse, N.Y., Simeral, J.D., Vogel, J., Haddadin, S., Liu, J., Cash, S.S., van der Smagt, P., Donoghue, J.P.: Reach and grasp by people with tetraplegia using a neurally controlled robotic arm. Nature 485, 372–375 (2012)
7. Abdulkader, S.N., Atia, A., Mostafa, M.-S.: Brain computer interfacing: applications and challenges. Egypt. Inf. J. 16(2), 213–230 (2015)
8. Gulati, T., Won, S.J., Ramanathan, D.S., Wong, C.C., Bodepudi, A., Swanson, R.A., Ganguly, K.: Robust Neuroprosthetic control from the stroke perilesional cortex. J. Neurosci. 35(22), 8653–8661 (2015)
9. Mao, X., Li, M., Li, W., Niu, L., Xian, B., Zeng, M., Chen, G.: Progress in EEG-based brain robot interaction systems. Comput. Intell. Neurosci. (2017)
10. Rosca, S., Leba, M., Ionica, A., Gamulescu, O.: Quadcopter control using a BCI. IOP Conference Series: Materials Science and Engineering, Vol. 294, International Conference on Applied Sciences (2017)
11. Remsik, A., Young,B., Vermilyea, R., Kiekoefer, L., Abrams, J., Elmore, S.E., Schultz, P., Nair, V., Edwards, D., Williams, J., Prabhakaran, V.: A review of the progression and future implications of brain-computer interface therapies for restoration of distal upper extremity motor function after stroke. Expert Rev. Med. Dev. 13(5), 445–454 (2016)
12. Monge-Pereira, E., Ibañez-Pereda, J., Alguacil-Diego, I.M., Serrano, J.I., Spottorno-Rubio, M.P., Molina-Rueda, F.: Use of electroencephalography brain-computer interface systems as a rehabilitative approach for upper limb function after a stroke: a systematic review. PM&R 9(9), 918–932 (2017)
13. Sabathiel, N., Irimia, D.C., Allison, B.Z., Guger, C., Edlinger, G.: Paired associative stimulation with brain-computer interfaces: a new paradigm for stroke rehabilitation. foundations of augmented cognition. Neuroergonomics and Operational Neuroscience. Lecture Notes in Computer Science, pp. 261–272 (2016)
14. Chatelle, C., Chennu, S., Noirhomme, Q., Cruse, D., Owen, A.M., Laureys, S.: Brain–computer interfacing in disorders of consciousness. Brain Inj. 26(12), 1510–1522 (2012)
15. Boly, M., Massimini, M., Garrido, M.I., Gosseries, O., Noirhomme, Q., Laureys, S., Soddu, A.: Brain connectivity in disorders of consciousness. Brain Connectivity 2(1), 1–10 (2012)
16. Gibson, R.M., Ferná¡Ndez-Espejo, D., Gonzalez-Lara, L.E., Kwan, B.Y., Lee, D.H., Owen, A.M., Cruse, D.: Multiple tasks and neuroimaging modalities increase the likelihood of detecting covert awareness in patients with disorders of consciousness. Front. Human Neurosci. 8, 950 (2014)
17. Risetti, M., Formisano, R., Toppi, J., Quitadamo, L.R., Bianchi, L., Astolfi, L., Cincotti, F., Mattia, D.: On ERPs detection in disorders of consciousness rehabilitation. Front. Hum. Neurosci. 7, 775 (2013)
18. Ritaccio, A., Brunner, P., Gunduz, A., Hermes, D., Hirsch, L.J., Jacobs, J., Kamada, K., Kastner, S., Knight, R.T., Lesser, R.P., Miller, K., Sejnowski, T., Worrell, G., Schalk, G.: Proceedings of the fifth international workshop on advances in Electrocorticography. Epilepsy Behav. 41, 183–192 (2014)

19. McCarthy, R.A., Warrington, E.K.: Cognitive neuropsychology: a clinical introduction. Academic Press, San Diego (1990)
20. Geschwind, N., Galaburda, A.M.: Cerebral Lateralization. The Genetical Theory of Natural Selection. Clarendon Press, Oxford (1987)
21. Parmer Computer, K., Hansen, Kringelbach, M.L., Holliday, I., Barnes, G., Hillebrand, A., Singh, K.H., Cornelissen, P.L.: Visual word recognition: the first half second. NeuroImage 22–4, 1819–1825 (2004)
22. Iwata, M., Kawamura, M., Otsuki, M. et al.: Mechanisms of writing, Neurogrammatology (in Japanese), IGAKU-SHOIN Ltd, pp. 179–220 (2007)
23. Yamazaki, T., Kamijo, K., Kiyuna, T., Takaki, Y., Kuroiwa, Y., Ochi, A., Otsubo, H.: PC-based multiple equivalent current dipole source localization system and its applications. Res. Adv. Biomedi. Eng. **2**, 97–109 (2001)
24. Yamanoi, T., Toyoshima, H., Ohnishi, S., Yamazaki, T.: Localization of brain activity to visual stimuli of linear movement of a circle by equivalent current dipole analysis (in Japanese). Proceeding of the 19th Symposium on Biological and Physical Engineering, pp. 271–272 (2004)
25. Yamanoi, T., Yamazaki, T., Vercher, J.-L., Sanchez, E., Sugeno, M.: Dominance of recognition of words presented on right or left eye-Comparison of Kanji and Hiragana-," Modern Information Processing from Theory to Applications, Bouchon-Meunier, B., Coletti, G., Yager, R.R. (Eds.), Elsevier Science B.V., pp. 407–416 (2006)
26. Anderson, T.W.: An introduction to multivariate statistical analysis. Wiley, New York (1984)
27. Yamanoi, T., Toyoshima, H., Yamazaki, T., Ohnishi, S., Sugeno, M., Sanchez, E.: Micro robot control by use of electroencephalograms from right frontal area. J. Adv. Computat. Intell. Intell. Inf. **13**(2), 68–75 (2009)
28. Yamanoi, T., Toyoshima, H., Yamazaki, T., Ohnishi, S.: Discrimination of playing cards by EEGs during recognizing and recalling: Playing cards estimation magic without a trick. J. Japan Soc. Fuzzy Theory Intell. Inf. **28**(3), 639–646 (2016)

Automatic Synthesis of Rule Bases of Fuzzy Control Systems Based on Genetic Algorithms

Oleksiy Kozlov and Yuriy Kondratenko

Abstract This chapter is dedicated to the method of the rule base (RB) automatic synthesis for the fuzzy automatic control systems (FACS) of Mamdani-type with the determination of optimal consequents based on the genetic algorithms. The obtained method allows to effectively synthesize the rule bases with optimal consequents for Mamdani-type FACSs at insufficient initial information (in conditions of a high degree of information uncertainty), at a sufficiently large number of rules when the compilation of the FACS rule base on the basis of expert knowledge is not always effective, and at the various levels of experts qualifications. The study of the efficiency of the proposed method is carried out at the development of the rule base of the fuzzy controller of Mamdani-type for the automation system of the multipurpose caterpillar mobile robot (MR) with the ability of moving on vertical and inclined ferromagnetic surfaces. The results of computer simulation confirm the high effectiveness of the designed method.

Keywords Fuzzy control system · Rule base · Synthesis method · Genetic algorithm · Fuzzy controller · Multipurpose mobile robot

1 Introduction

Artificial intelligence systems based on fuzzy logic and the fuzzy sets theory are widely used in various fields of science and technology to solve problems of control, identification, modeling of complex physical phenomena, classification, clustering, pattern recognition, etc. [1–3]. These systems allow to effectively use expert information, formalize the processes of human decision-making and thinking, as well as

O. Kozlov (✉) · Y. Kondratenko
Department of Intelligent Information Systems, Petro Mohyla Black Sea National University, Mykolaiv, Ukraine
e-mail: kozlov_ov@ukr.net

Y. Kondratenko
e-mail: yuriy.kondratenko@chmnu.edu.ua

© The Author(s), under exclusive license to Springer Nature Switzerland AG 2023
Sh. N. Shahbazova et al. (eds.), *Recent Developments and the New Directions of Research, Foundations, and Applications*, Studies in Fuzziness and Soft Computing 422,
https://doi.org/10.1007/978-3-031-20153-0_9

form linguistic models of complex objects and processes [4–6]. According to the fuzzy approximation theorem [7], any arbitrarily complex mathematical dependence can be approximated by a fuzzy system. This allows reproducing with high accuracy arbitrary input–output dependencies using natural language statements-rules of the type "IF ..., – THEN ...", with their subsequent formalization using algorithms of the theory of fuzzy sets, without the use of complex calculations based on differential and integral equations [8–10].

Most often, fuzzy systems (FS) of the Mamdani [11–13] and Takagi–Sugeno [14] types are used for automatic control of complex nonlinear and non-stationary plants as well as for decision making in conditions of uncertainty [11]. To formalize expert information, these FSs use various (a) linguistic variables (for example, "coolant temperature", "vehicle speed", "operator fatigue level"), (b) linguistic terms (for example, "Low", "Very big", "Above average"), (c) membership functions (for example, trapezoidal, triangular, π-like), (d) the rule base consisting of antecedents and consequents, as well as (e) a fuzzy logic inference engine that includes consecutive steps of fuzzification, aggregation, activation, accumulation and defuzzification [13]. Herewith, the main difference between the Mamdani and Takagi–Sugeno systems is the construction of rule consequents. So, for Mamdani-type FSs, rule consequents are specified in the form of linguistic terms (LT) of output variables with time-invariant membership functions (MF) of various types (triangular, trapezoidal, etc.) [15]. In turn, impulse-type membership functions are used as consequents of the rules of Takagi–Sugeno systems, which are sums of weighted instant values of all input variables [16–18].

The successful usage of fuzzy automatic control systems with inferences of Takagi–Sugeno and Mamdani types requires solving two main problems: (1) determining the optimal structure of the FACS; (2) finding the optimal parameters of the FACS [19]. The first problem includes the choice of: the number of inputs and outputs of the FACS; the numbers of linguistic terms for all input and output variables; the number of the RB rules; types of membership functions for each term; types of procedures of defuzzification, aggregation, activation and accumulation [20]. The second problem, in turn, consists of (a) determining the parameters of the LT membership functions for all inputs and outputs and normalizing coefficients to convert them to relative units, as well as (b) synthesis of rule consequents for FACS [13].

In many cases, the optimal structure and parameters of the FACS for automation of a particular plant are determined using some basic recommendations, the expert knowledge, and assessments [11]. Herewith, the subjective factor has a significant impact on the process of FACS developing. Under conditions of insufficient full amount of initial information and expert knowledge, as well as when making erroneous design decisions, the efficiency of the FACS can be significantly reduced or their functioning will be performed with understated, in terms of potential capabilities, indicators [21–23]. To increase the efficiency of the FACS functioning, as well as to reduce the negative impact of subjective factors on the design process, scientists from around the world have recently developed and implemented methods of FACS synthesis and training, which are based on certain optimization procedures [24–26]. In particular, the methods and algorithms of FACS structural optimization using the choice of optimal types of the LT membership functions, methods of defuzzification, RB reduction and interpolation are presented in [27–29]. In turn, the designing methods that include parametric optimization procedures for the LT membership functions of the systems of Mamdani type, as well as synthesis of weight coefficients for the rule consequents of Takagi–Sugeno FACSs are given in [30–32]. The results of published studies show that intelligent methods of multi-agent and evolutionary optimization [33–35], which include genetic methods [36, 37], evolutionary strategies [38] as well as methods, that model the interacting behavior of collective animals, insects and microorganisms, are quite promising for solving the problems of the FACS synthesis and training [39–41]. These methods are based of stochastic global optimization which have a number of advantages compared with the classical optimization methods: (a) do not impose any restrictions on the objective functions; (b) allow detailed study of large, nonsmooth, multimodal search space, eliminating the possibility of looping at local extremes; (c) effectively optimize fuzzy systems of large dimension [33, 37].

The problem of synthesis of the optimal RB consequents for the FACSs of the Mamdani type in the absence or insufficiently full amount of expert knowledge about real control processes [42, 43] is a complicated optimization problem, for the solution of which the usage of the effective stochastic methods of global optimization is most expedient.

This paper is devoted to the development and study of the method of RB automatic synthesis for the fuzzy control systems of the Mamdani-type with the determination of optimal consequents based on the use of algorithms of evolutionary optimization, in particular genetic. The paper is organized in the following way. Section 2 presents the statement of the research problem, the purpose of this work, and the analysis of publications in the researched area. Section 3 sets out in detail the theoretical approach to the RB automatic synthesis method developed by the authors, which is presented as a sequential computational algorithm. Section 4 presents the results of the effectiveness study of the designed method on a specific example of FACS for the speed of the multi-purpose vertical moving mobile robot, with a detailed discussion of the results of computer simulation. The conclusions and references list, in turn, are given at the end of this paper.

2 Problem Statement and Related Works

Currently, the advanced FACSs use fuzzy inference devices to solve various tasks, in particular, as control devices (controllers), identifiers, observers, modes setting blocks of upper control level, etc. [44]. The generalized SISO fuzzy automatic control system with the fuzzy controller (FC) and automated designing unit (ADU) is presented in Fig. 1 [30], where the following notations are adopted: EO is the expert-operator; OFCU is the objective function calculating unit; SPOU is the structural-parametric optimization unit; SD is the setting device; ISCU is the input signals calculating unit; FU is the fuzzification unit; FIE is the fuzzy inference engine;

AGG, ACT and ACC are the aggregation, activation and accumulation operations; DFU is the defuzzification unit; CP is the control plant; S is the sensor; y_S and y_R are the set and real values of the control variable; $\mathbf{X}_{OF}$ is parameters vector of the objective function J; $\mathbf{X}_0$ is the preliminary value (preliminary hypothesis) of the vector $\mathbf{X}$ that determines the FC structure and parameters; ε is the FACS control error; $\mathbf{X}_{IS}$ is the vector that defines the FC input signals vector $\mathbf{E}$, which is calculated on the basis of the error ε; $\mathbf{X}_{IT}$ is the vector that determines the number, types and parameters of linguistic terms of the FC input variables; $\mathbf{R}_X$ is the RB consequents vector; $\mathbf{X}_{FIE}$ is the vector that defines the operations of aggregation, activation and accumulation of the fuzzy inference engine; $\mathbf{X}_{DF}$ is the vector that determines the number, types and parameters of linguistic terms of the FC output variable u_{FC}, as well as the defuzzification method; $\mathbf{F}_D$ is the vector of disturbances that affect the control plant; u_{SD} and u_S are the output signals of the FACS setting device and the sensor respectively.

In the presented fuzzy control system (Fig. 1) the input signals calculating unit ISCU, based on the vector $\mathbf{X}_{IS}$ forms the vector $\mathbf{E}$, which can consist of various combinations of error ε signals, its derivatives of various orders, integral, etc. For example, the vector $\mathbf{E}$ can be represented by the following expression

$$E = \left\{ k_P \varepsilon, k_D \dot{\varepsilon}, k_I \int \varepsilon dt \right\}, \tag{1}$$

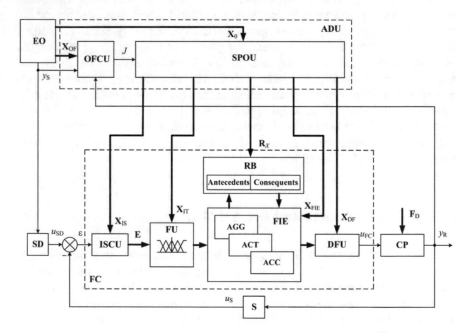

Fig. 1 Structure of the generalized FACS with FC and ADU

where k_P, k_D and k_I are the normalizing coefficients of the $\mathbf{X}_{IS}$ vector that are applied to convert the FC input signals into relative units from their maximum values.

The fuzzification unit FU defines the membership degrees of the numerical values of the input signals vector $\mathbf{E}$ to the corresponding FC input linguistic terms [16], the number, types and parameters of which are determined by the vector $\mathbf{X}_{IT}$. In turn, the fuzzy inference engine FIE, based on the fuzzified signals and received data from the RB, sequentially performs aggregation, activation, and accumulation operations, the types of which are specified by the $\mathbf{X}_{FIE}$ vector [28]. The FC rule base is a set of rules consisting of certain antecedents and consequents. Moreover, for the Mamdani-type FC with an input vector (1), the RB rules are set on the basis of the expression

$$\text{IF } k_P\varepsilon = A_1 \text{ AND } k_D\dot{\varepsilon} = A_2 \text{ AND } k_I \int \varepsilon dt = A_3 \text{ THEN } u_{FC} = A_4, \qquad (2)$$

where A_1, A_2, A_3 and A_4 are corresponding linguistic terms of the FC input and output signals.

The defuzzification unit DFU implements the conversion of a consolidated fuzzy inference into a crisp numerical FC output signal [19].

The problem of the fuzzy controller designing of this FACS (Fig. 1) is reduced to sequential or parallel finding the optimal values of the $\mathbf{X}_{IS}$, $\mathbf{X}_{IT}$, $\mathbf{R}_X$, $\mathbf{X}_{FIE}$, $\mathbf{X}_{DF}$ vectors by means of the ADU, which includes the calculation unit of the objective function OFCU, as well as the structural-parametric optimization unit SPOU [32]. Herewith, the expert-operator EO sets only the parameters vector of the objective function $\mathbf{X}_{OF}$ and the initial values (initial hypotheses) $\mathbf{X}_0$ vector of the FC structure and parameters.

In this case, one of the most important and complicated tasks is the rule base synthesis with the determination of the optimal consequents vector $\mathbf{R}_X$ [42, 43].

The rules antecedents are various combinations of linguistic terms of the FC input variables [19], and the consequent LT_r of each r-th RB rule is selected from the set of all possible rules consequents $\{LT^1, LT^2, \ldots, LT^\nu\}$, which includes ν linguistic terms of the FC output variable u_{FC}

$$LT_r \in \{LT^1, LT^2, \ldots, LT^\nu\}. \qquad (3)$$

For example, at choosing the number of linguistic terms of the FC output variable ν = 5 and $\nu = 7$, the following sets can be formed

$$LT_r \in \{LT^1, LT^2, LT^3, LT^4, LT^5\} = \{\text{BN, SN, Z, SP, BP}\}; \qquad (4)$$

$$LT_r \in \{LT^1, LT^2, LT^3, LT^4, LT^5, LT^6, LT^7\} = \{\text{BN, N, SN, Z, SP, P, BP}\}, \qquad (5)$$

where BN is big negative; N is negative; SN is small negative; Z is zero; SP is small positive; P is positive; BP is big positive.

The consequents vector $\mathbf{R}_X$ can be formed in different ways, herewith the task of the consequents optimizing is reduced to the task of finding the optimal vector of consequents $\mathbf{R}_{opt}$ from the set of all possible alternative variants, which provides optimal quality indicators of the fuzzy control system [42].

The consequents vector $\mathbf{R}_X$ for the x-th alternative variant of the rule base in general terms can be represented as follows [43]

$$\mathbf{R}_X = \{LT_{X1}, LT_{X2}, \ldots, LT_{Xr}, \ldots, LT_{Xs}\},$$
$$LT_{Xr} \in \{LT^1, LT^2, \ldots, LT^v\}, X \in \{1, 2, \ldots, v^s\}, \tag{6}$$

where v^s is total number of all possible variants of the vector $\mathbf{R}_X$, which is defined as a number of the output variable LT v, raised to the power of the total number of the RB rules s.

Thus, the task of synthesis of the fuzzy system RB consequents is reduced to finding such a vector of the RB consequents $\mathbf{R}_X = \mathbf{R}_{opt}$, at which the value of the objective function J of the FACS will be optimal ($J = J_{opt}$) [42].

This task is a complicated task of discrete optimization of large dimension, the solution of which requires an effective stochastic method of global search that takes into account the peculiarities of the consequents formation of the RB rules in the conditions of information uncertainty [32, 43]. At solving this task by exhaustive search of all possible vectors $\mathbf{R}_X$ ($X = 1, 2, \ldots, v^s$) the value of the objective function J will need to be calculated v^s times, which even with a small dimension of the RB will require significant computational and time costs [43]. For example, for the FC with a sufficiently simple structure with nine rules ($s = 9$) in the rule base and three linguistic terms of the output variable ($v = 3$) the value of the objective function J during the exhaustive search of the vector $\mathbf{R}_X$ need to be calculated $3^9 = 19{,}683$ times, which will take a lot of time when using significant computing resources.

Analysis of existing intelligent stochastic methods of global optimization shows that genetic methods (genetic algorithms) are quite effective for solving complex discrete optimization problems of large dimension [32, 33, 43]. A number of published works present examples of the successful application of various genetic algorithms for solving discrete optimization problems such as: traveling salesman problem [45–47], scheduling problem [48, 49], optimization of vehicle routes [50–52], graph coloring problem [53–55], etc. [56–58]. There are also many examples of the effective use of genetic algorithms for optimization of various fuzzy control systems [59–61], in particular, fuzzy control systems of industrial and mobile robots [62–64], heat power plants [65, 66], water and air vehicles [67, 68] and others [69, 70]. Thus, it is expedient to carry out the synthesis of the RB consequents of the fuzzy control systems on the basis of genetic algorithms, adapting them to the specifics of this task.

The main purpose of this work is development and study of the method of the rule base automatic synthesis for the Mamdani-type fuzzy control systems with the formation of optimal consequents based on the genetic algorithms. Application of the method will allow to synthesize the rule bases with optimal consequents for

Mamdani-type FACSs at insufficient initial information (at a high degree of uncertainty of information), if the number of rules is quite large, so that the compilation of the FACS rule base on the basis of expert knowledge is not always effective, and at the various levels of expert qualifications.

3 Method of Rule Bases Automatic Synthesis for Fuzzy Control Systems Based on GA

Genetic algorithms are the techniques of the heuristic search that allow solving optimization tasks using operations of random selecting, varying, and combining search parameters that are similar to natural selection [33, 35]. They relate to evolutionary computation algorithms that solves optimization problems using methods of natural evolution, such as inheritance, mutations, selection, and crossover. A distinctive feature of the genetic algorithm is the emphasis on the use of the "crossover" operator, which performs the operation of recombination of candidate solutions, whose role is similar to the role of crossing in wildlife [38, 51]. Genetic algorithms operate with populations consisting of a finite set of individuals at solving the optimization problem. The individuals included in the population are represented by the chromosomes with sets of task parameters encoded in them, i.e. solutions, that are also can be called points in the search space [51, 53]. In turn, each chromosome consists of a number of genes, each of which is the specific parameter of the optimization problem. To evaluate the chromosome fitness and correspondingly the goodness of the optimization task solution the fitness function is used.

The generalized genetic algorithm includes the next steps [58]:

(1) initialization or selection of the initial population of individuals;
(2) evaluation of the fitness of chromosomes in a population;
(3) checking the stopping condition of the algorithm;
(4) selection of chromosomes;
(5) implementation of genetic operators;
(6) formation of a new population;
(7) choice of the "best" chromosome.

In this work, to adapt the genetic algorithm for solving the problem of automatic synthesis of the rule base for the fuzzy control systems the RB consequents vector $\mathbf{R}_X$ is represented as a chromosome, each gene of which is the specific consequent LT_r ($r = 1, 2, \ldots, s$), that is selected from the set of all linguistic terms v of the fuzzy control system output variable. The fitness function calculation is implemented on the basis of the FACS objective function J. Since, genetic algorithms are function maximization algorithms, then to minimize the the objective function J value of the fuzzy control system, the fitness function f should be calculated as follows

$$f = \frac{1}{J}. \tag{7}$$

In turn, as the FACS objective function J, for example, the mean integral quadratic control error can be selected

$$J(t, \mathbf{R}_X) = \frac{1}{t_{\max}} \int\limits_0^{t_{\max}} \varepsilon^2 dt \rightarrow \min, \tag{8}$$

where $t_{\max}$ is the FACS total transient time.

Thus, the proposed method of the rule base automatic synthesis for the Mamdani-type fuzzy control systems with the determination of optimal consequents using the genetic algorithms consists of the following sequential steps.

Step 1. Method initialization. At this stage, the structure of the FACS's rule base is constructed, as well as parameters and genetic operators types of the genetic algorithm are selected. Also, structure, parameters and desired value for the FACS's objective function, as well as the optimization expiration criterion are selected. Herewith, the structure of the FACS rule base is constructed based on a preliminary selection of the LTs number for all input variables [35]. The rules total number s is defined by all the possible combinations of LTs of input variables and the rules consequents LT_r can be selected from the set of all possible terms v of the FACS output variable. In turn, for the selection genetic operator it is advisable to use the proportional selection, for the crossover is the one-point crossover operator and for the mutation operator—the simple mutation, for which a gene is randomly selected in the mutating chromosome, and then randomly replaced. Also, at this stage the crossover P_C and mutation P_M probabilities, as well as the size of initial population Z are specified.

The mean integral quadratic control error (8) or the integral deviation of the actual transient response of the FACS from the desired transient response of its reference model (RM) can be chosen as the objective function J. In, turn, the fitness function f is calculated based on the objective function J according to the Eq. (7).

As for the optimization expiration criterion, achievement of the following conditions can be used.

Condition 1. Achievement of the fitness function f_{opt} value optimal of the FACS. If the fitness function value f of some chromosomes reaches f_{opt} in the search process of GA, then the expiration criterion is reached.

Condition 2. Performing the previously set maximum iterations number $n_{\max}$. In this case, if the iterations number $n = n_{\max}$, then the expiration criterion is also reached.

Condition 3. Population degeneration. In this case, the improvement factor ρ for the population is calculated using the Eq. (9) based of the fitness functions of the best individual. If this improvement coefficient ρ is less than initially set threshold ρ_p, then the expiration criterion is achieved, and the genetic algorithm operation is stopped. Such checking is implemented at every iteration starting from n_p, which is also set at the initialization stage.

$$\rho = \frac{f_{best(n)} - f_{best(n-1)}}{f_{best(n-1)}}, \tag{9}$$

where $f_{best(n)}$ is the fitness function best value f at the iteration n; $f_{best(n-1)}$ is the fitness function best value f at the previous iteration $n-1$.

Step 2. Encoding of the RB consequents vector into the chromosome. At this stage the RB consequents vector $\mathbf{R}_X$ is encoded into the population chromosome that include specific genes. In turn, each gene corresponds to the specific consequent LT_r ($r = 1, 2, \ldots, s$). In this case, real encoding is used [33].

Step 3. Generation of the basic population that consists of individuals. Initially, the basic population is generated that include Z individuals, each of which corresponds to Z-th different variant (among all existing variants) of the RB consequents vector $\mathbf{R}_X$ for the FACS controller. The given population P_0 created at the first iteration is presented in the following form:

$$P_0 = \{H_1, H_2, \ldots, H_j, \ldots, H_Z\}, \tag{10}$$

where H_j is the j-th chromosome of the basic population; Z is the total number of individuals in the population.

In most cases the values of the chromosomes genes are set randomly for the initial population. In the presence of one or several preliminary hypotheses of the RB consequents values obtained on the basis of expert knowledge, the chromosomes with genes corresponding to these consequents are added to the created population P_0.

Step 4. Determination of the individuals' fitness for the current population. The individuals' fitness is defined using the next steps.

Step 4.1. Individuals decoding. The decoding of each population individual into the RB consequents vector $\mathbf{R}_X$ is implemented at this stage.

Step 4.2. Building of the current rule base for the FACS. At this stage the rule bases with the consequents vectors that corresponds to the population individuals are built.

Step 4.3. Calculation of the objective (fitness) function of the FACS. At this stage simulation of the FACS with developed rule base is carried out and the value of the fitness function f is obtained for all population individuals. Herewith, the FACS modeling for each j th individual ($j = 1,.., Z$) is performed considering all existing input and disturbing actions as well as other operating conditions. The fitness function f is calculated on the basis of the objective function J according to the Eq. (7).

Step 5. Checking of the optimization stopping criteria. The optimization process is stopped when one of the conditions of the expiration criterion, given in the *Step 1*, occurs. When the *Condition 1* occurs, it is necessary to move to *Step 10*. If the stopping criterion is defined by *Conditions 2* or *3*—go to *Step 1*.

Step 6. Individuals' selection for crossover. To create new variants of the rule base vector the individuals of the current population are selected at this step based of the fitness function value f for implementation the crossover operation. The operation of the proportional selection [38] is used, that consists of the next stages.

Step 6.1. Determination of the average fitness function value over the current population. The average fitness function f_M value over the current population is defined as the arithmetic mean value from all the chromosomes fitness functions [43],

$$f_M = \frac{1}{Z} \sum_{j=1}^{Z} f_j. \tag{11}$$

Step 6.2. Determination of the coefficient of selection. The coefficient $P_S(j)$ of the selection operation is defined for each chromosome of the current population by means the Eq. (12) [38]

$$P_S(j) = \frac{f_j}{f_M}. \tag{12}$$

Step 6.3. Creation of the individuals array for the crossover. If the coefficient of selection $P_S(j) > 1$, then this chromosome is sufficiently well adapted and can be allowed for implementation of the crossover operation. The array of such selected chromosomes is created at this stage.

Step 7. Individuals' crossover. The operation crossover is implemented at this step for all the individuals of the current population, that are previously selected and admitted. In turn, the parent couples are chosen randomly [51].

Step 7.1. Individuals' numbering. The selected chromosomes' numbering is performed arbitrarily.

Step 7.2. Parents' selection. The random number is selected in the interval [0, 1] for all the admitted individuals, starting from the first individual of the population. If this randomly selected number is greater than or equal to the crossover probability P_C (previously selected at *Step 1*), then the given individual is considered as the first parent in the parent couple. The next individual of the population, for which the crossover probability P_C is less than or equal to its randomly generated number will be the second parent. This stage with the generation of random numbers is performed cyclically until all the parent pairs of the current population are compiled.

Step 7.4. Implementation of the crossover operation for the admitted pairs. The one-point crossover operation is implemented at this step for all selected parent pairs [33]. The point of crossing for each chromosomes' pair is defined in a random way.

Step 8. Individuals' mutation. The operation mutation is implemented at this step for all the individuals of the current population, that are selected at *Step 6* on the basis of the mutation probability value P_M.

Step 8.1. Parent individuals' copying to the successor individuals. The parental individuals that are allowed for operation mutation are copied to the successor individuals.

Step 8.2. The mutant gene selection. The gene that will mutate in each individual (the consequent of the *r*-th rule), admitted to mutation, is chosen randomly [53].

Step 8.3. Determination of the mutant gene new value. The new value of the mutating gene is defined randomly from the set of all possible linguistic terms v of the FACS output variable.

Step 9. New generation creation. The new generation is created of the successor individuals that are obtained with the help of the crossover and mutation operations. The elite individuals are also included in this generation. Then go to *Step 4*.

Step 10. Implementation and experimental research of the FACS with the developed rule base. The implementation and researching of the FACS with the developed rule base is performed at this stage.

To conduct the efficiency research of the proposed method of the rule base automatic synthesis of the FACSs of Mamdani-type with creation of the optimal vector of consequents, the development of the RB of the speed FACS for the industrial mobile robot with caterpillar ferromagnetic movers [16] is performed in this paper.

4 Synthesis of the Rule Base of the Speed FACS for the Industrial Mobile Robot with Caterpillar Ferromagnetic Movers

The industrial mobile robots with caterpillar ferromagnetic movers at a proper automatic control can be quite effectively used for implementation of different technological tasks in various branches of industry (mechanical engineering, petroleum industry, shipbuilding, agriculture, ship and aircraft repair, and others) [16, 18]. They can move with different working equipment on inclined metal surfaces of a large area due to the magnetic caterpillar movers for solving tasks of inspection, ultrasonic diagnostics, staining, corrosion cleaning, spot welding of seams, etc. Moreover, the given MRs allow moving along complex trajectories with implementation of complicated hazardous types of work without human intervention, as well as in hard-to-reach places [16].

Herewith, the industrial mobile robots of such type must function in randomly changing conditions, as well as with incomplete initial information, since in the process of movement and performance of basic operations, various random disturbances can occur (surfaces of different types, coatings, thicknesses and roughness;

various irregularities and sharply arising obstacles; unevenly acting loads from technical operations, etc.) [18]. Thus, automatic control of the linear motion speed in conditions of various disturbances action is the quite important automation task of MRs of such type. The control system of the multi-purpose MR based on the fuzzy controller has the same structural organization as the generalized FACS shown in Fig. 1. As a controlled coordinate y in this system acts the movement speed of MR V_{MR}.

The development of the speed FACS is carried out for the industrial mobile robot that has the next basic technical specifications: the full equipped weight of the robot is 150 kg, the MR's width and length are 0.8 m and 1 m, the radius of the driving wheel of robot is 0.15 m. Also, the studied MR is equipped with 2 drive electric motors 2PB132MH which gear ratio is 105. The problem of automatic control and stabilization of the linear speed of movement along an inclined surface was solved for a set value of 0.2 m/s [18]. The computer model of the robot for all studies is built on the basis of the next main expressions

$$
\begin{aligned}
M_{EMa} = & \frac{D_M}{\eta_M} \frac{d\omega_M}{dt} + \frac{1}{k_R \eta_{MR}} \left[D_{\sum W} \frac{d\omega_W}{dt} + R_W \left(\zeta \frac{G \cos\gamma + F}{2} + G \sin\gamma \left(\frac{\cos\varphi_{MR}}{2} + \right. \right. \right. \\
& \left. + b \frac{x_O}{B} \sin\varphi_{MR} + b\zeta \frac{h_C}{B} \sin\varphi_{MR} \right) - b \frac{\mu_T L (G \cos\gamma + F)}{4B} \left(1 + \frac{4x_O^2}{L^2} \right) + \\
& + F_{TO} \left(\frac{1}{2} \cos\beta - \frac{(x_1 - x_O)}{B} \sin\beta \right) \\
& + \left(\frac{G \cos\gamma + F}{2g} + bG \sin\gamma \sin\varphi_{MR} \frac{h_C}{gB} \right) \frac{dV_{ca}}{dt} + \\
& + \left(m_{MR} + \frac{F}{g} \right) \frac{L^2 + B^2}{12B} \frac{d\omega_{MR}}{dt} \bigg) \bigg],
\end{aligned}
\tag{13}
$$

$$
F = \frac{F_{TO} + G(\zeta \cos\gamma + \sin\gamma - \xi \cos\gamma) + m_{MR} c\lambda}{\xi - \zeta};
\tag{14}
$$

$$
\omega_{MR} = V_{MR}/R_T = (V_{C2} - V_{C1})/B;
\tag{15}
$$

$$
V_{MR} = (V_{C2} + V_{C1})/2;
\tag{16}
$$

$$
V_{C1} = \omega_{MR}(R_T - 0.5B) = \omega_{W1} R_W;
\tag{17}
$$

$$
V_{C2} = \omega_{MR}(R_T + 0.5B) = \omega_{W2} R_W;
\tag{18}
$$

$$
R_T = \frac{0.5(V_{C2} + V_{C1})}{(V_{C2} - V_{C1})};
\tag{19}
$$

where F and F_{TO} are the specific forces values of the pressing of electromagnets and the load of technical operation; G is the full equipped weight of the robot; γ is the angle of the ferromagnetic surface inclination; ζ is the rolling friction factor; ξ is the adhesion factor; m_{MR} is the mass of MR; λ is the added mass coefficient, $\lambda = 1.15 + 0.001k_R^2$; k_R is the gear coefficient; c is the acceleration of MR; V_{C1}, V_{C2} are the values of speed for the running and lagging caterpillars; R_T is the radius of MR's turning; V_{MR}, φ_{MR} are the values of MR's moving speed and rotation angle, ω_{MR} is the robot's rotation speed, $\omega_{MR} = d\varphi_{MR}/dt$; B is the distance between the caterpillars and MR's center; ω_{W1} and ω_{W2} are rotation speed of running and lagging driving wheels; R_W is the driving wheel radius; a and b are the factors, which consider the direction of MR's rotation, for the running caterpillar $a = 2$ and $b = -1$; for the lagging caterpillar $a = 1$ and $b = 1$; M_{EMa} are the drive electromotors' torques; η_M, η_{MR} are the efficiency of the electromotor and the MR directly; D_M is the motor anchor's inertia moment; $D_{\Sigma W}$ is the aggregated inertia moment of the MR's caterpillar and its two wheels; L, h_C are the MR's length and height of the gravity center; x_O is the distance from the MR's transverse axis to the caterpillars' turning centers; x_1 is the distance from the MR's transverse axis to the fixing point of the technical equipment; β is the deviation angle of the force F_{TO} from the MR's longitudinal axis; μ_T is the resistance to turning, which depends on the radius of turning.

In this paper, to prove the proposed method efficiency, a synthesis of the RB for the speed fuzzy controller of the FACS is performed. In turn, the other parameters and structure of the FC are set using the knowledge of expert-operators.

The developed speed FC of the Mamdani-type implements the control law based on the dependence

$$u_{FC} = f_{FC}\left(k_P\varepsilon, k_D\dot{\varepsilon}, k_I \int \varepsilon dt\right). \tag{20}$$

The vector of FC input signals $\mathbf{E}$ in this case is represented by the expression (1). The normalizing coefficients that are used to convert the FC input signals into relative units, have the following values: $k_P = 5$; $k_D = 0.33$; $k_I = 60$.

Five linguistic terms of a triangular type are selected for the first input of the speed FC: BN is the big negative; SN is the small negative; Z is the zero; SP is the small positive; BP is the big positive. In turn, three LTs of the triangular type are selected for the second and third input variables: N is the negative; Z the zero; P is the positive. Also, seven terms are selected for the FC output variable u_{FC}: BN, N, SN, Z, SP, P, BP. The appearance of the selected linguistic terms and their parameters is shown in Fig. 2.

At the *Step 1* of the proposed method the total number of the RB rules s for the fuzzy controller is defined by all the existing combinations of LT of the input signals and is equal to 45, $s = 5 \times 3 \times 3 = 45$ [18]. Each r-th rule of the given RB ($r = 1$, ..., 45) is represented by the expression (2).

The consequents of each r-th rule, in this case, are selected from the set of possible rule consequents, which consists of 7 linguistic terms (big negative, negative, small

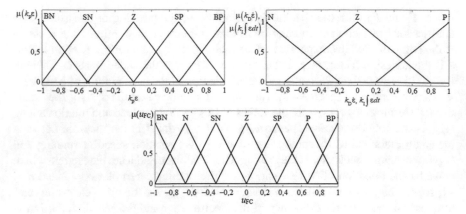

Fig. 2 FC linguistic terms and their parameters

negative, zero, small positive, positive, big positive) of the FC output variable u_{FC}. The consequent vector (6) $\mathbf{R}_X$ of this FC has the form

$$\mathbf{R}_X = \{LT_{X1}, LT_{X2}, \ldots, LT_{Xr}, \ldots, LT_{X45}\},$$
$$LT_{Xr} \in \{LT^1, LT^2, \ldots, LT^7\}, X \in \{1, 2, \ldots, 7^{45}\}.$$

The operation "min" is selected as an aggregation operation, the operation "max" is selected for both activation and accumulation operations. In turn, the gravity center method is chosen as the defuzzification method of the given FC.

For a comparative analysis and effectiveness study of the designed in this paper technique the development of the RB for the mobile robot FACS by means of the expert knowledge and using the RB synthesis method with determination of the proper vector of consequents based on GA is performed in this paper.

The fragment of the developed on the basis of the experts knowledge RB for the FC is presented in [35].

At the synthesis of the FC RB using the proposed method at the initialization stage (*Step 1*) the objective function J is chosen, which is the generalized integral deviation of the actual transient of the FACS of the MR speed $V_{MR}(t, \mathbf{R}_X)$ from the desired transient $V_D(t)$ of its reference model (RM) [42], that is represented by the transfer function

$$W_{RM}(s) = \frac{V_D(s)}{V_S(s)} = \frac{1}{(T_{RM}s + 1)^2}, \tag{21}$$

where T_{RM} is the time constant of the reference model, V_S is the set value of the MR speed.

In turn, the goal function is calculated based on the expression

$$J(t, \mathbf{R}_X) = \frac{1}{t_{\max}} \int\limits_{0}^{t_{\max}} [(E_V)^2 + k_{J1}(\dot{E}_V)^2 + k_{J2}(\ddot{E}_V)^2]dt, \qquad (22)$$

where k_{J1}, k_{J2} are weighting coefficients; E_V is the deviation of $V_{\mathrm{MR}}(t, \mathbf{R}_X)$ from $V_{\mathrm{D}}(t)$, $E_V = V_{\mathrm{D}}(t) - V_{\mathrm{MR}}(t, \mathbf{R}_X)$, $t_{\max}$ is the total time of the FACS transient for the robot movement speed;

In turn, the objective function optimal value J_{opt} is selected to be equal 0.1, at which the deviation of $V_{\mathrm{MR}}(t, \mathbf{R}_X)$ from the $V_{\mathrm{D}}(t)$ lies within acceptable limits.

At the initialization stage of the genetic algorithm parameters (*Step 1*), 4 separate populations with a different number of individuals Z were created: (1) $Z = 50$; (2) $Z = 100$; (3) $Z = 150$; (4) $Z = 200$. In turn, the synthesis of consequents of the FC rule base was carried out alternately with the help of each separate population. Also, these experiments for each separate population were carried out 10 times with subsequent averaging of the results. In addition, for each of the created populations, the same genetic operators and probabilities values P_C and P_M (for the crossover and mutation) were used. In turn, the probabilities for the crossover and mutation are equal to 0.25 and to 0.5.

At calculating the objective function (22) values for each j-th individual of population at *Step 4* of each n-th iteration of the method, the simulation of the mobile robot speed FACS transients was conducted for all existing conditions (under the action of various input and disturbing influences) to efficiently synthesize consequents of all the RB rules. In turn, checking of the optimization expiration criteria at *Step 5* was conducted using *Condition 1*, that is achievement the optimal value of the FACS objective function ($J(n) \leq J_{\mathrm{opt}}$).

The averaged experimental results obtained during the RB synthesis of the speed FC using each separate population are given in Table 1. In turn, the next abbreviations are adopted: $\upsilon_{J\mathrm{opt}}$ is the total number of calculations of the value of the objective function J required to achieve its optimal value J_{opt}; $n_{J\mathrm{opt}}$ is the total number of iterations of the method required to achieve the objective function J optimal value.

In turn, $\upsilon_{J\mathrm{opt}}$ is the main parameter characterizing the total time and computational costs of the method, and is determined as follows

$$\upsilon_{J\mathrm{opt}} = Z n_{J\mathrm{opt}}. \qquad (23)$$

Table 1 Averaged experimental results obtained at the RB synthesis	Population number	Experiment Parameters		
		Z	$n_{J\mathrm{opt}}$	$\upsilon_{J\mathrm{opt}}$
	1	50	98	4900
	2	100	47	4700
	3	150	29	4350
	4	200	23	4600

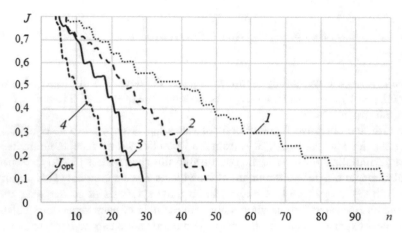

Fig. 3 Changing curves of the objective function (22) best values during the synthesis of consequents of the FC rule base

Figure 3 shows the changing curves of the objective function (22) best values during the synthesis of consequents of the FC rule base using the developed method using various populations: (1) $Z = 50$; (2) $Z = 100$; (3) $Z = 150$; (4) $Z = 200$.

As the Table 1 and Fig. 3 show, the larger the number of individuals Z in the population, the fewer iterations n are required for finding the optimal RB consequents vector $\mathbf{R}_{opt}$. So, for the 4-th population ($Z = 200$), the objective function J reached the optimal value J_{opt} at 75 iterations earlier than for the 1st ($Z = 50$). However, at the same time, the total number of calculations υ_{Jopt} of the objective function J value for the 4-th population is somewhat larger than for the 3rd. Thus, analyzing the above results, it can be concluded that 3rd population ($Z = 150$) is the most optimal for the synthesis of consequents of the FC rule base, in this case, since it has the smallest value of the parameter υ_{Jopt} ($\upsilon_{Jopt} = 4350$).

The RB fragment with the consequents synthesized using this 3rd population is presented in Table 2, and the full vector of consequents $\mathbf{R}_{opt}$ obtained as a result of the implementation of the developed method has the form:

$\mathbf{R}_{opt} =$ (BN, BN, BN, BN, BN, BN, BN, BN, BN, BN, BN, N, N, N, SN, SN, SN, SN, N, N, SN, SN, Z, Z, SP, SP, P, SP, SP, SP, SP, P, P, SP, P, BP, BP, BP, BP, BP, BP, BP, BP, BP, BP).

The Fig. 4 shows the characteristic surfaces $u_{FC} = f_{FC}(k_P\varepsilon, k_D\dot{\varepsilon})$ of the speed fuzzy controller at a fixed value $k_I \int \varepsilon dt = 0, 5$ with the developed RBs: a is for the FC using the expert knowledge; b is for the FC based on the synthesis method with using the 3rd population.

The Fig. 5 presents the MR's accelerating characteristics when it moves along an inclined ferromagnetic surface at the following parameters: angle of inclination of the working surface $\gamma = 60°$; set value of the MR speed $V_S = 0.2$ m/s; permanent disturbance in the form of the technical operation's load force that equals to 900 N ($F_{TO} = 900$ N).

Table 2 Fragment of the RB synthesized by means of the 3rd population

Number of Rule	FC's variables (input and output)			
	$k_P \varepsilon$	$k_D \dot{\varepsilon}$	$k_I \int \varepsilon dt$	u_{FC}
1	BN	N	N	BN
5	BN	Z	Z	BN
18	SN	P	P	SN
23	Z	Z	Z	Z
30	SP	N	P	SP
36	SP	P	P	BP
41	BP	Z	Z	BP
45	BP	P	P	BP

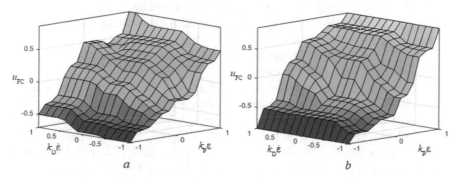

Fig. 4 Characteristic surfaces $u_{FC} = f_{FC}(k_P \varepsilon, k_D \dot{\varepsilon})$ of the speed fuzzy controller at a fixed value $k_I \int \varepsilon dt = 0,5$

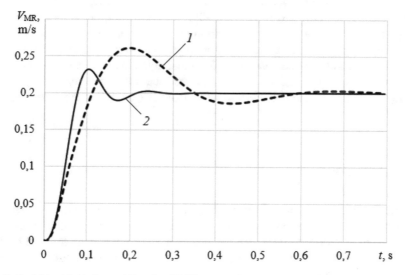

Fig. 5 Speed transients for mobile robot FACS

Table 3 Comparative analysis of the quality indicators for the speed FACS of the MR

Main quality indicators	Quality indicators values for the speed FACS	
	RB formed using the expert knowledge	RB based on the proposed method
σ, %	35.4	15.3
t_r, s	0.121	0.078
t_{max}, s	0.54	0.19
J	0.244	0.099

These graphs of transients of speed change are obtained for the speed FACS with the developed FC: *1* is for the RB using the expert knowledge; *2* is for the RB based on the proposed synthesis method using the 3rd population.

Also, Table 3 presents the quality indicators comparative analysis of the MR speed FACS with developed rule bases: using the expert knowledge and the synthesis method designed in this paper using the 3rd population.

In turn, the main abbreviations are adopted in Table 3: σ is an overshoot, $\sigma = \frac{V_{MR_{max}} - V_{MR}}{V_{MR}} \cdot 100\%$; t_r is the rise time.

The Table 3 and Fig. 5 prove that FACS of the mobile robot speed with the developed RB using the designed by the authors synthesis method have a significantly fewer value of the function J and better control quality indicators than FACS of the MR speed with the RB developed on the expert knowledge. In addition, the process of finding the optimal vector of the RB consequents $\mathbf{R}_{opt}$ using this method, required minor time and computational costs, that proves its rather high effectiveness.

In turn, if it is necessary to further increase the indicators of control quality of the given MR speed FACS and simplify its hardware and software realization, after synthesis of the consequents vector $\mathbf{R}_X$ of the rule base by means of the developed technique, it is advisable to carry out additional optimization of the FC's structure and parameters, in particular, optimization of: the vector $\mathbf{X}_{IS}$ that defines the FC input variables; the vector $\mathbf{X}_{IT}$ that determines the number, types and parameters of LTs of the controller inputs; the vector $\mathbf{X}_{FIE}$ that defines the main operations of the fuzzy inference engine and the vector $\mathbf{X}_{DF}$ that determines the number, types and parameters of LTs of the controller output variable u_{FC}, as well as the defuzzification method.

5 Conclusions

The development and study of the method of the rule base automatic synthesis for the FACSs of the Mamdani-type with the generation of optimal consequents by means of the genetic algorithms is presented in this work. Application of the method will allow to synthesize the RBs with best consequents for the FACSs of Mamdani-type at insufficient initial data (at a high degree of information uncertainty), if the number of rules is quite large, so that the compilation of the FACS rule base using knowledge of experts-operators is not always effective, and at the various levels of expert qualifications.

For studying the designed method's effectiveness, the RB development is performed in this paper for the mobile robot's speed FACS. Herewith, the synthesis of the RB consequents for the presented speed fuzzy controller is performed at different sizes of the population of individuals Z. Based on the analysis of the experiments results, the optimal configuration of the method population ($Z = 150$) is defined for which the RB consequents optimal vector $\mathbf{R}_{opt}$ is found at the lowest time and computational costs ($\upsilon_{Jopt} = 4350$). Also, the transients obtained during computer simulations show that the mobile robot speed FACS with the developed RB based on the presented by the authors technique with the proper adjustment of parameters gives better goal function J and higher indicators of control quality than FACS with the rule base developed on the basis of the knowledge of experts, that proves the high effectiveness of this method.

In turn, it is expedient to conduct further research of the proposed method of the RB automatic synthesis using various types of genetic operators, as well as changing other parameters of the genetic search.

References

1. Zadeh, L.A.: Fuzzy sets. Inf. Control **8**, 338–353 (1965)
2. Kacprzyk, J.: Multistage Fuzzy Control: A Prescriptive Approach. Wiley, New York, NY, USA (1997)
3. Solesvik, M., Kondratenko, Y., Kondratenko, G., Sidenko, I., Kharchenko, V., Boyarchuk, A.: Fuzzy decision support systems in marine practice. In: Fuzzy Systems (FUZZ-IEEE), 2017 IEEE International Conference on Fuzzy Systems, pp. 1–6 (2017). https://doi.org/10.1109/FUZZ-IEEE.2017.8015471
4. Zadeh, L.A.: The role of fuzzy logic in modeling, identification and control. MIC—Model. Identif. Control **15**(3), 191–203 (1994)
5. Mendel, J.M.: Uncertain Rule-Based Fuzzy Systems, Introduction and New Directions, Second Edition, Springer International Publishing, 684 p (2017)
6. Kondratenko, Y.P., Kondratenko, N.Y.: Synthesis of analytic models for subtraction of fuzzy numbers with various membership function's shapes. In: Gil-Lafuente, A., Merigó, J., Dass, B., Verma, R. (eds.), Applied Mathematics and Computational Intelligence, FIM 2015, Advances in Intelligent Systems and Computing, Vol. 730, pp. 87–100. Springer, Cham (2018). https://doi.org/10.1007/978-3-319-75792-6_8
7. Kosko, B.: Fuzzy systems as universal approximators. IEEE Trans. Comput. **43**(11), 1329–1333 (1994)
8. Kondratenko, Y., Kondratenko, N.: Real-time fuzzy data processing based on a computational library of analytic models. Data **3**(4), 59, 1–19 (2018). https://doi.org/10.3390/data3040059
9. Topalov, A. et al. Control processes of floating docks based on SCADA systems with wireless data transmission. In: Perspective Technologies and Methods in MEMS Design, MEMSTECH 2016—Proceedings of 12th International Conference, pp. 57–61 (2016). https://doi.org/10.1109/MEMSTECH.2016.7507520
10. Encheva, S., et al.: Decision support systems in logistics. AIP Conf. Proc. **1060**, 254–256 (2008). https://doi.org/10.1063/1.3037065
11. Zadeh, L.A., Abbasov, A.M., Yager, R.R., Shahbazova, S.N., Reformat, M.Z. (eds.): Recent Developments and New Directions in Soft Computing, STUDFUZ 317. Springer, Cham (2014)

12. Kondratenko, Y.P., Kozlov, O.V.: Mathematical model of ecopyrogenesis reactor with fuzzy parametrical identification. In: Zadeh, L.A. et al. (eds.), Recent Developments and New Direction in Soft-Computing Foundations and Applications, Studies in Fuzziness and Soft Computing 342, pp. 439–451. Springer, Berlin, Heidelberg (2016). https://doi.org/10.1007/978-3-319-32229-2_30
13. Kondratenko, Y., Korobko, V., Korobko, O., Kondratenko, G., Kozlov, O.: Green-IT approach to design and optimization of thermoacoustic waste heat utilization plant based on soft computing. Kharchenko, V., Kondratenko, Y., Kacprzyk, J. (eds.), Green IT Engineering: Components, Networks and Systems Implementation. Studies in Systems, Decision and Control, Vol. 105, pp. 287–311. Springer, Cham (2017). https://doi.org/10.1007/978-3-319-55595-9_14
14. Takagi, T., Sugeno, M.: Fuzzy Identification of systems and its applications to modeling and control. IEEE Trans. Syst. Man Cyberne. **SMC-15**(1), 116–132 (1985)
15. Pomanysochka, Y., Kondratenko, Y., Sidenko, I.: Noise filtration in the digital images using fuzzy sets and fuzzy logic. Zholtkevych, G. (ed.), ICT in Education, Research and Industrial Applications. Integration, Harmonization and Knowledge Transfer, Proceedings of ICTERI'2019. Volume III: Ph.D. Symposium, Kherson, Ukraine, CEUR Workshop Proceedings, pp. 63–72 (2019). CEUR-WS.org/Vol-2403/paper8.pdf
16. Kondratenko, Y.P., Zaporozhets, Y.M., Rudolph, J., Gerasin, O.S., Topalov, A.M., Kozlov, O.V.: Features of clamping electromagnets using in wheel mobile robots and modeling of their interaction with ferromagnetic plate. In: Proceedings of the 9th IEEE International Conference on Intelligent Data Acquisition and Advanced Computing Systems: Technology and Applications (IDAACS), Vol. 1, Bucharest, Romania, pp. 453–458 (2017)
17. Kondratenko, Y.P., Kozlov, O.V., Klymenko, L.P., Kondratenko, G.V.: Synthesis and research of neuro-fuzzy model of ecopyrogenesis multi-circuit circulatory system. In: Jamshidi, M., Kreinovich, V., Kazprzyk, J. (eds.), Advance Trends in Soft Computing, Series: Studies in Fuzziness and Soft Computing, Vol. 312, 1–14 (2014). https://doi.org/10.1007/978-3-319-03674-8_1
18. Kondratenko, Y.P., Kozlov, A.V.: Parametric optimization of fuzzy control systems based on hybrid particle swarm algorithms with elite strategy. J. Autom. Inf. Sci. **51**(12), 25–45 (2019). New York: Begel House Inc. https://doi.org/10.1615/JAutomatInfScien.v51.i12.40
19. Kozlov, O., Kondratenko, G., Gomolka, Z., Kondratenko, Y.: Synthesis and optimization of green fuzzy controllers for the reactors of the specialized pyrolysis plants. In: Kharchenko, V., Kondratenko, Y., Kacprzyk, J. (eds.), Green IT Engineering: Social, Business and Industrial Applications, Studies in Systems, Decision and Control, Vol. 171, pp. 373–396. Springer, Cham (2019). https://doi.org/10.1007/978-3-030-00253-4_16
20. Kozlov, O.: Optimal selection of membership functions types for fuzzy control and decision making systems. In: Proceedings of the 2nd International Workshop on Intelligent Information Technologies & Systems of Information Security with CEUR-WS, Khmelnytskyi, Ukraine, IntelITSIS 2021, CEUR-WS, Vol. 2853, pp. 238–247 (2021)
21. Kosheleva, O., Kreinovich, V.: Why Bellman-Zadeh approach to fuzzy optimization. Appl. Math. Sci. **12**, 517–522 (2018)
22. Kondratenko, Y.P., Altameem, T.A., Al Zubi, E.Y.M.: The optimisation of digital controllers for fuzzy systems design. Adv. Modell. Anal. AMSE Periodicals Series A **47**, 19–29 (2010)
23. Kondratenko, Y.P., Korobko, O.V., Kozlov, O.V.: Frequency tuning algorithm for loudspeaker driven thermoacoustic refrigerator optimization. In: Engemann, K.J., Gil-Lafuente, A.M., Merigo, J.M. (eds.), Lecture Notes in Business Information Processing: Modeling and Simulation in Engineering, Economics and Management, Vol. 115, pp. 270–279. Springer, Berlin, Heidelberg (2012). https://doi.org/10.1007/978-3-642-30433-0_27
24. Kondratenko, Y.P., Klymenko, L.P., Al Zu'bi E.Y.M.: Structural optimization of fuzzy systems' rules base and aggregation models. Kybernetes **42**(5), 831–843 (2013). https://doi.org/10.1108/K-03-2013-0053
25. Skakodub, O., Kozlov, O., Kondratenko, Y.: Optimization of linguistic terms' shapes and parameters: fuzzy control system of a quadrotor drone. In: Proceedings of the 11th IEEE International Conference on Intelligent Data Acquisition and Advanced Computing Systems: Technology

and Applications (IDAACS), Vol. 1, Cracow, Poland, pp. 566–571 (2021). https://doi.org/10. 1109/IDAACS53288.2021.9660926

26. Kondratenko, Y.P., Korobko, O.V., Kozlov, O.V.: Synthesis and optimization of fuzzy controller for thermoacoustic plant. In: Zadeh, L.A. et al. (eds.), Recent Developments and New Direction in Soft-Computing Foundations and Applications, Studies in Fuzziness and Soft Computing 342, pp. 453–467. Springer, Berlin, Heidelberg (2016). https://doi.org/10.1007/978-3-319-32229-2_31

27. Jayaram, B.: Rule reduction for efficient inferencing in similarity based reasoning. Int. J. Approximate Reason. **48**(1), 156–173 (2008)

28. Atamanyuk, I.P. et al.: The algorithm of optimal polynomial extrapolation of random processes. In: Engemann, K.J., Gil-Lafuente, A.M., Merigo, J.L. (eds.), Modeling and Simulation in Engineering, Economics and Management, International Conference MS 2012, New Rochelle, NY, USA, Proceedings. Lecture Notes in Business Information Processing, Volume 115, pp. 78–87. Springer (2012). https://doi.org/10.1007/978-3-642-30433-0_9

29. Kondratenko, Y.P., Klymenko, L.P., Sidenko, I.V.: Comparative analysis of evaluation algorithms for decision-making in transport logistics. In: Jamshidi, M., Kreinovich, V., Kazprzyk, J. (eds.), Advance Trends in Soft Computing, Series: Studies in Fuzziness and Soft Computing, Vol. 312, pp. 203–217 (2014). https://doi.org/10.1007/978-3-319-03674-8_20

30. Jamshidi, M., Kreinovich, V., Kacprzyk, J. (eds.): Advance Trends in Soft Computing. Springer, Cham (2013)

31. Kondratenko, Y., Khalaf, P., Richter, H., Simon, D.: Fuzzy real-time multiobjective optimization of a prosthesis test robot control system. In: Kondratenko, Y.P., Chikrii, A.A., Gubarev, V.F., Kacprzyk, J. (eds.), Advanced Control Techniques in Complex Engineering Systems: Theory and Applications, Dedicated to Professor Vsevolod M. Kuntsevich. Studies in Systems, Decision and Control, Vol. 203. Cham: Springer Nature Switzerland AG, pp. 165–185 (2019). https://doi.org/10.1007/978-3-030-21927-7_8

32. Kondratenko, Y.P., Simon, D.: Structural and parametric optimization of fuzzy control and decision making systems. In: Zadeh, L., Yager, R., Shahbazova, S., Reformat, M., Kreinovich, V. (eds.), Recent Developments and the New Direction in Soft-Computing Foundations and Applications, Studies in Fuzziness and Soft Computing, Vol. 361, pp. 273–289. Springer, Cham (2018). https://doi.org/10.1007/978-3-319-75408-6_22

33. Simon, D.: Evolutionary Optimization Algorithms: Biologically Inspired and Population-Based Approaches to Computer Intelligence. Wiley (2013)

34. Oh, S.K., Pedrycz, W.: The design of hybrid fuzzy controllers based on genetic algorithms and estimation techniques. J. Kybernetes **31**(6), 909–917 (2002)

35. Kozlov, O.V., Kondratenko, Y.P.: Bio-Inspired Algorithms for Optimization of Fuzzy Control Systems: Comparative Analysis. In: Kondratenko, Y.P., Kuntsevich, V.M., Chikrii, A.A., Gubarev, V.F. (eds.) Advanced Control Systems: Theory and Applications, pp. 83–128. Series in Automation, Control and Robotics, River Publishers, Denmark (2021)

36. Alcalá, R., Alcalá-Fdez, J., Gacto, M.J., Herrera, F.: Rule base reduction and genetic tuning of fuzzy systems based on the linguistic 3-tuples representation. Soft. Comput. **11**(5), 401–419 (2007)

37. Pedrycz, W., Li, K., Reformat, M.: Evolutionary Reduction of Fuzzy Rule-Based Models, Fifty Years of Fuzzy Logic and its Applications, STUDFUZ 326, pp. 459–481. Springer, Cham (2015)

38. Haupt, R., Haupt, S.: Practical Genetic Algorithms. Wiley, New Jersey 261 p. (2004)

39. Quijano, N., Passino, K.M.: Honey Bee Social Foraging Algorithms for Resource Allocation: Theory and Application. Columbus: Publishing house of the Ohio State University, 39 p. (2007)

40. Kim, D.H., Cho, C.H.: Bacterial foraging based neural network fuzzy learning. Proceedings of the 2nd Indian International Conference on Artificial Intelligence (IICAI – 2005), Pune: IICAI, pp. 2030–2036 (2005)

41. Engelbrecht, A.: A study of particle swarm optimization particle trajectories. Inf. Sci. **176**(8), 937–971 (2006)

42. Kondratenko, Y.P., Kozlov, O.V. Korobko, O.V.: Two modifications of the automatic rule base synthesis for fuzzy control and decision making systems. In: Medina, J. et al. (eds.), Information Processing and Management of Uncertainty in Knowledge-Based Systems: Theory and Foundations, 17th International Conference, IPMU 2018, Cadiz, Spain, Proceedings, Part II,, CCIS 854, Springer International Publishing AG, pp. 570–582 (2018). https://doi.org/10.1007/978-3-319-91476-3_47
43. Kondratenko, Y.P., Kozlov, A.V.: Generation of rule bases of fuzzy systems based on modified ant colony algorithms. J. Autom. Inf. Sci. 51(3), 4–25 (2019). New York: Begel House Inc. https://doi.org/10.1615/JAutomatInfScien.v51.i3.20
44. Kondratenko, Y.P., Kozlov, O.V.: Mathematic modeling of reactor's temperature mode of multi-loop pyrolysis plant. Modeling and Simulation in Engineering, Economics and Management, Lecture Notes in Business Information Processing 115, 178–187 (2012)
45. Potvin, J.-Y.: Genetic algorithms for the traveling salesman problem. Ann. Oper. Res. 63, 339–370 (1996)
46. Nagata, Y., Soler, D.: A new genetic algorithm for the asymmetric traveling salesman problem. Expert Syst. Appl. 39(10), 8947–8953 (2012)
47. Piwonska, A.: Genetic algorithm finds routes in travelling salesman problem with profits, Zeszyty Naukowe Politechniki Bia lostockiej, Informatyka, 51–65 (2010)
48. Sadegheih, A.: Scheduling problem using genetic algorithm, simulated annealing and the effects of parameter values on GA performance. Appl. Math. Model. 30(2), 147–154 (2006)
49. Sadegheih, A., Drake, P.R.: Network optimization using linear programming and genetic algorithm. Neural Network World Int. J. Non-Stand. Comput. Artif. Intell. 11(3), 223–233 (2001)
50. Sun, X., Wang, J.: Routing design and fleet allocation optimization of freeway service patrol: Improved results using genetic algorithm. Phys. A Stat. Mech. Appl. 501, 205–216 (2018)
51. Arakaki, R.K., Usberti, F.L.: Hybrid genetic algorithm for the open capacitated arc routing problem. Comput. Oper. Res. 90, 221–231 (2018)
52. Putha, R., Quadrifoglio, L., Zechman, E.: Comparing ant colony optimization and genetic algorithm approaches for solving traffic signal coordination under oversaturation conditions. Comput. Aided Civ. Infrastruct. Eng. 27, 14–28 (2012)
53. Abbasian, R., Mouhoub, M.: An efficient hierarchical parallel genetic algorithm for graph coloring problem. Proceedings of The 13th annual conference on Genetic and evolutionary computation, ACM, Dublin, Ireland, pp. 521–528 (2011)
54. Croitoru, C., Luchian, H., Gheorghies, O., Apetrei, A.: A new genetic graph coloring heuristic. Proceedings of The Computational Symposium on Graph Coloring and its Generalizations, Ithaca, New York, USA, pp. 63–74 (2002)
55. Glass, C.A., Prugel-Bennett, A.: Genetic algorithm for graph coloring: exploration of Galinier and Hao's algorithm. J. Comb. Optim. 3, 229–236 (2003)
56. Ashby, H., Yampolskiy, R.V.: Genetic algorithm and wisdom of artificial crowds algorithm applied to light up. Proceedings of 16th International Conference on Computer Games, AI, Animation, Mobile, Interactive Multimedia, Educational & Serious Games, Louisville, KY, USA, pp. 27–30 (2011)
57. Shen, J.W.: Solving the Graph Coloring Problem using Genetic Programming, Genetic Algorithms and Genetic Programming at Stanford 2003, Stanford Bookstore, pp. 187–196 (2003)
58. Singha, S., Bhattacharya, T., Chaudhuri, S.R.B.: An approach for reducing crosstalk in restricted channel routing using graph coloring problem and genetic algorithm. Proceedings of The International Conference on Computer and Electrical Engineering, Phuket Island, Thailand, pp. 807–811 (2008)
59. Khan, S., et al.: Design and implementation of an optimal fuzzy logic controller using genetic algorithm. J. Comput. Sci. 4(10), 799–806 (2008)
60. Cordon, O., Gomide, F., Herrera, F., Hoffmann, F., Magdalena, L.: Ten years of genetic fuzzy systems: current framework and new trends. Fuzzy Sets Syst. 141(1), 5–31 (2004)

61. Fan, L., Joo, E.M.: Design for auto-tuning PID controller based on genetic algorithms. Industrial Electronics and Applications, 2009, ICIEA 2009, 4th IEEE Conference on IEEE, pp. 1924–1928 (2009)
62. Zhao, J., Han, L., Wang, L., Yu, Z.: The fuzzy PID control optimized by genetic algorithm for trajectory tracking of robot arm. 2016 12th World Congress on Intelligent Control and Automation (WCICA), Guilin, China, 556–559 (2016)
63. Narvydas, G., Simutis, R., Raudonis, V.: Autonomous Mobile Robot Control Using Fuzzy Logic and Genetic Algorithm, IEEE International Workshop on Intelligent Data Acquisition and Advanced Computing Systems: Technology and Applications, pp. 460–464. Dortmund, Germany (2007)
64. Chen, Ch., Li, M., Sui, J., Wei, K., Pei, Q.: A genetic algorithm-optimized fuzzy logic controller to avoid rear-end collisions. J. Adv. Transp. **50**, 1735–1753 (2016)
65. Xinbin., L, Chunxia, D., Xiaohong, N.: Design of clustering adaptive fuzzy controller of drum boiler superheat temperature based on genetic algorithm. Proceedings of the 26th Chinese Control Conference, Zhangjiajie, Hunan, China, pp. 341–344 (2007)
66. Liu, X.-H., Kuai, R., Guan, P., Ye, X.-M., Wu, Z.-L.: Fuzzy-PID control for arc furnace electrode regulator system based on genetic algorithm. Proceedings of the Eighth International Conference on Machine Learning and Cybernetics, Baoding, pp. 683–689 (2009)
67. Li, L., Zhu, Q., Gao, Sh.: Design and realization of waterjet propelled craft autopilot based on fuzzy control and genetic algorithms. Proceedings of the 2006 IEEE International Conference on Mechatronics and Automation, Luoyang, China, pp. 1362–1366 (2006)
68. Cervantes, L., Castillo, O.: Automatic design of fuzzy systems for control of aircraft dynamic systems with genetic optimization. World Congress AFSS International Conference, pp. 4131–4137 (2011)
69. Cervantes, L., Castillo, O.: Statistical comparison of type-1 and type-2 fuzzy systems design with genetic algorithms in the case of three tank water control, Joint IFSA World Congress and NAFIPS Annual Meeting (IFSA/NAFIPS) 2013, Edmonton, AB, Canada, pp. 1056–1061 (2013)
70. Cervantes, L., Castillo, O.: Hierarchical genetic algorithms for optimal Type-2 fuzzy system design. Annual Meeting of the North American Fuzzy Information Processing Society, 324 329 (2011)

Fuzzy Logic and Its Applications

Creating a Database of Estimates of Noise Characteristics for Monitoring the Technical Condition of Industrial Facilities

Telman Aliev(iD) and Naila Musaeva(iD)

Abstract It is noted that at the moment of formation of damage in industrial facilities, additional noise appears, which correlates with the useful component of the noisy signal. As a result, the calculated estimates of the characteristics of the noisy signal do not allow assessing the technical condition of the facility adequately. It is shown that it is the characteristics of this noise that are the most informative when solving monitoring problems. Therefore, the algorithms for calculating the estimates of the characteristics of the correlated noise are developed. A database of informative attributes of the technical condition of an industrial facility is created. It is shown that the database consists of estimates of cross-correlation functions and correlation coefficients between the useful signal and the noise, as well as of high-order moments of the noise at different time instants. The technology for analyzing the technical condition of a facility by means of this database is developed. It is noted that the application of the proposed techniques in monitoring and control systems makes it possible to detect defects and malfunctions of industrial facilities in the latent period of formation, which allows preventing accidents with catastrophic consequences.

Keywords Noisy signal · Noise · Industrial facilities · Technical condition · Monitoring

1 Introduction

It is known that in the process of operation industrial facilities are subjected to various impacts, as a result of which deformations, cracks, wear, breakages, corrosion and other malfunctions and defects emerge. To prevent accidents in such situations, continuous monitoring systems are in place to detect dangerous defects as they

T. Aliev
Institute of Control Systems of ANAS, B. Vahabzade 68, Baku, Azerbaijan

N. Musaeva (✉)
Azerbaijan University of Architecture and Construction, A. Sultanova 11, Baku, Azerbaijan
e-mail: musanaila@gmail.com

© The Author(s), under exclusive license to Springer Nature Switzerland AG 2023
Sh. N. Shahbazova et al. (eds.), *Recent Developments and the New Directions of Research, Foundations, and Applications*, Studies in Fuzziness and Soft Computing 422,
https://doi.org/10.1007/978-3-031-20153-0_10

occur or develop and to warn the operating personnel in advance. Existing modern monitoring systems collect, transfer, process, analyze data in a qualified manner and make decisions about the condition of monitored facilities. This makes it possible to prevent major industrial accidents, prevent environmental damage, avoid human and material losses.

However, the existing monitoring and control systems do not allow identifying the early period of initiation of malfunctions in industrial facilities. This is especially important for countries in seismically active regions, when after frequent weak earthquakes or minor landslides, microscopic changes appear in the technical condition of the facility, which can subsequently lead to serious damage or destruction [1–4].

It is shown in [5–10] that in the normal technical condition of the industrial facility, the noisy signal that comes from the corresponding sensor contains only the noise caused by the influence of external factors. At the moment of formation of even the smallest damage, additional noise appears, which correlates with the useful component of the noisy signal. Therefore, the problem arises of determining the early stage of the initiation of a defect in the industrial facility and the dynamics of its development by creating a database of estimates of the characteristics of the noise correlated with the useful signal of the noisy signal.

2 Problem Statement

It is known that in practice, real signals are the sum of useful signals $X(t)$ and noises $E(t)$, i.e.

$$G(t) = X(t) + E(t). \tag{1}$$

Because useful signals $X(t)$ are contaminated with noise $E(t)$, tangible errors emerge in determining their characteristics. In this case, the characteristics of the noise are the variance, the mean square deviation, high-order moments, the correlation function, the correlation coefficient between the useful signal and the noise, and the distribution law. Let us consider the impact of the noise $E(t)$ on the results of the calculation of the above estimates.

Here, as shown in [1, 5–10], the overall noise $E(t)$ is made up of the noise $E_1(t)$ caused by external factors and the noise $E_2(t)$ caused by the initiation of a defect during the operation of objects, i.e.,

$$E(t) = E_1(t) + E_2(t). \tag{2}$$

Suppose that $G(t)$ is sampled stationary random signal with a normal distribution law, consisting of the useful signal $X(t)$ and the noise $E(t)$ with a zero mathematical expectation m_E.

The formula for calculating the estimate D_G of the variance of the noisy signal $G(t)$ here is written as [1, 5–10]:

$$D_G = R_{GG}(0) = \frac{1}{N} \sum_{i=1}^{N} G^2(i\,\Delta t)$$

$$= R_{XX}(0) + 2R_{XE}(0) + R_{EE}(0). \tag{3}$$

Therefore, the error of the obtained result is

$$\lambda_{GG}(\mu = 0) = 2R_{XE}(0) + R_{EE}(0) = D_E,$$

where $R_{XE}(0) = \frac{1}{N} \sum_{i=1}^{N} X(i\,\Delta t)E(i\,\Delta t)$ is the cross-correlation function; $R_{EE}(0) = \sum_{i=1}^{N} E(i\,\Delta t)E(i\,\Delta t)$ is the variance of the noise $E(t)$.

The formula for calculating the estimate of the correlation function $R_{GG}(\mu)$ at $\mu \neq 0$ can be also written as [1, 5–10]:

$$R_{GG}(\mu) = \frac{1}{N} \sum_{i=1}^{N} G(i\,\Delta t)G((i+\mu)\Delta t)$$

$$= R_{XX}(\mu) + R_{EX}(\mu) + R_{XE}(\mu) + R_{EE}(\mu) \tag{4}$$

Since

$$R_{EE}(\mu) = 0 \text{ when } \mu \neq 0,$$

the overall noise will be

$$\lambda_{GG}(\mu) \approx \begin{cases} 2R_{XE}(0) + R_{EE}(0) \ npu \ \mu = 0 \\ 2R_{XE}(\mu) \qquad npu \ \mu \neq 0 \end{cases} \tag{5}$$

hence the obvious inequality

$$R_{XX}(\mu) \neq R_{GG}(\mu).$$

For this reason, in practice, in many cases the noise is filtered. Therefore, it is not possible to ensure the adequacy of the results of the tasks to be solved using the estimate of $R_{GG}(\mu)$, because in this case, some of the valuable information contained in the noise $E(t)$ of the signals $G(t)$ is lost as a result of the filtering. In this regard, it is obvious that it is necessary to create algorithms and techniques for determining the estimate of the noise variance D_E, the cross-correlation function $R_{XE}(\mu)$ and the correlation coefficient between the useful signal and the noise.

Here, the main purpose of calculating these characteristics is to develop new techniques that allow the use of valuable information, which is often contained in the noise of the noisy signals $G(t)$, to solve the problems of monitoring and control of industrial facilities [1, 5–10]. At the same time, the algorithms being developed may also be of interest on their own, since they open up the possibility of improving the accuracy of traditional techniques for correlation analysis of noisy signals. For this reason, in the following sections, we propose algorithms for calculating the relay cross-correlation function and correlation coefficient between the useful signal and the noise, as well as techniques for solving the problem of monitoring the technical condition of industrial facilities using a database of estimates of characteristics of the noise, consisting of the variance, the high-order moments, the correlation function, the correlation coefficient between the useful signal and the noise, the distribution law.

3   Algorithms and Techniques for Determining the Cross-Correlation Function and the Correlation Coefficient Between the Useful Signal and the Noise

As shown above, during normal operation of the object, the noise $E(t) = E_1(t)$ emerges due to random external factors, which does not correlate with the useful signal $X(t)$. However, at the beginning of the latent period of changes in the technical condition of the object as a result of the initiation of various defects, the noise $E_2(t)$ emerges, which correlates with the useful signal. Therefore, from this moment on, the correlation between the useful signal $X(t)$ and the overall noise $E(t)$ is different from zero. Here, the initiation and development of malfunctions is essentially manifested in the estimates of the cross-correlation functions $R_{XE}(\mu)$ between $X(t)$ and $E(t)$ [1, 5–10]. Therefore, to control the beginning and dynamics of changes in the technical condition of industrial facilities, it is advisable to use the estimate of $R_{XE}(\mu)$.

We first adopt the following notation and conditions [1]:

$$sgnG(i\Delta t) = \begin{cases} +1 & when \ \ G(i\Delta t) > 0 \\ 0 & when \ \ G(i\Delta t) = 0 \ . \\ -1 & when \ \ G(i\Delta t) < 0 \end{cases}$$

It is known that the estimates of the relay correlation functions can be calculated from the formula [1]:

$$R_{XE}^r(\mu) = \frac{1}{N} \sum_{i=1}^{N} sgn(X(i\Delta t))E((i + \mu)\Delta t).$$

Obviously, to use this formula, it is necessary to determine the samples of the noise $E(i\Delta t)$ and the useful signal $X(i\Delta t)$, which cannot be measured directly or isolated from the noisy signal $G(t)$ [1].

In this regard, we will consider one of the possible options for the approximate calculation of estimates of the relay cross-correlation function $R_{XE}^{r*}(\mu)$ between the useful signal $X(t)$ and the noise $E(t)$ as a result of calculating the relay correlation function $R^r_{GG}(\mu)$ of the noisy signal $G(t)$.

Then the relay correlation function $R^r_{GG}(\mu)$ of the noisy signal $G(t)$ will be [1]:

$$R^r_{GG}(\mu) = \frac{1}{N} \sum_{i=1}^{N} sgn(G(i\,\Delta t))G((i+\mu)\Delta t).$$

Considering that [1]

$$sgnG(i\,\Delta t) = sgnX(i\,\Delta t), \tag{6}$$

the expression for calculating the relay correlation function $R^r_{GG}(\mu)$ of the noisy signal $G(t)$ for $\mu = 0$ can be written as:

$$R^{r*}_{GG}(\mu = 0) = \frac{1}{N} \sum_{i=1}^{N} sgn(G(i\,\Delta t))G(i\,\Delta t)$$

$$= \frac{1}{N} \sum_{i=1}^{N} sgn(X(i\,\Delta t))X(i\,\Delta t) + \frac{1}{N} \sum_{i=1}^{N} sgn(X(i\,\Delta t))E(i\,\Delta t)$$

$$= R^r_{XX}(0) + R^r_{XE}(0), \tag{7}$$

where $R^r_{XX}(\mu)$, $R^r_{XE}(\mu)$ are the relay correlation function of the useful signal and the cross-correlation function between the useful signal and the noise.

In the literature [1] it is shown that the estimates of the relay cross-correlation function $R^{r*}_{XE}(0)$ can be calculated from the expression

$$R^{r*}_{XE}(0) = R^r_{GG}(0) - 2R^r_{GG}(1) + R^r_{GG}(2)$$

$$= \frac{1}{N} \sum_{i=1}^{N} sgn(G(i\,\Delta t))G(i\,\Delta t)$$

$$- \frac{1}{N} \sum_{i=1}^{N} 2sgn(G(i\,\Delta t))G((i+1)\Delta t)$$

$$+ \frac{1}{N} \sum_{i=1}^{N} sgn(G(i\,\Delta t))G((i+2)\Delta t). \tag{8}$$

In this case, if the conditions of stationarity and normality of the distribution law of noisy signals are satisfied, then the following equalities will be valid for the controlled objects [1]

$$\begin{cases} R_{XE}^{r*}(0) \approx \dfrac{1}{N} \sum_{i=1}^{N} sgn(X(i\Delta t))E(i\Delta t) \neq 0 \\[2mm] R_{XX}^{r*}(0) + R_{XX}^{*}(2\Delta t) - 2R_{XX}^{*}(\Delta t) \approx 0 \\[2mm] R_{XE}^{r*}(\Delta t) \approx \dfrac{1}{N} \sum_{i=1}^{N} sgn(X(i\Delta t))E((i+1)\Delta t) \approx 0 \\[2mm] R_{XE}^{r*}(2\Delta t) \approx \dfrac{1}{N} \sum_{i=1}^{N} sgn(X(i\Delta t))E((i+2)\Delta t) \approx 0 \end{cases} \tag{9}$$

and due to this, the right-hand side of expression (8) takes the form

$$R_{XE}^{r*}(\mu = 0) = R_{XE}^{r}(\mu = 0).$$

Thus, we can consider that the estimate obtained by formula (8) is an estimate of the relay cross-correlation function $R_{XE}^{r*}(\mu \Delta t)$ between the useful signal $X(t)$ and the noise $E(t)$. The relationship between relay and normalized cross-correlation functions for normally distributed random processes $X(t)$, $E(t)$ here is expressed by the relation [1]:

$$R_{XE}^{r}(\mu) = \sqrt{\frac{2}{\pi}} \rho_{XE}(\mu)\sigma_{E}, \tag{10}$$

where $\rho_{XE}(\mu)$ is the normalized cross-correlation function between the useful signal $X(t)$ and the noise $E(t)$, σ_E is the mean square deviation of $E(t)$.

Hence the normalized cross-correlation function $\rho_{XE}(\mu)$ is calculated from the expression [1]:

$$\rho_{XE}(\mu) = \frac{R_{XE}^{r}(\mu)}{\sqrt{\frac{2}{\pi}} \cdot \sigma_{E}}. \tag{11}$$

It follows from this formula that to calculate the normalized cross-correlation function $\rho_{XE}(\mu)$ between the useful signal $X(t)$ and the noise $E(t)$, we need to know the relay cross-correlation function $R_{XE}^{r}(\mu)$ and the mean square deviation σ_E of the noise. Obviously, using formula (8), it is possible to calculate the relay cross-correlation function $R_{XE}^{r*}(\mu)$. At the same time, the mean square deviation σ_E of the noise can be calculated from the expression [1, 5–10]:

$$\sigma_{E}^{*} = \sqrt{R_{GG}(0) - 2R_{GG}(\Delta t) + R_{GG}(2\Delta t)}. \tag{12}$$

Therefore, the calculation of the normalized cross-correlation function can be reduced to the form:

$$\rho_{XE}^*(\mu) = \frac{R_{XE}^{r*}(\mu)}{\sqrt{\frac{2}{\pi}} \cdot \sigma_E^*}. \tag{13}$$

It is known here that the value of the normalized cross-correlation function $\rho_{XE}(0)$ at $\mu = 0$ is the correlation coefficient:

$$r_{XE} = \rho_{XE}(0) = \frac{R_{XE}^r(0)}{\sqrt{\frac{2}{\pi}} \cdot \sigma_E}. \tag{14}$$

Therefore, it is possible to calculate the value of the correlation coefficient r_{XE} between the useful signal $X(t)$ and the noise $E(t)$:

$$r_{XE}^* = \rho_{XE}^*(0) = \frac{R_{XE}^{r*}(0)}{\sqrt{\frac{2}{\pi}} \cdot \sigma_E^*}. \tag{15}$$

Thus, using formulas (8), (12), (13), (15), we can calculate the variance, the correlation function and the correlation coefficient between the useful signal $X(t)$ and the noise $E(t)$.

4   Algorithms for Calculating High-Order Moments of the Noise of the Noisy Signal

The noise $E_1(t)$ caused by external influences, as a rule, is known and its value varies within the range of $0 < D_{E_1} << 1$. At the same time, when deviations appear in the technical condition of industrial facilities, the variance of the overall noise $E(t)$ = $E_1(t) + E_2(t)$ in most cases is $D_E > 1$ and for unfiltered signal may vary within the range of $1 < D_E \leq 10$ [5–10]. If the signal is filtered, the value of overall noise variance can be in the range of $0 < D_E << 1$, but such values are not informative, and then the noise variance cannot be used as a carrier of information [5–10].

It is known [5–10] that expressions are used to calculate high-order moments of the noise $E(t)$:

$$\mu_{Eq} = \int\limits_{-\infty}^{+\infty} (\varepsilon - m_E)^q \frac{1}{\sigma_E \sqrt{2\pi}} e^{-\frac{(\varepsilon - m_E)^2}{2\sigma_E^2}} d\varepsilon. \tag{16}$$

where σ_E is the mean square deviation of the noise.

It is also known that the odd central moments of the normal distribution are zero [5–10] and there is a relation expressing the high-order moments through the variance $D_E = \mu_{E2}$ of the normal distribution [5–10]:

$$\mu_{Eq} = (q-1)\sigma_E^2 \mu_{E(q-2)} \text{ or } \mu_{E4} = 3\sigma_E^4, \ \mu_{E6} = 15\sigma_E^6, \ \mu_{E8} = 105\sigma_E^8, \dots$$

The mean square deviations here can be determined from expression (12).

It should be noted that these even high-order moments of the noise, which have the property of "sensitivity," can be used as characteristics that are diagnostic indicators for determining the dynamics of damage development.

This is due to the fact that, when the variance (the central second-order moment) changes by an insignificant value as compared to the initial value, which is greater than 1, fourth-, sixth-, eighth-order, etc. moments change by values which, depending on the moment order, change by a more significant value. Then, according to the order and value of the moment of the noise, it is possible to identify the initial latent period of defect occurrence and determine the degree and dynamics of damage development, which ultimately leads to a decrease in the strength of industrial facilities. Knowing the values of these indicators makes it possible to prevent premature destruction of the facility.

Given expressions (12), (16), the distribution density $N*(\varepsilon)$ of a normally distributed noise $E(t)$ with the mathematical expectation $m_E = 0$ can be calculated from the expression:

$$N^*(\varepsilon) = \frac{1}{\sigma_E^* \sqrt{2\pi}} e^{-\frac{\left(\varepsilon - m_E^*\right)^2}{2\left(\sigma_E^*\right)^2}}, \tag{17}$$

where the values of $N*(\varepsilon)$ are calculated within the range of $m_E^* - 3\sigma_E^* \leq E(t) \leq m_E^* + 3\sigma_E^*$.

Then the high-order moments of the noise $E(t)$ can be calculated from the formula:

$$\mu_{Eq}^* = \int\limits_{-\infty}^{+\infty} \left(\varepsilon - m_E^*\right)^q \frac{1}{\sigma_E^* \sqrt{2\pi}} e^{-\frac{\left(\varepsilon - m_E^*\right)^2}{2\left(\sigma_E^*\right)^2}} d\varepsilon. \tag{18}$$

In addition, the high-order moments of the normally distributed noise $E(t)$ can also be calculated from the recurrence relations

$$\mu_{Eq}^* = (q-1)\left(\sigma_E^*\right)^2 \mu_{E(q-2)}^* \text{ or}$$
$$\mu_{E4}^* = 3\left(\sigma_E^*\right)^4, \ \mu_{E6}^* = 15\left(\sigma_E^*\right)^6, \ \mu_{E8}^* = 105\left(\sigma_E^*\right)^8, \tag{19}$$

In the following sections, we consider the possibility of using such characteristics of noise as variance, mean square deviation, high-order moments, correlation function, correlation coefficient between the useful signal and the noise, the distribution law to create a database for early detection of the transition of an industrial facility from one technical state to another, and to determine the degree of defect development in the latent initial stage.

5   Creating a Database of Estimates of the Characteristics of the Noise for Monitoring the Technical Condition of Industrial Facilities

For early detection of defects in industrial facilities and monitoring the dynamics of malfunction development, the following technology is proposed to create a database of estimates of the characteristics of the noise.

1. At the initial time instant t_0, when the industrial facility is in an operational condition, i.e., there are no changes in the form of deformations, cracks, wear, breakages, corrosion and other faults and defects, we calculate by formulas (12), (13), (15), (17)–(19) the estimates of the mean square deviation $\sigma^*_{E,t0}$ of the noise, high-order moments $\mu^*_{E2,t0}, \mu^*_{E4,t0}, \mu^*_{E6,t0}, \mu^*_{E8,t0}, \ldots$, the normalized cross-correlation function $\rho^*_{XE}(\mu)_{t0}$ and correlation coefficient $r^*_{XE,t0}$ between the useful signal $X(t)$ and the noise $E(t)$. It is obvious that at this moment the equations $\rho^*_{XE}(\mu)_{t0} = 0$, $r^*_{XE,t0} = 0$ are valid. The values of estimates of the mean square deviation $\sigma^*_{E,t0}$ and high-order moments $\mu^*_{E2,t0}, \mu^*_{E4,t0}, \mu^*_{E6,t0}, \mu^*_{E8,t0}, \ldots$ are the result of the influence of the noise $E_1(t)$ caused by external factors. These data on the operational state of the industrial object and the characteristics of the noise with an indication of the type of external factors are recorded in the data bank as reference values of the operational state of the facility.

2. After a certain time interval at the instant t_1, when the industrial facility is subjected to all possible loads, the above characteristics are recalculated. If the equalities $\rho^*_{XE}(\mu)_{t1} = 0$, $\rho^*_{XE}(\mu)_{t1} \approx 0$, $\left| r^*_{XE,t1} \right| \approx 0$ are satisfied, and at the same time the estimates of normalized cross-correlation function and mean square deviation do not exceed some threshold value $\rho^*_{XE}(\mu)_{t1} \leq \Delta_1$, $\left| r^*_{XE,t1} \right| \leq \Delta_2$, then it means that the facility continues to remain in operational condition. But if at the same time the impact of external factors has changed, then the values of the estimates of the characteristics $\sigma^*_{E,t1}, \mu^*_{E2,t1}, \mu^*_{E4,t1}, \mu^*_{E6,t1}, \mu^*_{E8,t1}, \ldots$ of the noise $E_1(t)$ will also change. These values are also recorded in the data bank as reference values of the operational state of the facility with an indication of the type of external influence.

3. Then at the instant t_2 we recalculate the estimates of the noise characteristics $\sigma^*_{E,t2}$, $\mu^*_{E2,t2}, \mu^*_{E4,t2}, \mu^*_{E6,t2}, \mu^*_{E8,t2}, \ldots, \rho^*_{XE}(\mu)_{t2}, r^*_{XE,t2}$. If the inequalities $\rho^*_{XE}(\mu)_{t2} \neq 0$ and $r^*_{XE,t2} \neq 0$ are satisfied, then it indicates the appearance of a defect or malfunction in the technical condition of the industrial facility. Then we compare the high-order moments. Here, if $\mu^*_{E2,t2} \approx \mu^*_{E2,t1}$, then a fourth-order moment $\mu^*_{E4,t2}$ should be calculated for the noise. If $\mu^*_{E4,t2} \approx \mu^*_{E4,t1}$, we should calculate a sixth-order moment $\mu^*_{E6,t2}$. This process should be continued until the value of some moment differs significantly from the value at the previous time instant t_1. Then it is necessary to identify the fault, memorize the order of the moment and its numerical value, and establish a correspondence between the type and degree of the malfunction and the order of the moment.

4. The above calculation procedure is also repeated at time instants $t_3, t_4, \ldots, t_k$. The results of the calculations of the noise characteristics obtained at each stage

are entered into the data bank of informative attributes and are used to assess the current technical state of the industrial facility and to determine the dynamics of malfunction development. Here the matrix of informative attributes for the noise can be written as:

$$
N = \begin{pmatrix}
\rho^*_{XE}(\mu)_{t0} & r^*_{XE,t0} & \mu^*_{E2,t0} & \mu^*_{E4,t0} & \cdots & \mu^*_{En,t0} & 0 & 0 \\
\rho^*_{XE}(\mu)_{t1} & r^*_{XE,t1} & \mu^*_{E2,t1} & \mu^*_{E4,t1} & \cdots & \mu^*_{En,t1} & 1 & 1 \\
\rho^*_{XE}(\mu)_{t2} & r^*_{XE,t2} & \mu^*_{E2,t2} & \mu^*_{E4,t2} & \cdots & \mu^*_{En,t2} & 2 & 2 \\
\cdots & \cdots & \cdots & \cdots & \cdots & \cdots & \cdots & \cdots \\
\rho^*_{XE}(\mu)_{tk} & r^*_{XE,tk} & \mu^*_{E2,tk} & \mu^*_{E4,tk} & \cdots & \mu^*_{En,tk} & n & n
\end{pmatrix},
$$

where the penultimate column characterizes the malfunction that occurred and means: 0—fully operational; 1—operational without damage; 2—partially operational with minor damage; 3—limited operational; 4—non—operational; 5—pre-emergency; 6—emergency, etc. depending on the values of estimates of the characteristics; the last column characterizes the dynamics of malfunction development and mean: 0—facility is operational; 1—the defect exists but is not developing; 2—the defect is developing; 3—the defect is developing rapidly, …, n—the defect is developing very rapidly.

6　Conclusion

The study has shown that one of the possibilities to identify the initial latent period of occurrence of damage and defects of industrial facilities and to determine the dynamics of their development is by calculating, analyzing and creating a database of estimates of characteristics of the noisy signal's noise correlated with the useful component, at different instants of time of the facility's operation. The application of the proposed techniques in monitoring and control systems allows identifying defects and malfunctions of industrial facilities at an early stage and preventing accidents with tragic consequences.

Acknowledgements The authors thank the sponsors and organizers of the conference.

References

1. Aliev, T.A.: Noise control of the Beginning and Development Dynamics of Accidents. Springer, 201 p (2019). DOI:https://doi.org/10.1007/978-3-030-12512-7
2. Mao, J., Wang, H., Xu, Y., et al.: Deformation monitoring and analysis of a long-span cable-stayed bridge during strong typhoons. Adv. Bridge Eng. **1**(8), 1–19 (2020). https://doi.org/10.1186/s43251-020-00008-5

3. Chen, H., Ulianov, C., Shaltout, R.: 3D laser scanning technique for the inspection and moni-toring of railway tunnels. Transp. Problems **10**, 73–84 (2015). https://doi.org/10.21307/tp-2015-063

4. Guo, W., Jin J., Hu, S.J.: Profile monitoring and fault diagnosis via sensor fusionfor ultrasonic welding. J. Manuf. Sci. Eng. **141**(8), 081001–1–81001–13(2019). https://doi.org/10.1115/1.4043731

5. Aliev, T.A., Musaeva, N.F., Suleymanova, M.T., Gazizade, B.I.: Analytical representation of the density function of normal distribution of noise. J. Autom. Inf. Sci. **47**(8), 24–40 (2015). https://doi.org/10.1615/JAutomatInfScien.v47.i8.30

6. Aliev, T.A., Musaeva, N.F., Gazizade, B.I.: Calculation algorithms of the high order moments of interference of noisy signals. J. Autom. Inf. Sci. **50**(6), 1–13 (2018). https://doi.org/10.1615/JAutomatInfScien.v50.i6.10

7. Aliev, T.A., Musaeva, N.F., Suleymanova, M.T.: Algorithms for indicating the beginning of accidents based on the estimate of the density distribution function of the noise of technological parameters. Autom. Control Comput. Sci. **52**(3), 231–242 (2018). https://doi.org/10.3103/S0146411618030021

8. Aliev, T.A., Musaeva, N.F.: Technologies for early monitoring of technical objects using the estimates of noise distribution density. J. Autom. Inf. Sci. **51**(9), 12–23 (2019). https://doi.org/10.1615/JAutomatInfScien.v51.i9.20

9. Aliyev, T.A., Musaeva, N.F., Rzayeva, N.E., Mammadova, A.I.: Development of technologies for reducing the error of traditional algorithms of correlation analysis of noisy signals. Measur. Techn. Springer **6**, 421–430 (2020). https://doi.org/10.1007/s11018-020-01804-1

10. Aliev, T.A., Musaeva, N.F., Rzayeva, N.E., Mamedova, A.I.: Technologies for forming equiv-alent noises of noisy signals and their use. J. Autom. Inf. Sci. **52**(5), 1–12 (2020). https://doi.org/10.1615/JAutomatInfScien.v52.i5.10

On the Errors of the Theory of Safety of Environmentally Hazardous Facilities

Aminagha Sadigov

Abstract Provided the main methodological provisions of the optimal safety management of environmentally hazardous objects. An analysis of the methodological limitations of the concept of protection of the Environmentally Hazardous Facilities (EHF) is given. It is shown that the concept of protection is focused on the statistical regularity of a severe accident, which axiomatically excludes the possibility of managing the safety of an EHF. The methodology for managing environmental safety, taking into account economic efficiency, contains the concept of maximum safety at the minimum possible costs. The accident is considered statistically irregular, which is practically not obligatory, but which theoretically cannot be excluded due to safety management error. A virtual accident is used. It has no regularities and is virtually not required, but theoretically not excluded because of error prevention and mitigation of the accident. This position is qualitatively different from the accepted postulate of the probabilistic model of the accident. The proposed methodology is designed to optimally manage the safety of any EHF.

Keywords Accident · Environmental security · Security management · Mathematical model

1 Introduction

The development of the modern world depends on such important sectors of society as energy, transport, telecommunications, etc. The failure of any one sector threatens society with serious losses or destruction. In the world, there is now a trend of negative processes caused by natural, techno genic, environmental or socio-political cataclysms. The number of terrorist attacks and sabotage, military operations and conflicts, natural disasters is increasing in the world, the number of cyber-attacks,

A. Sadigov (✉)
Institute of Control Systems, Ministry of Science and Education of the Republic of Azerbijan, Baku, Azerbaijan
e-mail: aminaga.sadigov@gmail.com

© The Author(s), under exclusive license to Springer Nature Switzerland AG 2023   139
Sh. N. Shahbazova et al. (eds.), *Recent Developments and the New Directions of Research, Foundations, and Applications*, Studies in Fuzziness and Soft Computing 422, https://doi.org/10.1007/978-3-031-20153-0_11

which poses a threat to the stable functioning of any state. Given the situation, security and implementation issues, critical infrastructure protection concepts are becoming increasingly relevant.

In order to solve the problem of minimizing and managing risks arising from the operation of thermal power systems of critical infrastructure facilities, it is necessary to study and analyze the patterns of occurrence and the effects of environmental and techno genic risks arising from the operation of thermal power systems, which are the cause of the deterioration of environmental, techno genic and socio-political security.

The problem of the environmental safety is particularly acute in technology related to the extraction of energy sources, the production of electricity and waste disposal. The negative impact on the environment as a result of energy consumption is increasing dramatically. In the absence of protection, the costs of eliminating negative consequences exceed the cost of the generated electricity [1, 2]. The problem of environmental safety management is complex. Therefore, it cannot be formulated within the framework of any one theory. Any of the theories, due to its axiomatic construction, allows us to formulate only a limited range of tasks, which is a small part of the totality of the tasks of the problem.

In the process of solving problems, various theories are used, such as cybernetics, automatic control theory, diagnostics, strength, reliability, safety, etc. Each of these theories has its axiomatic, methodology, mathematical apparatus and application area, different from the other theories. Therefore, in essence, each of these theories is local and solves a certain part of the problems related to the problem (without taking into account the connection with the solutions of other problems) with the help of the corresponding local theories. The unification of solutions to particular problems into a general algorithm for solving the problem cannot be performed by any local theory, due to a partial approach to solving the problem. Typically, a general algorithm for solving a problem is developed by a highly qualified practitioner who combines local solutions into a system solution at a logical-expert level [1–18].

The consequence of the logical-expert approach to solving the problem is the impossibility to build a mathematical model for managing complex complex objects. This leads to the impossibility of optimization and, as a result, to the danger of large economic losses.

This is completely true in relation to both the classical theory of safety and the classical theory of reliability. They do not consider the theory of safety and reliability management. They are limited to analysis only.

The influence of the human factor in the technology of managing the safety of human–machine systems is not limited to the assessment of the operator. As noted, this influence should be considered only as one, but by no means the most important of the factors in the problem of the effectiveness of the technological process of managing the safety of human–machine systems, which is a nuclear power plant.

2 Basic Terms and Definitions

An accident—an event exceeding the permissible rate of emissions of harmful substances outside an environmentally hazardous facility.

A virtual accident—an accident for which it is impossible to establish regularities (recurrence of causes and effects), which is hypothetically possible (not excluded due to an error in safety technology and incomplete knowledge about the accident), but is practically not necessary.

Maximum system (element) safety—a complex property that includes nuclear, radiation and environmental safety of the system.

The possibility of an accident—a sequence of events of the alleged causes of the occurrence of an accident, which does not have a priori stable statistics of occurrence.

A random variable is a mathematical prototype of a variable in probability theory.

Dose—the amount of radiation energy transmitted to the human body.

Total safety costs—the sum of the costs of preventing, mitigating and insuring the consequences of an accident.

The interval of the safety controllability of an object—a zone of potential danger of the transition of an object from an operable state to an emergency state due to an error in the accident prevention technology and incomplete knowledge.

The control loop of a system (element)—a closed system consisting (consisting) of a control object (element)—direct connection (transfer of information from input to output) and a control subsystem—feedback (transfer of information from output to input) of the system (element).

The reliability of a system (element)—a complex property that includes the reliability, durability and maintainability of the system (element).

The flow of information (information flow)—an information variable indicating the direction of information transfer.

Safety (reliability) management system—a set of managed environmentally hazardous (unreliable) objects with a safe (reliable) subsystem for managing its safety (reliability).

Randomness—a statistical regularity that takes into account the property (patterns) of the mass of the same type of elements (phenomena) under conditions of statistical stability.

Security management of a system (element)—a sequence of measures aimed at ensuring the security of a system (element).

Optimal system security management—ensuring the technical and economic efficiency of system security management.

Statistical stability—conditions for the repetition of randomness according to experimentally determined physical laws (reasons).

Technical efficiency—the determination of the optimal values of the reliability indicators of the elements of the safety management system according to the given optimal value of the system accident risk indicator.

Economic efficiency—the determination of the optimal value of the risk indicator of a system failure with a minimum of possible total costs for maximum safety.

3   Analysis of the Concept of Protection of Environmentally Hazardous Facilities

The five levels of technical and organizational measures to ensure the protection of the EHF described in [3] are purely expert and qualitative. There is no methodology of economic efficiency on their basis. It must be emphasized that safety is the goal of optimal management, and costs determine the practical possibilities of ensuring safety. Minimization of safety costs is a necessary condition for the competitiveness of nuclear power. Taken by itself, the thesis "maximum security" has neither theoretical nor practical meaning. Taking into account competition, nuclear energy can exist only if the necessary profit and cost of energy are ensured. The cost also includes security costs, including insurance. Otherwise, a necessary condition for the practical provision of security is to minimize the cost of it, as one of the elements of profit maximization in the common science-intensive technology of control on the quality base of the logic-expert methods. This practically excludes the uniqueness of safety control models and their adequacy to the technological processes. The possibility of optimization of the whole control process is excluded. In particular, the theory of the safety analysis and reliability theory as well as local ones. The risk analysis itself cannot be considered independently from the control, as it is its constituent part and can have positive practical importance only in interconnection with other control procedures provided they correspond to the required validity.

Figure 1 shows the curves that correspond to the principles of prevention and mitigation of accidents [3].

These principles are not sufficient for optimal security management. Curve 1 shows that a decrease in the risk of an accident requires an increase in costs. According

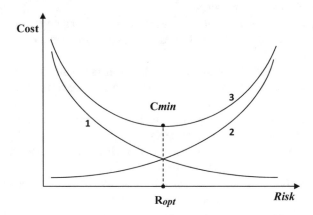

Fig. 1 Dependence of the total safety costs on the accident risk indicator. *Curve 1*—costs for the technology of prevention and mitigation of the accident; *Curve 2*—insurance costs for the consequences of a possible accident; *Curve 3*—total costs for both technologies (1 and 2); C_{min}—minimum total security costs; R_{opt}—the optimal value of the virtual accident risk index

to it, it is impossible to substantiate the optimal value of the accident risk indicator, as well as the corresponding minimum safety costs.

Thus, the concept of safety of EHF is purely expert. It is based on the methodology of economic efficiency, as well as the principles of its implementation by methods of optimal safety management of nuclear power plants.

4 Modeling Peculiarities of the Accident Time

The principal concept of the safety control theory is the choice of the mathematical model of the time of accident occurrence at the nuclear power plant (NPP) [4–7]. The concept of the unlimited random continuous variable of the time of occurrence of a severe accident η belongs to the semi-infinite interval: $y \in [0; \infty]$. As is known, the probability

$$P\{\eta \in [0; \infty]\} = \int_0^\infty \psi(y)dy = 1 \tag{1}$$

The probability of a severe accident at the interval of the operation life $[0, \tau]$ equals

$$P\{\eta \in [0, \tau]\} = \int_0^\tau \psi(y)dy = 1 - \int_\tau^\infty \psi(y)dy \tag{2}$$

With the generally normalized assumed value of severe accident risk probability 10^{-7} reactor/year and operation life $\tau = 30$ years, the probability of the accident during the period of the operation life

$$P\{\eta \in [0.\tau]\} = \int_0^\tau \psi(y)dy \cong 3.10^{-6} \text{ reactor/operation life} \tag{3}$$

is insignificant. This, according to rate-setters, makes it impossible for any severe accident with the emission of radioactive substances materials to happen. However, the error of approximation of the limited operation life of the object is not taken into account by the unlimited random value which leads to the next incorrectness of the mathematic model. Indeed, for the unlimited random value of the time of the severe accident η, the value of the accident probability at the interval after removing NPP from service $[\tau; \infty]$ equals

$$P\{\eta \in [\tau; \infty]\} = \int_{\tau}^{\infty} \psi(y)dy = 0,999997 \tag{4}$$

From the expression (4) it is clear that the greatest (almost one hundred percent) accident probability, for example, melting of the active zone, will come after removing NPP from service. Naturally, such a conclusion contradicts the main concepts of the NPP properties. It is obvious that after removing from service (provided there is no nuclear fuel in the reactor) melting of the active zone is impossible.

Let's dwell upon the correctness of the model of the limited random value of severe accident ξ with the probability density $\varphi(x)$, when the value x of the random value ξ belongs to the final interval: $x \in [0; \tau]$. According to the theory, probability of the severe accident occurrence during the operation life

$$P\{\xi \in [0; \tau]\} = \int_{0}^{\tau} \psi(y)dy = 1 \tag{5}$$

Consequently, according to the model of the limited random value ξ, severe accident at the NPP during the operation life is inevitable. According to the above mentioned, postulation of the model both as unlimited η, and limited ξ random values of time of the severe accident occurrence is not suited for the optimal safety management of the NPP to prevent the accident as with such postulation accidents become statistically regular (theoretically inevitable) and, thus, any protection is theoretically impossible.

In the classical theory of reliability, the event of failure-free operation of the system A consisting of an object and a safety management subsystem (the health events of which respectively ε_1 and ε_2) is equal to the product of the health events respectively ε_1 and ε_2: $A = \varepsilon_1\varepsilon_2$.

The main concept for ensuring the environmental safety of a system whose object is potentially environmentally hazardous is to ensure the reliability of the control subsystem in the form of defense in depth. Per the concept, the failure event of the system $\overline{B}$ occurs when the failure event of the object $\overline{\varepsilon}_1$ and the failure event of defense in depth subsystems $\overline{\varepsilon}_2$ occur together, i. e. event $\overline{B} = \overline{\varepsilon}_1 \overline{\varepsilon}_2$. Accordingly, in the event of the absence of an accident $B = \varepsilon_1 + \varepsilon_2$.

The accepted model of the absence of an accident—an event B and the model corresponding to it—an event $\overline{B}$ are correct from the standpoint of the accident analysis. However, these models are not consistent with reliability models. Indeed, as already noted, the event of failure-free operation of the system A, consisting of their object and protection subsystem, according to modern reliability theory is equal to $A = \varepsilon_1\varepsilon_2$ and the failure event $\overline{A} = \overline{\varepsilon}_1 \overline{\varepsilon}_2 = \overline{\varepsilon}_1 + \overline{\varepsilon}_2$.

As you can see, event A is not equal to event B. It follows that for a given value of the probability of acceptable risk of an accident $P(B)$ and consequently, the probability of the absence of an accident $P(\overline{B})$, it is impossible to determine the requirements for the probability of failure-free operation of the system $P(A)$.

The contradiction between models A and B is fundamental. It indicates the lack of general methodological foundations necessary for the analysis of the risk of an accident and the reliability of NPPs in the safety management algorithm according to the concept of defense in depth.

In classical structural theories (theories of transport systems, energy systems and communication systems, electrical circuits, etc.), an object model is built based on the laws of connection between the material flows of elements and object flows. These laws could not be directly used to build models of the object, taking into account restoration and protection, since there are no material (energy) flows of the object. In this regard, in these classical theories, a logical-mathematical approach was applied. It is based on the construction of a mathematical model for the analysis of the technological process of safety management based on the algebra of logic, probability theory and mathematical statistics. A model, for example, of an object's security is represented as a set of enumeration of all possible emergency sequences—combinations of operable and inoperable states of elements. For an object of n elements, the number of all possible emergency sequences—combinations of operable and inoperable states of elements—the dimension of the model is $z = 2^n$. For example, for $n = 18$, the value is $>260,000$.

A qualitative increase in the model dimension with an increase in the number of elements leads to a dimension problem that is unsolvable in the framework of classical reliability and safety theories.

Due to the lack of consideration of the laws of connection of elements in the structure of an object, in the classical theories of reliability and safety, the feedback theory is impossible, without which, accordingly, the theory of reliability and safety control is impossible.

5   The Principle of Maximum Safety of EHF at a Minimum of Possible Costs

Optimal safety management of EHF by combining technical efficiency (ensuring maximum safety) with the practical possibility of its implementation based on economic efficiency (minimizing the total costs of safety and their optimal distribution) is the basis of the Concept of maximum (nuclear, radiation and environmental) safety at a minimum of costs for it [8].

The problem of control with incomplete knowledge has already been considered, for example, in cybernetics, according to the entropy estimate of the probability of an accident. However, the statistical stability of the accident was postulated. The need to combine the analysis of probabilistic regularities with the analysis of the absence of such regularities (with an analysis of possibilities) is explained by the lack of statistically stable data on severe accidents and the unacceptability of the postulate of statistical stability (patterns) of an accident. In this regard, it is advisable to introduce the concept of a virtual accident as an antipode to the concept of a probabilistic accident.

The safety manageability margin interval of the EHF is one of the basic concepts related to ensuring safety under the virtual accident hypothesis. This stock is on the frontier of sustainable security management. It defines the zone of unstable control and the potential danger of the transition of the EHF from an operable to an emergency state, caused by an error in preventing an accident.

Therefore, the EHF safety manageability margin interval can be determined through the values of z_1 and z_2 controlled value τ, for example, the temperature of the coolant for NPPs, in the following form:

$$\Delta_1 = [z_1, z_2]$$

where z_1 is the lower limit, and z_2 is the upper limit of the value of the controlled value τ.

6 Conclusion

The classical methods of reliability analysis are considered, both for non-restorable elements and systems, and for elements and subsystems with restoration. The analysis of logical-probabilistic methods and methods using Markov processes has been carried out. It was revealed that the classical methods of reliability analysis do not take into account the events of information flows of elements and subsystems, being limited to state events, as well as health (failure) events of elements and subsystems. This basically rules out the possibility of developing a feedback theory. As a result, these methods are limited only by the ability to analyze the reliability of recoverable systems and systems with redundancy. They do not contain a reliability management methodology. In particular, using these methods, it is impossible to build correct models of system reliability without redundancy, taking into account the events of information flows and control and protection systems. These methods are not suitable for optimization.

In general, reliability analysis methods are widely used in practice. It is expedient to use them in the EHF safety control algorithm as local methods for analyzing the reliability of recoverable systems and systems with redundancy.

Errors of safety analysis methods are considered on the examples of the logical-probabilistic method and the event tree method. It is shown that it is advisable to use the logical-probabilistic method to build an object safety model when a probabilistic accident model is acceptable and it is impossible to determine information flows, and the connection between the individual components of the object can be established only on the basis of a logical analysis of the object's functioning.

Particular attention is paid to the analysis of the methodological limitations of the event tree method. This is due to the fact that this method is widely used in practice. It has dimensionality and optimization limitations that have been noted for reliability analysis methods. The used hypothesis of the probabilistic model of the accident has disadvantages, which were mentioned in the presentation of the main provisions of the first chapter. It is shown that this method has fatal methodological limitations due to the fact that the dependence of failure events (operability) of safety and protection controls is not taken into account. As a result, there are unremovable theoretical errors in the analysis of the safety of EHF and, in particular, nuclear power plants. It is proved that the method allows to find a rough upper estimate of the accident risk index. The essential advantages of the event method include the fact that it divides the general task of security analysis into incompatible parts. For each part, the NPP specialist performs a safety analysis, which makes it possible to avoid the errors of a purely mathematical approach that does not take into account the technological features of the NPP.

Optimal safety control with the use of physical models of NPP regimes is considered. An example of a combination of physical models and the theory of NPP safety management with the help of several technological control loops is given.

An analysis of the effectiveness of the human factor in the safety management technology of the EHF was carried out using the example of a nuclear power plant. Two aspects of this problem are considered. The fundamental limitations of the control structure are shown, in which the operator directly enters as an element, without which it is impossible to close the control loop.

The proposed methodology is designed for optimal security management of any EHF.

References

1. Balat M.: Usage of energy sources and environmental problems. Energy Exploration & Exploitation **23**(2), 141–168 (2005). https://doi.org/10.1260/0144598054530011
2. Sorkin, L.R. et al.: Environmental Security Management: Industry and Settlements of the Future. Moscow (2022). http://www.researchgate.net/publication/357701189_upravlenie_ekol ogiceskoj_bezopasnostu_industria_i_poselenia_budusego
3. Pampuro, V.I.: A system of indicators of environmental safety of objects. Reports Nat. Acad. Sci. Ukraine **12**, 186–193 (2012). http://nbuv.gov.ua/UJRN/dnanu_2012_12_31
4. Sadigov, A.B.: Methodological restrictions of the theory of ensuring the safety of ecologically dangerous objects. https://www.arpapress.com/Volumes/Vol13Issue1/IJRRAS_13_1_19.pdf
5. Uryadnikova, I.A. et al.: Methodological bases of management of man-made risks at the heat power facilities of the critical infrastructures (2014). https://cyberleninka.ru/article/n/metodo logicheskie-osnovy-upravleniya-tehnogennymi-riskami-na-teploenergeticheskih-obektah-kri ticheskoy-infrastruktury/viewer
6. Deterministic Safety Analysis for Nuclear Power Plants. IAEA Safety Standards. No. SSG-2 (Rev. 1), 85 p (2019). https://www.iaea.org/publications/12335/deterministic-safety-analysis-for-nuclear-power-plants
7. Raso, A.: Use of reliability engineering tools in safety and risk assessment of nuclear facilities. International Nuclear Atlantic Conference—INAC 2017. Brazil, October 22–27, (2017). https://inis.iaea.org/collection/NCLCollectionStore/_Public/49/018/49018251.pdf
8. Sadigov, A.B., Zeynalov, R.M.: Optimal control in the problems of calculating the benefit/cost ratio in emergency response. Inf. Control Problems **40**(1), 47–56 (2020). https://icp.az/2020/1-06.pdf

9. Sadigov, A.B.: Creation of mathematical models and methods for solving tasks of operational control in emergency situations. Comput. Math. **1**, 37–45 (2011). http://nbuv.gov.ua/UJRN/Koma_2011_1_6

10. Sadigov, A.B. et al.: Formulation of the problem and general requirements for optimization methods of operational and organizational control systems. International Conference on Problems of Logistics, Management and Operation in the East-West Transport Corridor (PLMO) (2021). https://conf.cyber.az/files/papers/217-225.pdf

11. Duffy, V.G. (Ed).: Digital human modeling. Applications in Health, Safety, Ergonomics and Risk Management. Springer. DHM (2014). https://books.google.az/books?id=7ee5BQAAQBAJ&pg=PA134&lpg=PA134&dq=human+factor+in+the+technology+of+optimal+control+of+environmental+safety+of+human-machine+systems

12. Sadigov, A.B. et al.: Modeling human actions and psychophysical reactions in emergencies. Inf. Control Problems **42**(1) (2022). https://doi.org/10.54381/icp.2022.1.01

13. Sadigov, A.B., Zeynalov, R.M.: Mathematical modeling of environmental processes at military facilities. Inf. Control Problems **40**(2), 31–37 (2020). https://icp.az/2020/2-04.pdf

14. Sadigov, A.B., Mustafayev, I.I., Hajimatov, G.N., Gafarov, E.K.: Description of a high-risk region as a control object. Inf. Control Problems **41**(2), 3–7 (2021). https://doi.org/10.54381/icp.2021.2.01

15. Haddow, G.D. et al.: Natural and technological hazards and risk assessment. In book: Introduction to Emergency Management, pp.33–77 (2017). https://www.researchgate.net/publication/313826615_Natural_and_Technological_Hazards_and_Risk_Assessment

16. Twenty-five Years after Chornobyl Accident: Safety for the Future. Kiev, 328 p (2011). https://inis.iaea.org/collection/NCLCollectionStore/_Public/52/029/52029712.pdf

17. Sadigov, A.B.: Methods of risks assessment of life activity in the environment. In: 1st International Turkish World Engineering and Science Congress in Antalya. Turkey (2017). https://drive.google.com/file/d/0B2qIyCqQOx80UDRlbFltaFMyN2hvclZQREFLbF9BaWFhdy1z/view?resourcekey=0-U9qOBKo0S9xk3zDvqjI1uQ

18. Sadigov, A.B.: Models and technologies for solving management problems in emergencies. Baku, Elm, (2017) 372 p. (In Russian). https://www.researchgate.net/publication/315893563

Benefits Realization Approach to Project Management Using Social Fuzzy Consensus Group Decision Support

Ru. D Perera, A. Ghildyal, and E. Chang

Abstract In recent years there is an upsurge in the adoption of the Benefit Approach to Project and PPP (Project, Program, and Portfolio) management in the Government sector and large business particularly in the area of procurement, commercial grade contracts, IT governance, innovation investment space. However, our studies show that 80% of project benefits laid out in the procurement contracts are intangible benefits and benefits comprise an unclassified broad range of indirect achievable, and unquantifiable tangible and intangible benefits that require theoretical interpretation and empirical weightage so that such a Benefit Approach to project or PPP approach can contribute to the long-term project success. Setting measurable achievable benefits is not an accident as the nature of benefits is fuzzy, evolving, and dynamic and rarely supports benefit monitoring and evaluation. In this paper, we address two substantial issues. Firstly, the nature and scope of intangibility cause vagueness in the definition of the Benefits. Therefore, a consensus on the agreed achievable tangible and intangible benefits and their measurement metrics are accepted among stakeholders. These are underpinned by theoretical interpretation of the achievable benefits. Secondly, the measurability and quantification of achievable benefits particularly intangible benefits is new and emerging. The lack of a systematic approach may lead to accountability and governance failure as the project evolves over time. In this study, we use consensus-reaching group decision support techniques to quantify the achievable benefits which are a prerequisite to long-term project success.

Ru. D Perera
Defence University, Colombo, Sri Lanka
e-mail: dhinesha_drp@kdu.ac.lk

A. Ghildyal
Department of Defence, NNSO, Navy, Canberra, ACT, Australia
e-mail: amit.ghildyal@defence.edu.au

E. Chang (✉)
Defence Logistics Innovation Group, IIIS, School of ICT, Griffith University, Queensland, QLD, Australia
e-mail: e.chang@griffith.edu.au

Keywords Benefit approach · Project benefits realization measures · Fuzzy consensus · Group decision support

1 Introduction

The recent emergence of high demand on enterprise consulting for project management has resulted in the emergence of new academic discipline development in Project, program & portfolio (PPP) in many institutions worldwide [1, 2]. This is due to the fact that customers such as government departments are not aware of the benefit that has been identified or presented in a project contract or as part of PPP offered by the providers. This lack of achievable to quantifiable has led to only 2% of digital transformation projects succeeding (CIO group, 2022). It is generally viewed that the performance of the portfolio is based on its programs' performance and each program's performance is contributed by the performance of its projects. Therefore, in terms of benefits identification in PPP, the benefits are transferable to the upper layers of the PPP, where the benefit in PPP is the composite of the projects and programs' benefits and must be in line with the Organization or Enterprise objectives or goals [3].

Our recent field studies and practical PPP evaluation of the benefit approaches in PPP showed that although tangible benefits such as cost saving, Return on Investment, etc can be identified and measured, approximately 80% of benefits identified in the project contracts or projects are Intangible benefits, such as capability, efficiency, etc and no significant measurement or techniques or tools are available to date to deal with them in the Benefit Approach to PPP. This paper provides our studies and proposals on how to measure those intangible benefits in PPP, to achieve "what gets measured gets delivered". The benefits (Tangible and intangible) are largely identified in the context of strategic intent, social economic drivers, industrial inventive steps, or futuristics [4]. Due to the vagueness and imprecision of benefit concepts, particularly intangible benefits, there are no solid measurement techniques and tools available to help customers or large government organizations to **realize** the benefit resulting in repeated high costs in procurement, contracts, IT and investment. We have previously [5, 6] drawn upon the methodologies in Stratification (CST), Fuzzy inference system (FIS), and Computing with Words an Evidence-based approach to developing a scientific analytics and measurement framework to quantify intangible benefits in order to overcome the vagueness, fuzziness, and imprecision of intangible benefits in Project or PPP management. The output of intangible benefits measures is presented as a realization value that is mapped to the direct and indirect project value with financial and non-financial values. Our previous work relies on the opinions of experts. However, an important aspect of providing an acceptance of the results is agreement between the different stakeholders about the realized benefits. To provide this important dimension in our benefits realization approach, we utilize the fuzzy consensus-based approach in this paper.

2 Reaching Consensus on Benefit Approach in PPP

In technology related IT Projects the benefits are usually captured easily but its measurement is vague and fuzzy. The study focused on IT related projects used in companies or the public service to make efficient functions in the organization at different strategic functional units namely, HR, finance, sales, marketing, compliance, process automation etc. (Table 1).

The public sector explains that benefits management in terms of intangible benefits is counter intuitive and needs to be effectively governed and regulated. Only 23% of identified benefits are measurable and realizable from a quantifiable perspective [7]. The Department of Prisons & innovative industries in unison explains that the benefit approach to measure intangibles is fuzzy, weak and uncertain. It is a timely and pertinent to approach project benefit measures without compromising the utility value emerging from IT driven projects [8]. The specific project under study is a supply chain automation project which integrates e-governance into its operational process optimization. The supply chain that caters to the Defence industry from an enterprise viewpoint comprises of sourcing, processing and final product manufacturing from an upstream and customer delivery, customer development and after sales services from a downstream point of view. Each stage of the supply chain is computerized, and automation is crucial as a business process reengineered application. The system has periodical user testing and upgrades by the intervention of domain experts, project managers and customers or system end users. Thereby the project can be classified as a system user project of medium scale and the target benefits are stated as; short term deliverables at initiation and growth phases (3–6 months from system development), medium termed deliverables at maturity phases (6 months 2 years) from system implementation and long-term outcomes at stability phase (2–5 years).

Whilst financial consideration was the baseline for most project managers in the past, today the project benefit managers are not the only ones who are interested in the project but there are other parties namely, the domain experts and the larger interest groups like the stakeholders. Benefit realization is usually measured at individual level of each of the interested groups and the aggregate consensus leads to the project level (Table 2).

In the light of fuzzy set theory and related principles we approach the representation of our case study. Most(Q) of the relevant(B) project contract managers (y's) agree (F) as to almost all of the decision support process. The general notation and expression can be exemplified as:

$$Qy's \text{ are } F \tag{1}$$

where, Q is the linguistic quantifier (eg: Almost), $Y = \{y\}$ is a set of objects (eg: projects) and F is a property (eg: merchantable quality)

In developing the essence we may assign a different importance such as a specific characteristic B, to the particular y set of objects yielding an expression

Table 1 Operationalization

Variable	Dimension	Mathematical expression	Social Fuzzy Preference statement by StakeH
Product Benefit (PC VAR PRODUCT BENEFIT)	Product Requirement 1	$\mu Q = \left(\dfrac{\sum_{i=1}^{p} \left[\mu B(psi)^{\wedge} \mu F(psi) \right]}{\sum_{i=1}^{p} \mu B(psi)} \right)$	The most of the relevant project sponsors (ps) agree as to almost all of the product function 1
	Product Requirement 2	$\mu Q = \left(\dfrac{\sum_{i=1}^{p} \left[\mu B(ci)^{\wedge} \mu F(ci) \right]}{\sum_{i=1}^{p} \mu B(ci)} \right)$	The most of the relevant project clients (c) agree as to almost all of the product function 2
	Product Requirement 3	$\mu Q = \left(\dfrac{\sum_{i=1}^{p} \left[\mu B(ei)^{\wedge} \mu F(ei) \right]}{\sum_{i=1}^{p} \mu B(ei)} \right)$	The most of the relevant project experts (e) agree as to almost all of the product function 3
	Product Requirement 4	$\mu Q = \left(\dfrac{\sum_{i=1}^{p} \left[\mu B(ei)^{\wedge} \mu F(ei) \right]}{\sum_{i=1}^{p} \mu B(ei)} \right)$	The most of the relevant project experts (e) agree as to almost all of the product function 4
	Product Requirement 5	$\mu Q = \left(\dfrac{\sum_{i=1}^{p} \left[\mu B(ei)^{\wedge} \mu F(ei) \right]}{\sum_{i=1}^{p} \mu B(ei)} \right)$	The most of the relevant project experts agree as to almost all of the product function 5
	Product Requirement 6	$\mu Q = \left(\dfrac{\sum_{i=1}^{p} \left[\mu B(ei)^{\wedge} \mu F(ei) \right]}{\sum_{i=1}^{p} \mu B(ei)} \right)$	The most of the relevant project experts agree as to almost all of the product function 6
	Product Requirement 7	$\mu Q = \left(\dfrac{\sum_{i=1}^{p} \left[\mu B(ei)^{\wedge} \mu F(ei) \right]}{\sum_{i=1}^{p} \mu B(ei)} \right)$	The most of the relevant project experts agree as to almost all of the product function 7

(continued)

Table 1 (continued)

Variable	Dimension	Mathematical expression	Social Fuzzy Preference statement by StakeH
	Product Requirement 8	$\mu Q = \left(\dfrac{\sum_{i=1}^{p} \left[\mu B(ei)^{\wedge} \mu F(ei) \right]}{\sum_{i=1}^{p} \mu B(ei)} \right)$	The most of the relevant project experts agree as to almost all of the product function 8
Time to Market (PC VAR TIME TO MARKET)	Budget plan	$\mu Q = \left(\dfrac{\sum_{i=1}^{p} \left[\mu B(psi)^{\wedge} \mu F(psi) \right]}{\sum_{i=1}^{p} \mu B(psi)} \right)$	The most of the relevant project sponsors (ps) agree as to almost all of budget plan
	Contingency Budget plan	$\mu Q = \left(\dfrac{\sum_{i=1}^{p} \left[\mu B(ei)^{\wedge} \mu F(ei) \right]}{\sum_{i=1}^{p} \mu B(ei)} \right)$	The most of the relevant project experts agree as to almost all of contingency budget plan
	Time overrun and projection report	$\mu Q = \left(\dfrac{\sum_{i=1}^{p} \left[\mu B(cmi)^{\wedge} \mu F(cmi) \right]}{\sum_{i=1}^{p} \mu B(cmi)} \right)$	Most of the relevant project contract managers (cm) agree as to almost all of the time overrun and projection report
	Delivering the project within the budget	$\mu Q = \left(\dfrac{\sum_{i=1}^{p} \left[\mu B(ei)^{\wedge} \mu F(ei) \right]}{\sum_{i=1}^{p} \mu B(ei)} \right)$	The most of the relevant project experts agree as to almost all of the project is delivered within the budget
Project Cost benefit (PC VAR PROJECT COST BENEFIT)	Overall Budget planning report	$\mu Q = \left(\dfrac{\sum_{i=1}^{p} \left[\mu B(cmi)^{\wedge} \mu F(cmi) \right]}{\sum_{i=1}^{p} \mu B(cmi)} \right)$	Most of the relevant project contract managers (cm) agree as to almost all of the overall budget planning report

(continued)

Table 1 (continued)

Variable	Dimension	Mathematical expression	Social Fuzzy Preference statement by StakeH
	Budget overrun report	$\mu Q = \left(\dfrac{\sum_{i=1}^{p} \left[\mu B(psi)^{\wedge} \mu F(psi) \right]}{\sum_{i=1}^{p} \mu B(psi)} \right)$	Most of the relevant project contract managers agree as to almost all of the overall budgets overrun report
	Resource overrun report	$\mu Q = \left(\dfrac{\sum_{i=1}^{p} \left[\mu B(pmi)^{\wedge} \mu F(pmi) \right]}{\sum_{i=1}^{p} \mu B(pmi)} \right)$	Most relevant project managers agree to most of the resource overrun report
Business or operation benefit (PC VAR BUSINESS OR OPE BENEFIT)	Return on investment	$\mu Q = \left(\dfrac{\sum_{i=1}^{p} \left[\mu B(psi)^{\wedge} \mu F(psi) \right]}{\sum_{i=1}^{p} \mu B(psi)} \right)$	Most relevant project sponsors agree to almost all of the return on investment
	Business Cost Saving	$\mu Q = \left(\dfrac{\sum_{i=1}^{p} \left[\mu B(sci)^{\wedge} \mu F(sci) \right]}{\sum_{i=1}^{p} \mu B(sci)} \right)$	Most relevant project subcontractors (sc) agree to most of the business cost saving
Organization Benefit (PC VAR ORG BENEFIT)	Risk Control and Management	$\mu Q = \left(\dfrac{\sum_{i=1}^{p} \left[\mu B(sci)^{\wedge} \mu F(sci) \right]}{\sum_{i=1}^{p} \mu B(sci)} \right)$	Most relevant project subcontractors agree to most of Risk control and mgmt
	Decision Support Process	$\mu Q = \left(\dfrac{\sum_{i=1}^{p} \left[\mu B(cmi)^{\wedge} \mu F(cmi) \right]}{\sum_{i=1}^{p} \mu B(cmi)} \right)$	Most relevant project contract managers agree to most of decision support process
	Enterprise Culture and Unity	$\mu Q = \left(\dfrac{\sum_{i=1}^{p} \left[\mu B(cmi)^{\wedge} \mu F(cmi) \right]}{\sum_{i=1}^{p} \mu B(cmi)} \right)$	Most relevant project contract managers agree to most of Enterprise Culture and Unity

Table 2 The social fuzzy consensus approach for individual benefit groups

Variable	Dimension	Mathematical expression	Social Fuzzy Preference statement by StakeH
Product Benefit (PC VAR PRODUCT BENEFIT)	Product Requirement 1	$\mu Q = \left(\dfrac{\sum_{i=1}^{p} \left[\mu B(psi)^{\Lambda} \mu F(psi) \right]}{\sum_{i=1}^{p} \mu B(psi)} \right)$	Most relevant project sponsors (ps) agree to most of the product function 1
	Product Requirement 2	$\mu Q = \left(\dfrac{\sum_{i=1}^{p} \left[\mu B(ci)^{\Lambda} \mu F(ci) \right]}{\sum_{i=1}^{p} \mu B(ci)} \right)$	Most relevant project clients (c) agree to almost all of the product function 2
	Product Requirement 3	$\mu Q = \left(\dfrac{\sum_{i=1}^{p} \left[\mu B(ei)^{\Lambda} \mu F(ei) \right]}{\sum_{i=1}^{p} \mu B(ei)} \right)$	Most relevant project experts (e) agree to almost all of product function 3
	Product Requirement 4	$\mu Q = \left(\dfrac{\sum_{i=1}^{p} \left[\mu B(ei)^{\Lambda} \mu F(ei) \right]}{\sum_{i=1}^{p} \mu B(ei)} \right)$	Most relevant project experts (e) agree to almost all of product function 4
	Product Requirement 5	$\mu Q = \left(\dfrac{\sum_{i=1}^{p} \left[\mu B(ei)^{\Lambda} \mu F(ei) \right]}{\sum_{i=1}^{p} \mu B(ei)} \right)$	Most relevant project experts agree to almost all of product function 5
	Product Requirement 6	$\mu Q = \left(\dfrac{\sum_{i=1}^{p} \left[\mu B(ei)^{\Lambda} \mu F(ei) \right]}{\sum_{i=1}^{p} \mu B(ei)} \right)$	Most relevant project experts agree to almost all of product function 6
	Product Requirement 7	$\mu Q = \left(\dfrac{\sum_{i=1}^{p} \left[\mu B(ei)^{\Lambda} \mu F(ei) \right]}{\sum_{i=1}^{p} \mu B(ei)} \right)$	Most relevant project experts agree to almost all product function 7
	Product Requirement 8	$\mu Q = \left(\dfrac{\sum_{i=1}^{p} \left[\mu B(ei)^{\Lambda} \mu F(ei) \right]}{\sum_{i=1}^{p} \mu B(ei)} \right)$	Most relevant project experts agree to almost all product function 8

(continued)

Table 2 (continued)

Variable	Dimension	Mathematical expression	Social Fuzzy Preference statement by StakeH
Time to Market (PC VAR TIME TO MARKET)	Budget plan	$\mu Q = \left(\dfrac{\sum_{i=1}^{p} [\mu B(psi)^A \mu F(psi)]}{\sum_{i=1}^{p} \mu B(psi)} \right)$	Most relevant project sponsors (ps) agree to almost all of budget plan
	Contingency Budget plan	$\mu Q = \left(\dfrac{\sum_{i=1}^{p} [\mu B(ei)^A \mu F(ei)]}{\sum_{i=1}^{p} \mu B(ei)} \right)$	Most relevant project experts agree to almost all of contingency budget plan
	Time overrun and projection report	$\mu Q = \left(\dfrac{\sum_{i=1}^{p} [\mu B(cmi)^A \mu F(cmi)]}{\sum_{i=1}^{p} \mu B(cmi)} \right)$	Most relevant project contract managers (cm) agree to almost all of the time overrun and projection report
	Delivering the project within the budget	$\mu Q = \left(\dfrac{\sum_{i=1}^{p} [\mu B(ei)^A \mu F(ei)]}{\sum_{i=1}^{p} \mu B(ei)} \right)$	Most relevant project experts agree to almost all of the project is delivered within the budget
Project Cost benefit (PC VAR PROJECT COST BENEFIT)	Overall Budget planning report	$\mu Q = \left(\dfrac{\sum_{i=1}^{p} [\mu B(cmi)^A \mu F(cmi)]}{\sum_{i=1}^{p} \mu B(cmi)} \right)$	Most relevant project contract managers (cm) agree to almost all of the overall budget planning report
	Budget overrun report	$\mu Q = \left(\dfrac{\sum_{i=1}^{p} [\mu B(psi)^A \mu F(psi)]}{\sum_{i=1}^{p} \mu B(psi)} \right)$	Most relevant project contract managers agree to almost all of the overall budgets overrun report

(continued)

Table 2 (continued)

Variable	Dimension	Mathematical expression	Social Fuzzy Preference statement by StakeH
Business or operation benefit (PC VAR BUSINESS OR OPE BENEFIT)	Resource overrun report	$\mu Q = \left(\dfrac{\sum_{i=1}^{p} \left[\mu B(pmi)^{\wedge} \mu F(pmi) \right]}{\sum_{i=1}^{p} \mu B(pmi)} \right)$	Most relevant project managers (pm) agree to almost all of the resource overrun report
	Return on investment	$\mu Q = \left(\dfrac{\sum_{i=1}^{p} \left[\mu B(psi)^{\wedge} \mu F(psi) \right]}{\sum_{i=1}^{p} \mu B(psi)} \right)$	Most relevant project sponsors agree to almost all of the return on investment
	Business Cost Saving	$\mu Q = \left(\dfrac{\sum_{i=1}^{p} \left[\mu B(sci)^{\wedge} \mu F(sci) \right]}{\sum_{i=1}^{p} \mu B(sci)} \right)$	Most relevant project subcontractors (sc) agree as to almost all of the business cost saving
Organization Benefit (PC VAR ORG BENEFIT)	Risk Control and Management	$\mu Q = \left(\dfrac{\sum_{i=1}^{p} \left[\mu B(sci)^{\wedge} \mu F(sci) \right]}{\sum_{i=1}^{p} \mu B(sci)} \right)$	Most relevant project subcontractors agree as to almost all of the Risk control and management
	Decision Support Process	$\mu Q = \left(\dfrac{\sum_{i=1}^{p} \left[\mu B(cmi)^{\wedge} \mu F(cmi) \right]}{\sum_{i=1}^{p} \mu B(cmi)} \right)$	Most relevant project contract managers agree as to almost all of the decision support process
	Enterprise Culture and Unity	$\mu Q = \left(\dfrac{\sum_{i=1}^{p} \left[\mu B(cmi)^{\wedge} \mu F(cmi) \right]}{\sum_{i=1}^{p} \mu B(cmi)} \right)$	Most relevant project contract managers agree as to almost all of Enterprise Culture and Unity

$$QB\,y's \text{ are } F \tag{2}$$

"Almost (Q) all 'ICT' (B) project benefit evaluators (y's) are convinced (F), that almost all enterprise culture that emerge with time" or

$$\mu Q = \left(\frac{\sum_{i=1}^{p}\left[\mu B(yi)^{\wedge}\mu F(yi)\right]}{\sum_{i=1}^{p}\mu B(yi)} \right)$$

Most (Q) of the important (B) project evaluators (we need stakeholders here, not just expert) (y's) are convinced (F) about the emerging project benefit.

Both Q y's are F&QB y's are F are rather fuzzy and contain doubt and requires simplicity while justifying its truthfulness. However, underpinned by the epistle of prominent authors in seminal literature we can support the known truth (y is F) meaning stakeholders of the project who impact the project are also being impacted by the outcome of the project. Stakeholders of the project are convinced, or project yields merchantable quality benefits [9, 10]. The manifestation of the mathematical model is explainable through Zadeh's method, a fuzzy linguistic quantifier Q is assumed to be a fuzzy set defined in [0,1]. For instance Q= "most" may be given as:

$$\mu\,most(x)\begin{cases} 1\,for\,x \geq 0.8, \\ 2x - 0.6\,for\,0.3 < x < 0.8, \\ 0\,for\,x \leq 0.3, \end{cases} \tag{3}$$

With another quantifier of 'important experts' we can extend the belief to an analogous discussion that project or program or portfolio managers with the experience of 15 years or more in the company will satisfy the condition of important experts by means of competencies, knowledge and conceptual thinking.

In the case of *"Almost (Q) all 'ICT' (B) project benefit evaluators (y's) are convinced (F) that almost all enterprise culture that emerge with time"; the fuzzy representation for Q= "almost all"* could be:

$$\mu\,almost\,all(x)\begin{cases} 1\,for\,x \geq 0.8, \\ 2x - 0.6\,for\,0.3 < x < 0.8, \\ 0\,for\,x \leq 0.3, \end{cases}$$

Further representing the fuzzy perspectives Property F is defined as a fuzzy set in Y. For instance, if Y = {X,Y,Z} is a set of experts and F is the property 'satisfied', then F may be explicitly written as

F = 'satisfied' = 0.1/X+0.6/Y+0.8/Z; Which means expert X: the project manager, expert Y: the program manager and expert Z: the portfolio manager are satisfied to degree 0.1, 0.6 & 0.8 respectively that the social benefit will be yielded at the end of the project lifecycle. In the light of project cost, quality, time constraints the project triangle explains, the experts are also satisfied to varied degrees how the project benefits, program benefits and portfolio benefits will be satisfied in the foreground of social project value. If the different types of benefit experts are satisfied about the realization of benefits then, $Y = \{y_1,...y_p\}$ it is assumed that the truthful (y_i is F) = $\mu_F(Y_i)$, I = 1, ...,p. The value associated for truth (Q y's are F) is determined by two devised steps of Zadeh [9].

3 Social Fuzzy Preference Relation for Group Decision Support for Benefit Approach

In this area we focus on defining agreed benefit particularly intangible benefit, measure and realization. In the continuum comprising of different stakeholders from client, supplier and third party individuals (need to be stakeholders of the project, not just experts, as here we deal with consensus and the agreed project "Benefits" and that is measurable when the 'important' expert criteria and parameter is added, B is defined as a fuzzy set in Y, and $\mu B(y_i) \in [0,1]$ is a degree of importance of y_i: 1 is important to 0 is unimportant, through all intermediate values. For instance, B='important' = 0.2/X+0.5/Y+0.6Z means that Expert X (Project manager) is important (relevant) to degree 0.2; Expert Y (Program manager) is important to degree 0.5 and Expert Z (Portfolio manager) is important to degree 0.6. In the case study, the Macro level project-level decision making by a Senior Manager aligning with the broader perspectives of organizational objectives like innovation, creative thinking by implementing the e-governance system shall be 'important' for project managers but unimportant or rarely important for program managers and portfolio-level benefit managers. In the instance of a related or non-related projects that are logically combined such as disseminating technology infrastructure to 75% or more rural citizens that will aim at achieving the objective of strengthening rural engagement to procure and acquire raw materials to perform supply chain operations through binding contracts and implementing digital strategy are 'important' for program managers like middle managers, but unimportant for other experts in the benefit management process. In the case of portfolio, the discipline-based subdivision for hierarchical management of departments like document filing, e- documentation processing, money payments are under the purview of portfolio managers like the line managers or supervising project officer. The expert decision is thereby important for portfolio-level managers but unimportant for project and program managers. Supply chain quality (o1), customer equity and fairness (o2), public enterprise good-will (o3) and stakeholder justice (o4) are a few of the intangible outcomes that a project manager must specifically recognize. The senior manager is engaged in a process of identifying and choosing an option [11] as to which intangible benefit in the collection is more essential for the success of the project. Moreover, the benefit manager can position these o1-o4 as a set of alternatives yielding different outcomes (eg: quality can yield financial goals, customer equity can yield customer satisfaction, good will can yield enterprise trade name value, stakeholder justice yields stakeholder satisfaction etc.,) and among these a best option (good, feasible or acceptable) is to be found.

The decision-making process entails a final choice which may be implemented in this case if customer equity and fairness is to be desired the challenge is to implement the stated, agreed benefit which includes its measurement and realization. As a non-financial- intangible benefit it is a benefit with indirect monetary value. The decision process is underpinned by various decision styles used by the benefit manager. The rational and collaborative style of decision-making is crucial for the senior manager in this context to identify, measure and later realize the 'agreed', 'target' or 'stated' benefit with intangible properties in nature. In vertical collaboration the strategic (T1), tactical (T2) and operational (T3) levels of the hierarchy takes centric value in Project Benefit as the projects are designed in PPP management in accordance with the structure. In the T1, T2 & T3 levels, the Benefit can be shared benefits at T1 level, at T2 level and at T3 level, that are synergized toward the broader objectives and vision of the organization. With shared or collaborative benefits, deliverables and project outcomes there is a need for managers to think about collaborative decision making to systematically approach benefit management. Let us assume that the e-governance (If there are 2 case studies, each needs to be described and discussed before using them) project focus is on 'accountability and transparency' which are the broader objectives of the organization at T1 tier. The strategic level project manager can adopt a decision-making style that is conceptual that implies the decision is creative and long-term oriented. This broader project goal can be achieved through customer equity and fairness which is a defined agreed project-level benefit (o2). There are related or non-related projects at T2 tier which emphasize on "customer awareness systems like 'know your customer better' " and 'customer trends and like analysis' which are both projects that are well-coordinated and logically combined to achieve the organizational objectives. In this T2 tier the program managers are crucial decision-makers as their collaboration among themselves and with the project managers at T1 is useful and directly impacts the organizational benefit management. With the essence of portfolio, at T3 tier the faculties such as 'document filing', 'document processing', 'vetting of customers', 'customer services', 'information providing' are few of the championing portfolio of the e-governance project that are cross-functional in nature. Each of the arms ideally has a portfolio benefit manager and their collaboration among one another and the other tier managers are important and crucial for the success of the project from an overall perspective. The understanding from an industrial perspective rooted in Kacprzyk's scholarship explains that at the lower-levels of the organization, tier T3, the decisions made are structured in character and tenor as the decisions are routine-based, operational and decision makers are bound by definite procedure. On the other paradigm, at higher levels the decision making is unstructured, tier T1, where the decision maker must provide judgement, evaluation, and insights into the problem definition and then the solution. Hence, the project overview is in the perspective of the senior project manager who can comment on the issues, problems and outcomes.

While multiple criteria decision analyses, or group decision making analyses are decision-making techniques involving the novel approach of several individual experts raising their voice or voting towards a common understanding or consensus; the process is a collection of logically united steps that lead to a project decision outcome when followed ideally. The process is a coarse classification of four stages namely; intelligence, design, choice and implementation In this range of rational decision-making the real decision making is done collectively, interdependently and persuasive jointly. Hence, the group decision making adheres to a behavioral style of decision-making characterized by multiple goals, multiple decision makers and multiple stages. In the case of meeting of minds of the project benefit stakeholders or their consensus the decision by consensus takes a definite path of evaluation.

A set of m agents $E = \{e_1...e_m\}$ comprising of $E = \{$Proj_mgr, Prog_mgr,Portf_mgr,Enterprise_mgr$\}$ I believe "Agent" here are project stakeholders, including customers. who are competent and experienced to provide testimonies over a set of n options, $O = \{o_1...o_n\}$ to say $O = \{$quality$[b_1]$, investment$[b_2]$, goodwill$[b_3]$, justice$[b_4]$, equity and fairness$[b_5]\}$, as individual fuzzy preference relations, $R_1......R_m$. At $t = 0$, the agent's initial fuzzy preference relation, R_{KK} $= 1,2,..._m$ on the set of options, may differ to a large extent so that the degree of consensus is likely to be lower. The project, program, portfolio and enterprise managers views and thoughts about the benefit b_2 at the point of investment may differ to a larger extent as their individual perception is different and contextual factors like the project interest is diverse. In relation to financial investment the insights of strategic benefit level is at a higher-order while the tactical level is relatively lower. In the case of goodwill, b_3, will be perceived at divergent levels like in case of top level management seek higher about the project's reputation compared with lower level management as it aligns with the strategic outcome of the organizational project.

The individual benefit's such as time, HR cost etc (should be at the benefit var level, the aggregate these is for the project, and that can be automated computed) functionalities are multi-faceted and the mediating parties are crucial for consensus at the project level. The project decision is ideally championed by a significant industrial party/ stakeholder like the organizational leader. In most occasions it is the project owner or funder who has the overview, brainchild and insights of the project, its benefits and after-effects or consequences. Thus, a majority prefers the intervention of a pioneering project expert to streamline the decision-making process, playing the role of a mediator. A super-agent takes the role of the moderator who initializes the exchange of ideas, facilitates the arguments in a networked setting. An expert like a Chief Project Officer with the authorization of the board of directors or majority shareholders facilitate effective decision-making. In the light of project failures, unsuccessful programs and project disbenefits [12, 1] it is a loss for stakeholders.

The process of consensus driven decision making is a focused, unbiased, respected ideology of giving equal share of concern to the views of all experts. In the case of stakeholder **justice** [o4] and approaching to decide if it is practically attained; the decision making process is gauged and substantiated by a degree of 'soft consensus' which is equated with the truth value [13–16]. "Most of the competent {knowledgeable, reliable} project experts agree upon as to almost all of the dimensions of stakeholder justice is observable in the project at different intervals of the lifecycle".

Zadeh's approach is to inspire the logic that explains each project expert agrees to give the consent to the decision reached even if it is not in oneness with his/her individual perspective. Moreover, the individual expert is free to modify his/ her testimony while each intellectual view, insight and idea is valued, heard and considered. The quantifier 'most' is approximately analogous to a unanimous decision among the experts.

3.1 The Derivation of the Degree of Consensus

The first step is to compute a degree of agreement as to the preferences between all relevant pairs of options for the pairs of agents (Tables 3 and 4).

In the scientific approach of voting system of variables, one should apply the social fuzzy consensus methodology. For social consensus, a step wise approach in the consensus reaching process uses a fuzzy preference relation and reaches a social fuzzy preference relation, at consensus. In compliance with democratic voting standards, if more than a half or 2/3 ^{rds} of the experts group votes in favor of a preference option [PC VAR PRODUCT BENEFIT; PC VAR TIME TO MARKET; PC VAR PROJECT COST BENEFIT; PC VAR BUSINESS OR OPE BENEFIT; PC VAR ORG BENEFIT; PC VAR SOC-CUL POL BENE] the majority rule is applied with provision for bad outcome of the decision as well. If the majority project experts vote in favor of two but abstains on one variable, then it may undermine the benefit process as procedure of allocation rewards is ignored deliberately by the group of experts or stakeholders of the project. Hence, rational decision-making must consider the balance of probabilities. The benefit measurement using social fuzzy preference relates to the characteristic, explanation, measures and the fuzzy relation of each expert. Accordingly, the variables in the process of benefit realization it is necessary to quantify and evaluate the values from a linguistic perspective. Therefore, we need to quantify some values that cannot be directly quantified. The Project Component variables (PC var) are hybrid as it has both tangible and intangible aspects. The intangible benefit realization (y) cannot be

Table 3 Degree of consensus

Pairwise stakeholders	Options in pairwise combination	Degree of consensus
P_Sup & P_client of Product benefit and time to market benefit	PC VAR PRODUCT BENEFIT & PC VAR TIME TO MARKET	Full {strict agreement}
P_client & P_exp of project cost benefit & product benefit	PC VAR PROJECT COST BENEFIT & PC VAR PRODUCT BENEFIT	Partial {sufficient agreement}
P_Sup & P_exp of product benefit and soci-cul benefit	PC VAR PRODUCT BENEFIT & PC VAR SOC-CUL POL BENE	Partial {sufficient agreement}
P_exp & P_client of orga benefit and time to market	PC VAR ORG BENEFIT & PC VAR TIME TO MARKET	Full {strict agreement}
P_client & P_sup	PC VAR PRODUCT BENEFIT & PC VAR SOC-CUL POL BENE	Partial {sufficient agreement}

Table 4 Degree of agreement

Stakeholder pairwise	Option pairwise	Linguistic quantifier	Aggregate degree of consensus
P_Sup & P_client of *Product benefit & time to market benefit*	PC VAR PRODUCT BENEFIT & PC VAR TIME TO MARKET	Always	Full {strict agreement}
P_client & P_exp of *project cost benefit and product benefit*	PC VAR PROJECT COST BENEFIT & PC VAR PRODUCT BENEFIT	Often	Partial {sufficient agreement}
P_Sup & P_exp of *product benefit and socio cul benefit*	PC VAR PRODUCT BENEFIT & PC VAR SOC-CUL POL BENE	Sometimes	Partial {sufficient agreement}
P_exp & P_client of *org' benefit and time to market benefit*	PC VAR ORG BENEFIT & PC VAR TIME TO MARKET	Rarely	Full {strict agreement}
P_client & P_sup	PC VAR org BENEFIT & PC VAR SOC-CUL POL BENE	Not	Partial {sufficient agreement}
P_client& P_EXP			

directly measured and hence we need to estimate y based on available information and known values of related quantities $x_1,...,x_n$, and the estimating algorithm is expressed $y = f(x_1,...,x_n)$ (Table 5).

Table 5 Benefit measurement using social fuzzy consensus measurement

Characteristic	Explanation	Measures
Set of n; n ≥ 2 options	$O = \{o_1...o_n\}$	PC VAR PRODUCT BENEFIT PC VAR TIME TO MARKET PC VAR PROJECT COST BENEFIT PC VAR BUSINESS OR OPE BENEFIT PC VAR ORG BENEFIT PC VAR SOC-CUL POL BENE
Set of m; m ≥ 2 agents	$E = \{e_1...e_m\}$	supplier_side (P_Sup), client_side (P_client), expert_side (P_exp)
Individual fuzzy relation of each expert e_k k = 1,2,...m	R_k μRk	– if PC VAR PRODUCT BENEFIT is definitely preferred to PC VAR SOC-CUL POL BENE – if PC VAR PRODUCT BENEFIT is slightly preferred to PC VAR SOC-CUL POL BENE in the case of indifference – if PC VAR PRODUCT BENEFIT is slightly preferred to PC VAR SOC-CUL POL BENE – if PC VAR PRODUCT BENEFIT is definitely preferred to PC VAR SOC-CUL POL BENE

Realization of the benefit to the stakeholders = f(average of product benefit[×2], average of time to market benefit[x_3], average of project cost benefit[x_4], average of business or operation benefit[x_5], average of organization benefit[x_6], average of social, cultural and political benefit[×7]) Does it become the most important consideration for benefit realization. The relevance of options, relevance of pairs of options, importance of experts, importance of pair of individuals all contributes towards the degree of agreement and determination of options [17]. Strict agreement and sufficient agreement are the fundamental degrees of agreement among the experts with reference to options and the outcome or project decision. Accordingly the relevance of options with norm values s_i: from 0 standing for 'Not' to 1 for 'always' through all intermediate values. When measuring 'stakeholder BR' the PRODUCT BENEFIT could be definitely relevant/Always Product benefit = '1' & product cost benefit = '1' with Product social benefit = '0'. The relevance of pairs of options notational are measuring the compatibility among the options $b^B_{prod_{ben}/proj_{cost}ben} = b^B_{proj_{cost}ben/prod\_ben}$ are straightforward relevant and $b^B_{prod_{ben}/prod\_ben}$ s is irrelevant as the notation explains that it is the same option.

One should establish benefit realization decision making in terms of accurately estimating between actual and desired variable values and understand the distinction between benefit and intangible benefit at higher order levels of projects as the benefits are rather fuzzy and uncertain. In our case the project component variables are the actual values as perceived by the project stakeholders to be the realism of benefit approach. The estimated achieved values are the desires of the stakeholders which are the ultimately achievable benefit realization. It is further describable that the actual benefit realization is the possibility or probability of attaining the said benefits to reach the end realization of the target. On the other hand, desired value is the extent to which the benefit is actually realized and estimated as a target reachability at mid-point and final review of the benefit realization. The following is the fuzzy representation of fuzzy control case variables (Table 6).

The ultimate purpose of concepts and degrees of consensus models based on fuzzy majority is to ensure group decision-making and consensus models fits closer to reality and is human consistent [8, 15]. This thereby aims at accounting for a fuzzy majority represented by a fuzzy linguistic quantifier [9, 10, 15, 16]. The three types of benefit STAKEHOLDERS {k = supplier(1), client(2), pro_expert(3), enterpriseM(4)} express the fuzzy preference relation for a pairwise options relating to the realization of stakeholder benefits of projects levels of the public and corporate enterprise project namely, product_benefit; time to market, project cost benefit, project operation, organizational benefit and socio-benefit. For the same 'individual fuzzy preference relation' we can impute & determine the decision-making using the 'social fuzzy preference relations'. The core of this approach is useful to obtain a social fuzzy preference relation based on Nurmi [18]. It is also has an intuitive appeal to conceptualize and apply the fuzzy consensus winner extended further as a fuzzy majority expressed by a fuzzy linguistic quantifier (Fig. 1).

The application of social consensus can be reflected and recalled by the social fuzzy preference relation: In the context of a common manager known as a benefit

Table 6 Application of fuzzy uncertainty to solve the benefit realization of estimated PROJ COMP v values

Quantity of interest	Actual value	Available value	Accuracy in Estimation to determine the realization of intangible benefit
PC $_{VAR}$ PRODUCT BENEFIT	$y = f \{ PC_{VAR \ PRODUCT \ BENEFIT1}, \ldots, PC_{VAR \ PRODUCT \ BENEFITn} \}$	$\tilde{y} = f \{ PCVARPRODUCTBENEFIT1, \ldots, PCVARPRODUCTBENEFITn \}$	"and"– operation for "and" "or"– operation for "some OPTIONAL" benefits
PC $_{VAR}$ TIME TO MARKET BENEFIT	$y = f \{ PC_{VAR \ TIME \ TO \ MARKET1}, \ldots, PC_{VAR \ TIME \ TO \ MARKETn} \}$	$\tilde{y} = f \{ PCVARTIMETOMARKET1, \ldots, PCVARTIMETOMARKETn \}$	
PC $_{VAR}$ PRODUCT COST BENEFIT	$y = f \{ PC_{VAR \ PROJECT \ COST \ BENEFIT1}, \ldots, PC_{VAR \ PROJECT \ COST \ BENEFITn} \}$	$\tilde{y} = f \{ PCVARPROJECTCOSTBENEFIT1, \ldots, PCVARPROJECTCOSTBENEFITn \}$	
PC $_{VAR}$ BUSINESS OR OPERATION BENEFIT	$y = f \{ PC_{VAR \ BUSINESS \ OR \ OPERATION \ BENEFIT1}, \ldots, PC_{VAR \ BUSINESS \ OR \ OP \ BENEFITn} \}$	$\tilde{y} = f \{ PCVARBUSINESSOROPERATIONBENEFIT1, \ldots, PCVARBUSINESSOROPBENEFITn \}$	
PC $_{VAR}$ ORG BENEFIT	$y = f \{ PC_{VAR \ ORG \ BENEFIT1}, \ldots, PC_{VAR \ ORG \ BENEFITn} \}$	$\tilde{y} = f \{ PCVARORGBENEFIT1, \ldots, PCVARORGBENEFITn \}$	
PC $_{VAR}$ SOCIAL BENEFIT	$y = f \{ PC_{VAR \ SOCIAL \ BENEFIT1}, \ldots, PC_{VAR \ SOCIAL \ BENEFITn} \}$	$\tilde{y} = f \{ PCVARPROJECTSOCIALBENEFIT1, \ldots, PCVARPROJECTSOCIALBENEFITn \}$	

$$\text{Supplier\_M}=\begin{matrix} 0 & 0.3 & 0.7 & 0.1 \\ 0.7 & 0 & 0.6 & 0.6 \\ 0.3 & 0.4 & 0 & 0.2 \\ 0.9 & 0.4 & 0.8 & 0 \end{matrix}$$

$$\text{Client\_M}=\begin{matrix} 0 & 0.4 & 0.6 & 0.2 \\ 0.6 & 0 & 0.7 & 0.4 \\ 0.4 & 0.3 & 0 & 0.1 \\ 0.8 & 0.6 & 0.9 & 0 \end{matrix}$$

$$\text{Expert\_M}=\begin{matrix} 0 & 0.5 & 0.7 & 0 \\ 0.5 & 0 & 0.8 & 0.4 \\ 0.3 & 0.2 & 0 & 0.2 \\ 1 & 0.6 & 0.8 & 0 \end{matrix}$$

$$\text{BENEFIT MANAGER}=\begin{matrix} 0 & 0 & 1 & 0 \\ 3/4 & 0 & 3/4 & 1/4 \\ 0 & 1/4 & 0 & 0 \\ 1 & 3/4 & 1 & 0 \end{matrix}$$

Fig. 1 Fuzzy consensus winner

manager, the social fuzzy preference relation abiding the group decision making by the experts and the fuzzy preference consensus of individual experts of project managers is considered. Hence, the implication of the method is to obtain a 'social fuzzy preference' & 'consensus winner' in decision making over the options of benefits that are determined by the experts. If we presume that Q= 'always', and S= then we obtain;

$$W_{always} = 1/15/s1 + 11/15/s2 + 1/s4 :$$
$$W_{0.8/always} = 1/15/s1 + 11/15/s4 :$$
$$W_{S/always} = 2/15/s1 + 11/15/s2 + 1/s4$$

In case of W_{always}, benefit option s1 belongs to the fuzzy Q core to the extent of 1/15, option s2 to the extent of 11/15 and benefit option s4 to the extent of 1. Examining of individual fuzzy preference relations also exemplified and deconstructed into the analogous aspects of $W_{0.8/\ 'always'}$ and $W_{S/\ always}$, relating to a benefit. The aspect will be acquired by the different stakeholder-oriented managers or experts of project benefit realization.

The results of the data analysis can infer the following findings FROM THE GENERATED Tables. The actual vs target benefits of the six project component variables can be summarized into six situational occurrences below.

(1) Proj component variables 01 & 02 always achieved with a full agreement by expert stakeholders
(2) Proj component variables 01 & 03 is often achieved with partial agreement
(3) Proj component variables 01 & 06 is sometimes achieved with partial agreement
(4) Proj component variables 02 & 05 rarely achieved with full agreement
(5) Proj component variables 05 & 06 not achieve with partial agreement
(6) Proj component variable 01 & 04 sometimes achieved with partial agreement.

Benefit realization in the benefit approach at the consensus levels will be dynamic at the stages of mid review and the final review phases. The degree of consensus and aggregate degree of consensus are both regarded as diverse opinions at a social fuzzy consensus level. The go- and no-go decisions are that project components 01, 02, 05 are to proceed with value from product benefit, time to market benefit and org' benefit are regarded to be 'go" decisions without a schedule overrun, and the project busi-operation benefit, social value and proj cost benefit related indicators are contributory to 'no-go' projects that will lead to failure. Therefore, these are operationally non-functional, ignorable projects with a probability of failure [12, 16]. Failure means not merely the tangible profit value but also the non-tangibles, that account for malfunctioning, misfortune, limitation of benefits, unaccountable and liability driven projects.

4 Conclusion

We proposed a consensus-based approach to dealing with the input of different stakeholders in the realization of both tangible and the fuzzy intangible benefits. Although social cultural and political aspect of benefit is targeted and desired, the experts affirm that the benefit managers are in a transition stage yet to desire the actual realization of the project benefit. Business operation benefits are also a similar aspect with partial agreement and hence regarded to be a much-needed foreseeable focus for future consideration.

References

1. Zwikael, O., Chih, Y., Meredith, J.R.: Project Management Benefits: setting effective target benefits. Int. J. Project Manage. **36**, 650–658 (2018)
2. Aubry, M., Sergi, V., El Boukri, S.: Opening the black box of benefits management in the context of projects. In IRNOP. Boston, MA (2017)
3. Ghildyal, A., Ru D., Yapa, S., Uthumange, A., Chang, E., Joiner K., Despande A.: "Conceptual Framework for Measuring Project Benefits using Belief—Plausibility and Type1 Fuzzy Inference System" (Chapter 20) in Studies in Fuzziness, Vol. 393 (2020)
4. Fedrizzi, M., Pasi, G.: Fuzzy logic approaches to consensus modeling in group decision making. In: Ruan, D., Hardeman, F., Meer, K.V.D. (eds.) Intelligent Decision and Policy Making Support Systems, pp. 19–37. Springer, Berlin-Heidelberg (2008)
5. Ruwanthi, D., Yapa, S., Uthumange, A., Deshpande, A., Chang, E.: Stratification and fuzzy inference system for measuring project benefits. 10th International Conference on Soft Computing and Intelligent Systems & 19th International Symposium on advanced intelligent systems. Japan (2018)
6. Ruwanthi, D., Yapa, S., Uthumange, A., Ghildyal, A., Chang, E., Despande, A.: Measuring project benefit using an integrated fuzzy-shafer algorithm, "soft computing: new directions in foundations and applications". J. Bus. Res. (JBR), Elsevier publications (2020)
7. Shahbazova, S.N. et al. (Eds.): Recent Developments and the New Direction in Soft-Computing Foundations and Applications, 978-3-030-47123-1. (2020). https://doi.org/10.1007/978-3-030-47124-8_20
8. Fedrizzi, M., Kacprzyk, J., Nurmi, H.: Consensus degrees under fuzzy majorities and fuzzy preferences using OWA (ordered weighted average) operators. Control. Cybern. **22**, 71–80 (1993)
9. Zadeh, L., Chang, E., Deshpande, A.: Fuzzy inference approach. J. Bus. Res. **23**, 567–546 (2017)
10. Zadeh, L.A.: Stratification, target set reachability, and incremental enlargement principle. Inf. Sci. 354, https://doi.org/10.1016/j.ins.2016.02.047 (2016)

11. Sahakian, B., LaBuzetta, J. N.: Bad moves: How decision making goes wrong, and the ethics of smart drugs. New York, NY: Oxford University Press. (2013)
12. Dwivedi, Y.K., Ravichandran, K., Kartik, M.: IS/IT project failures: a review of the extant literature for deriving a taxonomy of failure factors. Int. Fed. Inf. Process.**402**, 73–88 (2013)
13. Herrera-Viedma, E., Cabrerizo, F.J., Kacprzyk, J., Pedrycz, W.: Review of soft consensus models in fuzzy environments. Inf. Fusion. **17**, 4–13 (2014). https://doi.org/10.1016/j.inffus.2013.04.002
14. Herrera-Viedma, E., Alonso, S., Chiclana, F., Herrera, F.: A consensus model for group decision making with incomplete fuzzy preference relations. IEEE Trans. Fuzzy Syst. **15**(5), 863–877 (2007)
15. Kacprzyk, J., Fedrizzi, M.: A 'human-consistent' degree of consensus based on fuzzy logic with linguistic quantifiers. Math. Soc. Sci. **18**(3) 275–290 (1992)
16. Zadeh, L.A.: Fuzzy sets. Info. Control **8**(3), 338–353; Zadeh, L.A.: Probability measures of fuzzy events. J. Math. Anal. Appl. **23**(2), 421–427 (1965)
17. Kacprzyk, J., Fedrizzi, M., Nurmi, H.: Group decision making and consensus under fuzzy preferences and fuzzy majority. Fuzzy Sets Syst. **49**(1) 21–31 (1992). https://doi.org/10.1016/0165-0114(92)90107-F
18. Nurmi, H.: Approaches to collective decision making with fuzzy preference relations. Fuzzy Sets Syst. **6**, 249–259. (1986)
19. Cabreizo F.J.: Building consensus in group decision making with an allocation of information, granularity, Fuzzy sets and systems, **255**, 115–127 (2014)

Intuitionistic Fuzzy Tools in Evaluation of Macroeconomic Stability

G. Imanov⬤ and **A. Aliyev**⬤

Abstract In this paper an intuitionistic fuzzy technique is proposed to evaluate macroeconomic stability index (MSI), which ensures the possibility to determine not only the quantitative, but also the qualitative level of the macroeconomic condition in the country. Intuitionistic fuzzy instruments are suitable in this regard to deal with economic problems having uncertainty. As the generalization of fuzzy sets, intuitionistic fuzzy sets are more effective to deal with vagueness and uncertainty of economic information analysis. In comparison with classical fuzzy sets, intuitionistic fuzzy sets express information by membership and non-membership functions (which convey positive and negative association) and hesitation margin. From this perspective, intuitionistic fuzzy tools come to help to transform imprecise information into knowledge and employ in multiple criteria decision-making (MCDM) problems. In the paper MSIs are obtained using intuitionistic MCDM methods. The proposed methodology for estimation of MSI of the country is carried out on the basis of statistical annuals of IMF for 2016–2020 years.

Keywords Intuitionistic fuzzy weighted average · Linguistic terms · Macroeconomic stability index

1 Introduction

The concept of "Macroeconomic Stability" defines a national economy minimizing its susceptibility to external shocks, which in turn increases its prospects for sustainable development. Macroeconomic stability functions as a bumper against currency and interest wavering in the global market. It is an essential, but unsatisfactory

G. Imanov · A. Aliyev (✉)
Institute of Control Systems, Ministry of Science and Education Republic of Azerbaijan, 68 B.Vahabzadeh str., Baku AZ1141, Azerbaijan
e-mail: msc.aaliyev@gmail.com

© The Author(s), under exclusive license to Springer Nature Switzerland AG 2023
Sh. N. Shahbazova et al. (eds.), *Recent Developments and the New Directions of Research, Foundations, and Applications*, Studies in Fuzziness and Soft Computing 422,
https://doi.org/10.1007/978-3-031-20153-0_13

requirement for development of economy. Exposure to currency fluctuations, considerable debt burdens, and unguided inflation can cause economic crises and downfall in GDP.

Both the IMF and the EU put an emphasis on macroeconomic stability. Referring to the Maastricht criteria, macroeconomic stability is measured by five variables. World Bank defines macroeconomic stability as follows:

- when the inflation rate is low and predictable,
- real interest rates are opportune,
- the real exchange rate is competitive and forecastable,
- public sector saving rates are compliant with the resource mobilization requirements of the program,
- the state of payments balance situation is accepted as variable [1].

According to the Maastricht Treaty [2] the European Union has defined macroeconomic stability as an index measured with the help of five variables:

- Low and stable inflation (the Maastricht criteria capped at 3%);
- Low long-term interest rate (the Maastricht criteria restricted to the range of 9%);
- Low debt to Gross Domestic Product ratio (the Maastricht criteria capped at 60% of GDP);
- Low deficit (the Maastricht criteria capped at 3% of GDP);
- Monetary stability (the Maastricht criteria permitted fluctuation of at most 2.5%).

Since originally introduced by Atanassov [3], intuitionistic fuzzy sets (IFSs) is generalization of fuzzy sets, and can be used as an all-inclusive and full tool for describing uncertainty. IFSs have attracted much attention due to its power in expressing vagueness and ambiguity. As elements of IFSs, IFNs that is composed of three parameters: membership and non-membership functions and hesitation margin. IFNs are highly effective to convert and express imprecise data, and to deal with ambivalence.

If a problem under study relates to multiple criteria decision making problem, then in obtaining a decision result, an important step is the aggregation of IFNs. Xu [4] developed some aggregation operators, such as the intuitionistic fuzzy weighted averaging operator (IFWA) and established various properties of these operators. Later Huchang and Xu [5] developed intuitionistic fuzzy hybrid weighted average operators (IFHWA). Aggregation operators are very effective since they can be employed to fuse multidimensional values into an overall value.

Measuring of Macroeconomic Stability Index as an intuitionistic fuzzy MCDM problem, especially as an intuitionistic fuzzy weighted linear combination (WLC) problem was first introduced by Imanov [6], in which intuitionistic fuzzy linguistic aggregation considering argument weights was employed. G. Imanov and A. Aliyev investigated quality of life [7] with the application of IFHWA operator, in which aggregation of IFNs is carried out taking into account data and priority weights.

In this paper the purpose is to compute MSI employing intuitionistic fuzzy weighted average and taking into account priority weights of criteria.

The paper is organized as follows: Sect. 2 reviews basic operations on IFNs. In Sect. 3, crisp data on macroeconomic indicators are converted into intuitionistic fuzzy values (IFVs). Section 4 proposes methodology of assessment of priority weights of criteria. In last section IFWA operator is applied incorporating priority weights, and fused IFVs are obtained.

2 Preliminaries

An intuitionistic fuzzy set A in a finite set X is defined [4] as following:

$$A = \{\langle x, \mu_A(x), \nu_A(x)\rangle | x \in X\} \tag{1}$$

where $\mu_A(x) : X \to [0, 1]$ and $\nu_A(x) : X \to [0, 1]$ represent the membership and non-membership functions respectively, such that

$$0 \le \mu_A(x) + \nu_A(x) \le 1 \tag{2}$$

for all x $\in$ X.

The third parameter of the IFSs A is:

$$\pi_A(x) = 1 - \mu_A(x) - \nu_A(x), \forall x \in X \tag{3}$$

then $\pi_{A(x)}$ is called the indeterminacy degree or hesitation degree of whether x belongs to A or not.

Let $\alpha_1, \alpha_2, \alpha_3$ be three IFVs, where $\mu_\alpha \in [0, 1]$, $\mu_\alpha \in [0, 1]$, $\nu_\alpha \in [0, 1]$, $\mu_\alpha + \nu_\alpha \le 1$. The following operations are hold [5]:

$$1. \quad \lambda\alpha = \left(1 - (1 - \mu_\alpha)^\lambda, \nu_\alpha^\lambda\right), \lambda > 0, \tag{4}$$

$$2. \quad \alpha^\lambda = \left(\mu_\alpha^\lambda, 1 - (1 - \nu_\alpha)^\lambda\right), \lambda > 0, \tag{5}$$

$$3. \quad \sum_{j=1}^n \alpha_j = \left(1 - \prod_{j=1}^n (1 - \mu_{\alpha j}), \prod_{j=1}^n \nu_{\alpha j}\right), \tag{6}$$

$$4. \quad \prod_{j=1}^n \alpha_j = \left(\prod_{j=1}^n \mu_{\alpha j}, 1 - \prod_{j=1}^n 1 - \nu_{\alpha j}\right) \tag{7}$$

3 Macroeconomic Stability Index

In order to estimate MSI following indicators are investigated in this paper, which were proposed by International Monetary Fund (IMF):

- Real GDP growth (in percent)—GDP;
- Unemployment rate (in percent)—UNE;
- Consumer price index (period average)—CPI;
- Revenue (including grants, in percent of GDP)—REV;
- Expenditure (in percent of GDP)—EXP;
- General government gross debt (in percent of GDP)—GGD;
- Bank credit to the private sector (in percent of GDP)—BCP;
- Current account balance (in percent of GDP)—CAB;
- Foreign direct investment net inflows (in % of GDP)—FDI;
- Gross international reserves (in months of non-oil imports)—GIR;
- Real Effective Exchange Rate (average, percentage change)—REER.

Based on IMF Executive Board information and online sources [8–10] the data on macroeconomic indicators proposed by IMF are collected in Table 1.

In order to convert crisp data given in Table 1 into IFNs, first the data have to be normalized. For this purpose, max–min normalization method is applied that is expressed by the formula below (Table 2):

$$Y = \frac{x - x_{min}}{x_{max} - x_{min}} \tag{8}$$

The obtained results are given in Table 2.

Table 1 Macroeconomic stability indicators

Indicators	Periods				
	2016	2017	2018	2019	2020
GDP	−3.1	0.2	1.5	2.5	−4.3
UNE	5.0	5.0	4.9	4.8	6.5
CPI	12.4	12.9	2.3	2.7	2.8
REV	34.3	34.2	38.6	41.5	33.8
EXP	35.4	35.6	33.1	33.3	40.4
GGD	20.6	22.5	18.7	17.7	21.4
BCP	32.9	22.1	20.8	22.2	25.1
CAB	−3.6	4.1	12.8	9.1	−0.5
FDI	11.9	7.02	2.98	3.12	3.35
GIR	4.2	5.1	4.7	5.4	6.8
REER	−27.0	3.3	3.1	3.9	4.5

Table 2 Normalized macroeconomic stability indicators

Indicators	Periods				
	2016	2017	2018	2019	2020
GDP	0.18	0.66	0.85	1.00	0.00
UNE	0.12	0.12	0.06	0.00	1.00
CPI	0.95	1.00	0.00	0.04	0.05
REV	0.06	0.05	0.62	1.00	0.00
EXP	0.32	0.34	0.00	0.03	1.00
GGD	0.60	1.00	0.21	0.00	0.77
BCP	1.00	0.11	0.00	0.12	0.36
CAB	0.00	0.47	1.00	0.77	0.19
FDI	1.00	0.45	0.00	0.02	0.04
GIR	0.00	0.35	0.19	0.46	1.00
REER	0.00	0.96	0.96	0.98	1.00

Normalized indicators are converted into IFNs using the intuitionistic fuzzy triangular functions *iftrif* [11].

A definite intuitionistic fuzzy triangular membership and non-membership function of A is introduced as following:

$$\mu_A(x) = \begin{cases} 0 & ; \\ \left(\frac{x-a}{b-a}\right) - \varepsilon & ; \\ \left(\frac{c-x}{c-b}\right) - \varepsilon & ; \\ 0 & ; \end{cases} \quad v_A(x) = \begin{cases} 1- & ; \quad x \leq a \\ 1 - \left(\frac{x-a}{b-a}\right) & ; \ a < x \leq b \\ 1 - \varepsilon & ; \quad x \geq c \end{cases} \quad (9)$$

For the sake of simplicity isosceles triangles have been used here, and hesitation margin is preset as $\varepsilon = 0.01$. The results are illustrated in Table 3.

4 Assessment of Priority Weights of Criteria

In this section priority weights of criteria are obtained based on methodology in which weighting vector is obtained from intuitionistic preference relation matrix.

Fuzzy preference degree within 0.1–0.9 scale was developed including their meanings in work [12]. This scale is a basis for constructing intuitionistic fuzzy preference scale.

Referring to fuzzy scales between 0.1–0.9 given in Table 4, the intuitionistic fuzzy preference table of macroeconomic stability indicators expressed in linguistic terms with their IFNs counterparts is developed, that is illustrated in Table 5.

Let $\mathscr{R}$ be an intuitionistic fuzzy preference relation (IFPR), then the following conditions must hold [13]:

Table 3 Intuitionistic fuzzy macroeconomic stability indicators

Indicators	IFN	Periods				
		2016	2017	2018	2019	2020
GDP	μ	0.35	0.67	0.29	0.00	0.00
	ν	0.64	0.32	0.70	0.99	0.99
UNE	μ	0.23	0.23	0.11	0.00	0.00
	ν	0.76	0.76	0.88	0.99	0.99
CPI	μ	0.09	0.00	0.00	0.07	0.03
	ν	0.90	0.99	0.99	0.92	0.96
REV	μ	0.11	0.09	0.73	0.00	0.00
	ν	0.88	0.90	0.26	0.99	0.99
EXP	μ	0.63	0.63	0.00	0.05	0.00
	ν	0.36	0.36	0.99	0.94	0.99
GGD	μ	0.79	0.00	0.41	0.00	0.21
	ν	0.20	0.99	0.58	0.99	0.78
BCP	μ	0.00	0.21	0.00	0.23	0.33
	ν	0.99	0.78	0.99	0.76	0.66
CAB	μ	0.00	0.93	0.00	0.45	0.17
	ν	0.99	0.06	0.99	0.54	0.82
FDI	μ	0.00	0.89	0.00	0.03	0.01
	ν	0.99	0.10	0.99	0.96	0.99
GIR	μ	0.00	0.69	0.37	0.91	0.00
	ν	0.99	0.30	0.62	0.08	0.99
REER	μ	0	0.95	0.07	0.03	0
	ν	0.99	0.04	0.92	0.96	0.99

Table 4 Fuzzy scales between 0.1–0.9

0.1–0.9 Scales	Meanings
0.1	x_j is extremely preferred to x_i
0.2	x_j is strongly preferred to x_i
0.3	x_j is definitely preferred to x_i
0.4	x_j is slightly preferred to x_i
0.5	x_j is the same as x_i
0.6	x_i is slightly preferred to x_j
0.7	*x_i is definitely preferred to x_j*
0.8	*x_i is strongly preferred to x_j*
0.9	*x_i is extremely preferred to x_j*
Complementary number	If the preference degree or intensity of alternative x_j over x_i is r_{ij}, then the preference degree or intensity of alternative x_j over x_i is r_{ji} $= 1 - r_{ij}$

Table 5 Linguistic terms for rating the preference of criteria

Linguistic terms	IF preference numbers
Exactly equal (EE)	(050, 0.50)
Slightly preferred (SP)	(0.60, 0.30)
Definitely preferred (DP)	(0.70, 0.20)
Strongly preferred (STP)	(0.80, 0.10)
Extremely preferred (EP)	(0.90, 0.10)
Other useful midterms	*(0.65, 0.25) (0.75, 0.15)*

$$r_{ij} = (0.50, 0.50), \ \mu_{ij} = v_{ji}, \ v_{ij} = \mu_{ji}, \ \pi_{ij} = \pi_{ji}, \ \mu_{ij} + v_{ij} + \pi_{ij} = 1 \qquad (10)$$

The last condition provides the additive consistency of IFPR. The additive consistency is unsatisfactory in modeling consistency, so all attention has to be concentrated on multiplicative consistency of IFPR [14]. In the intuitionistic fuzzy analytic hierarchy process, in order to obtain a feasible solution, before deriving the priorities of the alternatives and criteria, we need to check whether IFPR is consistent or not.

Consistency is an important issue in preference relations and the lack of consistency of preference relations may lead to misleading solutions [15].

Based on Table 5, and considering additive consistency condition the IFPR matrix $\mathcal{R}$ is composed (Table 6).

Referring to Genç et al. [16, 17] the following method is proposed to adjust the given matrix into multiplicative consistent IFPR matrix, where, $(r_{ij})^* = (\mu_{ij}, v_{ij})^*$ $(i = 1, 2, \ldots, n; \ j = 1, 2, \ldots, n)$ and satisfies the following condition:

$$\left(\mu_{ij}\right)^* = \max\left\{\mu_{ij}, \max_p\left\{\frac{\mu_{ip}\mu_{pj}}{\mu_{ip}\mu_{pj} + (1 - \mu_{ip})(1 - \mu_{pj})}\right\}\right\} \qquad (11)$$

$$\left(v_{ij}\right)^* = \max\left\{v_{ij}, \max_p\left\{\frac{v_{ip}v_{pj}}{v_{ip}v_{pj} + \left(1 - v_{ip}\right)\left(1 - v_{pj}\right)}\right\}\right\} \qquad (12)$$

where, $\left(\mu_{ij}\right)^*$ and $\left(v_{ij}\right)^*$, the element of $(\mathcal{R})^*$ matrix, are the membership degree and the non-membership degree of the alternative x_i over x_j, respectively, and $0 \leq \left(\mu_{ij}\right)^* + \left(v_{ij}\right)^* \leq 1$ for all i, j, k = 1,2,...,n, then we call $(\mathcal{R})^*$ a multiplicative consistent IFPR. If $(\mathcal{R})^* = \left(r_{ij}\right)^*_{n \times n}$ does not satisfy the condition $0 \leq \left(\mu_{ij}\right)^* + \left(v_{ij}\right)^* \leq 1$ for any i, j, k = 1,2,...,n, then $(\mathcal{R})^*$ is called an inconsistent IFPR.

Applying formulas (11) and (12) the following multiplicative consistent IFPR matrix $(\mathcal{R})^*$ is composed that is illustrated in Table 7.

After obtaining multiplicative consistent aggregated IFPR matrix, the priority vector of criteria $w = (w_1, w_2, \ldots, w_n)^T$ can be estimated with the following equation proposed in papers [16, 17]:

Table 6 IFPR matrix

	GDP	UNE	CPI	REV	EXP	GGD	BCP	CAB	FDI	GIR	REER
GDP	(0.50,0.50)	(0.60,0.30)	(0.60,0.30)	(0.65,0.25)	(0.65,0.25)	(0.70,0.20)	(0.70,0.20)	(0.75,0.15)	(0.80,0.10)	(0.80,0.10)	(0.90,0.10)
UNE	(0.30,0.60)	(0.50,0.50)	(0.50,0.50)	(0.60,0.30)	(0.60,0.30)	(0.65,0.25)	(0.65,0.25)	(0.70,0.20)	(0.75,0.15)	(0.80,0.10)	(0.90,0.10)
CPI	(0.30,0.60)	(0.50,0.50)	(0.50,0.50)	(0.60,0.30)	(0.60,0.30)	(0.65,0.25)	(0.65,0.25)	(0.70,0.20)	(0.75,0.15)	(0.80,0.10)	(0.90,0.10)
REV	(0.25,0.65)	(0.30,0.60)	(0.30,0.60)	(0.50,0.50)	(0.50,0.50)	(0.60,0.30)	(0.60,0.30)	(0.65,0.25)	(0.70,0.20)	(0.75,0.15)	(0.80,0.10)
EXP	(0.25,0.65)	(0.30,0.60)	(0.30,0.60)	(0.50,0.50)	(0.50,0.50)	(0.60,0.30)	(0.60,0.30)	(0.65,0.25)	(0.70,0.20)	(0.75,0.15)	(0.80,0.10)
GGD	(0.20,0.70)	(0.25,0.65)	(0.25,0.65)	(0.30,0.60)	(0.30,0.60)	(0.50,0.50)	(0.50,0.50)	(0.60,0.30)	(0.65,0.25)	(0.70,0.20)	(0.75,0.15)
BCP	(0.20,0.70)	(0.25,0.65)	(0.25,0.65)	(0.30,0.60)	(0.30,0.60)	(0.50,0.50)	(0.50,0.50)	(0.60,0.30)	(0.65,0.25)	(0.70,0.20)	(0.75,0.15)
CAB	(0.15,0.75)	(0.20,0.70)	(0.20,0.70)	(0.25,0.65)	(0.25,0.65)	(0.30,0.60)	(0.30,0.60)	(0.50,0.50)	(0.60,0.30)	(0.65,0.25)	(0.70,0.20)
FDI	(0.10,0.80)	(0.15,0.75)	(0.15,0.75)	(0.20,0.70)	(0.20,0.70)	(0.25,0.65)	(0.25,0.65)	(0.30,0.60)	(0.50,0.50)	(0.60,0.30)	(0.65,0.25)
GIR	(0.10,0.80)	(0.10,0.80)	(0.10,0.80)	(0.15,0.75)	(0.15,0.75)	(0.20,0.70)	(0.20,0.70)	(0.25,0.65)	(0.30,0.60)	(0.50,0.50)	(0.60,0.30)
REER	(0.10,0.90)	(0.10,0.90)	(0.10,0.90)	(0.10,0.80)	(0.10,0.80)	(0.15,0.75)	(0.15,0.75)	(0.20,0.70)	(0.25,0.65)	(0.30,0.60)	(0.50,0.50)

Table 7 Multiplicative consistent IFPR matrix

	GDP	UNE	CPI	REV	EXP	GGD	BCP	CAB	FDI	GIR	REER
GDP	(0.50,0.50)	(0.60,0.33)	(0.60,0.33)	(0.69,0.31)	(0.69,0.31)	(0.74,0.25)	(0.74,0.25)	(0.78,0.21)	(0.81,0.17)	(0.86,0.14)	(0.90,0.10)
UNE	(0.33,0.60)	(0.50,0.50)	(0.50,0.50)	(0.60,0.33)	(0.60,0.33)	(0.69,0.27)	(0.69,0.27)	(0.74,0.21)	(0.78,0.17)	(0.82,0.14)	(0.90,0.10)
CPI	(0.33,0.60)	(0.50,0.50)	(0.50,0.50)	(0.60,0.33)	(0.60,0.33)	(0.69,0.27)	(0.69,0.27)	(0.74,0.21)	(0.78,0.17)	(0.82,0.14)	(0.90,0.10)
REV	(0.31,0.69)	(0.33,0.60)	(0.33,0.60)	(0.50,0.50)	(0.50,0.50)	(0.60,0.33)	(0.60,0.33)	(0.69,0.27)	(0.74,0.21)	(0.78,0.17)	(0.82,0.17)
EXP	(0.31,0.69)	(0.33,0.60)	(0.33,0.60)	(0.50,0.50)	(0.50,0.50)	(0.60,0.33)	(0.60,0.33)	(0.69,0.27)	(0.74,0.21)	(0.78,0.17)	(0.82,0.17)
GGD	(0.25,0.74)	(0.27,0.69)	(0.27,0.69)	(0.33,0.60)	(0.33,0.60)	(0.50,0.50)	(0.50,0.50)	(0.60,0.33)	(0.69,0.27)	(0.74,0.21)	(0.77,0.21)
BCP	(0.25,0.74)	(0.27,0.69)	(0.27,0.69)	(0.33,0.60)	(0.33,0.60)	(0.50,0.50)	(0.50,0.50)	(0.60,0.33)	(0.69,0.27)	(0.74,0.21)	(0.77,0.21)
CAB	(0.21,0.78)	(0.21,0.74)	(0.21,0.74)	(0.27,0.69)	(0.27,0.69)	(0.33,0.60)	(0.33,0.60)	(0.50,0.50)	(0.60,0.33)	(0.69,0.27)	(0.74,0.25)
FDI	(0.17,0.81)	(0.17,0.78)	(0.17,0.78)	(0.21,0.74)	(0.21,0.74)	(0.27,0.69)	(0.27,0.69)	(0.33,0.60)	(0.50,0.50)	(0.60,0.33)	(0.69,0.31)
GIR	(0.14,0.86)	(0.14,0.82)	(0.14,0.82)	(0.17,0.78)	(0.17,0.78)	(0.21,0.74)	(0.21,0.74)	(0.27,0.69)	(0.33,0.60)	(0.50,0.50)	(0.60,0.33)
REER	(0.10,0.90)	(0.10,0.90)	(0.10,0.90)	(0.17,0.82)	(0.17,0.82)	(0.21,0.77)	(0.21,0.77)	(0.25,0.74)	(0.31,0.69)	(0.33,0.60)	(0.50,0.50)

$$w_j = \left[w_j^L, w_j^U\right] = \left(\cfrac{1}{\sum_{j=1}^{n}\left(\cfrac{\left(1-\tilde{\mu}_{ij}^*\right)}{\tilde{\mu}_{ij}^*}\right)}, \cfrac{1}{\sum_{=1}^{n}\left(\cfrac{\tilde{v}_{ij}^*}{(1-\tilde{v}_{ij}^*)}\right)}\right) \quad (13)$$

Consequently, priority weight intervals are found:

$$
\begin{aligned}
w_1 &= [0.212, 0.233], & w_8 &= [0.043, 0.053] \\
w_2 &= w_3 = [0.138, 0.168], & w_9 &= [0.032, 0.040] \\
w_4 &= w_5 = [0.092, 0.108], & w_{10} &= [0.024, 0.030] \\
w_6 &= w_7 = [0.062, 0.073], & w_{11} &= [0.019, 0.020]
\end{aligned}
$$

Following the way proposed in works [18, 19] in the next step, reasonable weight vector of criteria $w^* = (w^*, w^*, \ldots, w^*)$ should be obtained so that all closeness coefficients of criteria could be as big as possible. In order to do that, the following multi-objective optimization model is applied:

Maximize

$$\sum_{j=1}^{n}\sum_{i=1}^{m} w_j \frac{d(r_{ij}, \alpha_j^-)}{d\left(r_{ij}, \alpha_j^+\right) + d(r_{ij}, \alpha_j^-)} \quad (14)$$

Subject to

$$w_1^L \le w_1 \le w_1^U$$
$$w_2^L \le w_2 \le w_2^U$$
$$\vdots$$
$$w_n^L \le w_n \le w_n^U$$
$$\sum_{j=1}^{n} w_j = 1$$
$$w_j \ge 0 \quad \text{for} \quad j = 1, 2, 3, \ldots, n$$

In order to apply formula (14), the following operations on IFSs and IFNs will be used.

Hamming distance between IFSs A and B is given as follows:

$$d(A, B) = \frac{1}{2}[|\mu_A(x) - \mu_B(x)| + |v_A(x) - v_B(x)| + |\pi_A(x) - \pi_B(x)|] \quad (15)$$

Let $\alpha_i^+ = (1, 0, 0)(i = 1, 2, 3, \ldots m)$ be m largest IFNs, then

$$A^+ = \left(\alpha_1^+, \alpha_2^+, \ldots, \alpha_m^+\right) \tag{16}$$

Is called as intuitionistic fuzzy positive ideal solution ($I\ F\ I\ S^+$).
Let $\alpha_i^- = (0, 1, 0)(i = 1, 2, 3, \ldots m)$ be m smallest IFNs, then

$$A^- = \left(\alpha_1^-, \alpha_2^-, \ldots, \alpha_m^-\right) \tag{17}$$

Is called as intuitionistic fuzzy positive ideal solution ($I\ F\ I\ S^-$).

Afterwards of application formulas (14) and (15) on data from the multiplicative consistent IFPR matrix, the below shown multi-objective optimization problem is obtained:

Maximize

$$3.03w_1 + 3.70w_2 + 3.70w_3 + 4.62w_4 + 4.62w_5 + 5.54w_6 + 5.54w_7 + 6.37w_8$$
$$+ 7.11w_9 + 7.80w_{10} + 8.45w_{11}$$

Subject to

$$0.212 \leq w_1 \leq 0.233$$
$$0.138 \leq w_2 \leq 0.168$$
$$0.138 \leq w_3 \leq 0.168 \qquad 0.062 \leq w_6 \leq 0.073$$
$$0.092 \leq w_4 \leq 0.108 \qquad 0.062 \leq w_7 \leq 0.073$$
$$0.092 \leq w_5 \leq 0.108 \qquad 0.043 \leq w_8 \leq 0.053$$
$$0.024 \leq w_{10} \leq 0.030 \qquad 0.032 \leq w_9 \leq 0.040$$
$$0.019 \leq w_{11} \leq 0.020$$

$$w_1 + w_2 + w_3 + w_4 + w_5 + w_6 + w_7 + w_8 + w_9 + w_{10} + w_{11} = 1$$

$$w_1, w_2, w_3, w_4, w_5, w_6, w_7, w_8, w_9, w_{10}, w_{11} \geq 0$$

Solving the linear optimization problem by the application of *linprog*–linear objective function of "Matlab Optimization Toolbox", the optimal weights are obtained as:

$$w^* = \left(w_1^*, w_2^*, w_3^*, w_4^*, w_5^*, w_6^*, w_7^*, w_8^*, w_9^*, w_{10}^*, w_{11}^*\right)$$
$$= 0.212, 0.138, 0.145, 0.108, 0.108, 0.073, 0.073, 0.053, 0.04, 0.03, 0.02$$

Table 8 IFWA results

№		2016	2017	2018	2019	2020
1	IFWA	(0.31, 0.68)	(0.52, 0.47)	(0.25, 0.74)	(0.13, 0.86)	(0.06, 0.93)

5   Evaluation of MSI by the Application of IFWA Operator

In this section, IFVs of the macroeconomic indicators given in Table 3, taking into account their priority weights obtained in section IV are aggregated according to the following IFWA formula [4, 5]:

$$IFWA = \left(1 - \prod_{j=1}^{n}\left(1 - \mu_j\right)^{w_j}, \prod_{j=1}^{n} v_j^{w_j}\right) \tag{18}$$

As an example, MSI for the year 2017 can be estimated as follows:

$$IFWA(2017) = \left(\left(1 - \left((1 - 0.67)^{0.212}\right) * (1 - 0.23)^{0.138} * (1 - 0)^{0.145}\right.\right.$$
$$* (1 - 0.09)^{0.108} * (1 - 0.63)^{0.108} * (1 - 0)^{0.073} * (1 - 0.21)^{0.073}$$
$$* (1 - 0.93)^{0.053} * (1 - 0.89)^{0.04} * (1 - 0.69)^{0.03} * (1 - 0.95)^{0.02}\right), \left(0.32^{0.212}\right.$$
$$* 0.76^{0.138} * 0.99^{0.145} * 0.9^{0.108} * 0.36^{0.108} * 0.99^{0.073} * 0.78^{0.073} * 0.06^{0.053}$$
$$\left.\left.* 0.1^{0.04} * 0.3^{0.03} * 0.04^{0.02}\right)\right) = 0.51, 0.47$$

The obtained results as aggregated intuitionistic values are not quite satisfactory to comprehend and infer reasonable conclusions. To this end, modified linguistic term equivalents (Table 8) for intuitionistic fuzzy scale that could contemplate and reflect humanistic perception for rating the MSI has been developed referring to [20].

IFVs of IFWA results, corresponding to intuitionistic fuzzy scale given in Table 9 are converted into linguistic terms which represent MSI for the analyzed years:

$$MSI(2016) = \mathbf{L - ML}$$
$$MSI(2017) = \mathbf{M - MH} \quad MSI(2019) = \mathbf{VL}$$
$$MSI(2018) = \mathbf{L} \quad MSI(2020) = \mathbf{VVL}$$

6   Conclusion

In this work the MSI is estimated using intuitionistic fuzzy weighted aggregation operator by considering priority weights. In the final stage MSI as an aggregated intuitionistic fuzzy value converted into crisp value to make it perceptional. One of

Table 9 Modified linguistic term equivalents for intuitionistic fuzzy scale

Linguistic terms	IF number for aggregated values
	(μ, v)
Very Very High—(VVH)	(0.9, 0.1)
Very High—(VH)	(0.8, 0.1)
High—(H)	(0.7, 0.2)
Medium High—(MH)	(0.6, 0.3)
Medium—(M)	(0.5, 0.4)
Medium Low—(ML)	(0.4, 0.5)
Low—(L)	(0.25, 0.6)
Very Low—(VL)	(0.2, 0.75)
Very Very Low—(VVL)	(0.1, 0.9)

the advantages of the proposed method might be that it clearly delineates the trend of advancement of phenomenon under study. But, the main restraint is the impact of time span covering time series data on conversion crisp data into IFVs, that can cause deviation from real values. Obtained results reflect macro-economic development level in the country. The proposed approach can be useful for decision makers in estimation of parameters of economy from different macro-economic perspectives.

References

1. Siegel, P.B., Johnson, T.G., Alwang, J.: Regional economic diversity and diversification. Growth Change **26**(2), 261–284 (1995)
2. Maastricht Treaty 17.45. https://eur-lex.europa.eu/legal-content/EN/TXT/?uri=celex%3A1 1992M%2FTXT
3. Atanassov, K.T.: Intuitionistic fuzzy sets. Fuzzy Sets Syst. **20**, 87–96 (1986)
4. Xu, Z.S.: Intuitionistic fuzzy aggregation operators. IEEE Trans. Fuzzy Syst. **15**, 1179–1187 (2007)
5. Huchang, L., Xu, Z.S.: Intuitionistic fuzzy hybrid weighted aggregation operators. Int. J. Intell. Syst. **29**, 971–993 (2014)
6. Imanov, G.J.: Fuzzy models in Economy. Springer, Berlin (2021)
7. Imanov, G.J., Aliyev. A.Z.: Intuitionistic fuzzy assessment of quality of life. In: 14th International Conference on Theory and Application of Fuzzy Systems and Soft Computing – ICAFS, pp. 174–182 (2020)
8. IMF Executive Board Concludes 2021 Article IV Consultation with Republic of Azerbaijan, June 18, 2021
9. https://www.ceicdata.com/en/azerbaijan/bank-loans/az-domestic-credit-to-private-sector-by-banks--of-gdp
10. https://www.oecd-ilibrary.org/sites/42efd1f2-en/index.html?itemId=/content/component/42e fd1f2-en
11. Radhika, C., Parvathi, R.: Intuitionistic fuzzification functions. Glob. J. Pure Appl. Math. **12**(2), 1211–1227 (2016)
12. Gong, Z.W., Lin, Y., Yao, T.: Uncertain Fuzzy Preference Relations and Their Applications, p. 210 (2012)

13. Gong, Z.W., Li, L.S., Zou, F.X.: Goal programming approaches to obtain the priority vectors from the intuitionistic fuzzy preference relations. Comput. Ind. Eng. **57**, 1187–2113 (2009)
14. Xu, Z.S., Huchang, L.: Intuitionistic fuzzy analytic hierarchy process. IEEE Trans. Fuzzy Syst. **22**(4) (2014). https://doi.org/10.1109/TFUZZ.2013.2272585
15. Huchang, L., Xu, Z.S.: Priorities of intuitionistic fuzzy preference relation based on multiplicative consistency. IEEE Trans. Fuzzy Syst. **22**(6), 1669–1681 (2014)
16. Genç, S., Boran, F.E., Akay, D.: Some approaches on estimating criteria weights from intuitionistic fuzzy preference relations under group decision making. J. Multiple-Valued Logic Soft Comput. (2010)
17. Genç, S., Boran, F.E., Akay, D., Xu, Z.S.: Interval multiplicative transitivity for consistency, missing values and priority weights of interval fuzzy preference relations. Inf. Sci. **180**, 4877–4891 (2010)
18. Boran, F.E.: An integrated intuitionistic fuzzy multi criteria decision making method for facility location selection. Math. Comput. Appl. **16**(2), 487–496 (2011)
19. Boran, F.E., Genc, S., Akay, D.: A method for solving multi criteria intuitionistic fuzzy decision making problems. In: 1st International Fuzzy Systems Symposium, Ankara, July 2019, pp. 47–51 (2019)
20. Kahraman, C., Öztayşi, B., Onar, S.Ç.: An integrated intuitionistic fuzzy AHP and TOPSİS aproach to evaluation of outsource manufacturers. J. Intell. Syst. **29**(1), 283–297 (2020)

The Use of Cognitive Modeling in Solving Problems of Ecological Sustainability of the Region

A. B. Hasanov⊙

Abstract The paper shows the prospects for combining two modeling platforms in a hybrid system of difficult to formalize tasks, the description of which is possible both at the cognitive and functional levels. One of the most difficult problems for the management involved in developing a strategy for the sustainable development of a region is understanding the complex causal chains that determine the impact of external and internal conditions on the environment. Today, this problem is compounded by the growing complexity and instability of the economic environment, leading to numerous uncertainties and risks.

Keywords The theory of dynamic analysis · Cognitive maps · Modeling

1 Introduction

Under these conditions, the use of well-known strategic management support tools, such as Total Quality Management (TQM), Business Process Reengineering (BPR), Balanced Scorecard (BSC), Six Sigma, performance management business (Business Performance Management, BPM), business intelligence systems (Business Intelligence, BI), ecision support systems (Decision Support Systems, DSS), strategic planning systems, etc., is faced with serious difficulties and limitations everywhere. New tools are needed that correspond to the creative nature of modern management [1], based on a research approach and long-term dynamic analysis of strategic decisions under various scenarios of future development.

Ideas and methods of cognitive modeling, presented, for example, in [3, 4] Recently, the methods of cognitive modeling are becoming more and more widespread for solving various kinds of analytical problems, especially those that are difficult to formalize [1]. In the traditional setting, the cognitive apparatus is based

A. B. Hasanov (✉)
Institute of Control Systems, Ministry Science and Education of the Republic of Azerbaijan, 68 B.Vahabzade, Baku AZ1141, Azerbaijan
e-mail: hesenli_ab@mail.ru

© The Author(s), under exclusive license to Springer Nature Switzerland AG 2023   183
Sh. N. Shahbazova et al. (eds.), *Recent Developments and the New Directions of Research, Foundations, and Applications*, Studies in Fuzziness and Soft Computing 422, https://doi.org/10.1007/978-3-031-20153-0_14

Fig. 1 The structure of the functional cognitive map

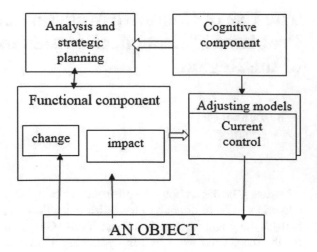

on expert assessments that characterize some conceptual qualitative parameters of an object and, as a rule, are not related, at least not directly, with the real physical parameters of the modeled system (object, process). In accordance with this, it can be indicated that there are independently two models of object description: cognitive, reflecting the qualitative level of object analysis, and functional, reflecting the quantitative level of object description. This determines the expediency of working out the unification of the two modeling platforms within a single hybrid system, which we will define as a functional cognitive map—FKK. Functional cognitive map.

A functional cognitive map should be presented as a system containing two components in accordance with the structure shown in Fig. 1.

If we consider the hierarchy of the system from the point of view of modeling goals, then it can be indicated that the functional component provides the current control of the object based on the measurement of its parameters and the provision of control actions. In technical objects, for example, for a boiler unit, the measured parameters are the readings of the sensors, and the signals to the actuators are the influences. In organizational systems, the measured objects and impacts are determined by the specifics of the object and are selected or formed on the basis of the relevant documentation.

The parameters obtained as part of the functional component are used for:

(a) calculating the values of the concept within the framework of the model itself, together with cognitive indicators;
(b) adjusting cognitive models.

In the cognitive map of the analysis of the functioning of the ecology of a particular region, the specified concept has causal relationships with a number of related concepts, for example, such as "personnel qualifications—"$K\_\phi n$"", "Tension of the production program—"K_nn"", Level

$$K_{CO} = \omega_{w/\text{кп}} * K_{\text{кп}} + \omega_{w/\text{кп}} * K_{\text{пп}} + \omega_{w/\text{оз}} * K_{\text{оз}} + \dots$$

Here ω are empirical coefficients that determine the level of influence of the corresponding concept. At the same time, within the framework of the functional component, the parameter "Equipment availability factor FGO" is used, which is subject to numerical measurement, the definition of which is based on the appropriate techniques and standards [3]. The values of this functional parameter, as mentioned above, can be used in two ways.

First, a function parameter can be used directly to compute the value of a concept:

$$K_{CO} = F_{co}\{\{K_{\text{кп}}, K_{\text{пп}}, K_{\text{оз}}\}, \Phi_{\text{го}}\}$$
$$K_{CO} = k_K\{K_{\text{кп}}, K_{\text{пп}}, K_{\text{оз}}\} + k_\Phi\{\Phi_{\text{го}}\}.$$

Here the coefficients k_K, k_Φ are the share of the contribution of the cognitive and functional components.

Secondly, the functional parameter can be used to analyze the correctness and make adjustments to the conceptual model according to the following scheme:

$$F'_{\text{го}} = P_{\text{го}}(F_{\text{го}}).$$
$$e = f(K_{CO} - F'_{\text{го}}).$$

Here: $F'_{\text{го}}$—phased value of the functional parameter;

$P_{\text{го}}$— fuzzification function;

e—residual (difference between cognitive and functional values).

In accordance with relation, the value of the functional parameter is phased out. In accordance with the relation, the discrepancy is determined, i.e. the difference between the meaning of the concept, formed within the framework of the cognitive environment, and the value of the corresponding functional parameter. The resulting discrepancy is the basis for correcting the cognitive model. The methods of such adjustment require a separate discussion.

The joint use of cognitive and formal processes within the framework of a single model requires an appropriate description of information objects—variables and procedures for their coordination. The classification of variables in the FCC is carried out in accordance with several, the following features and methods of application.

1. The main types of variables: cognitive/functional.
2. Presentation method: phased/defasified.

Taking into account the specifics of the complex model under consideration and the features of such elements defined in it, two types of representations are considered as part of the FCC: phased and defasified [4]. It is provided that any variable, if appropriate, can have both representations, for which a corresponding conversion table is entered. Based on the meaning and role of the variable, in the models under study, the primary and secondary process are distinguished, the variable can be converted to a secondary representation.

The use of FKK, in which, along with cognitive, formal descriptions are present, allow solving a number of new problems that are not characteristic of traditional methods. We will designate one of these tasks as the "Input Converter Task". When solving this problem, it is assumed that the object is represented in the form of a classical model of the control object from the theory of automatic control [5]. The main task of the transformation involves the analysis of the influence of inputs, which are natural from an applied point of view for each object, on the outputs.

Within this view, input, output and intermediate variables are introduced. The purpose of the output and intermediate variables is obvious. A separate description is required for input variables. In this context, we will single out control actions and independent actions as part of the input variables. Control variables describe targeted actions on an object; the independent variables describe the effects on the object that are not subject to targeted influence.

In a similar context, the inverse problem can be presented, the solution of which is associated with the analysis of the values of the inputs to achieve the desired values of the outputs.

The formal description of the FKK contains two components:

(a)   topological model of the FKK;
(b)   concept model.

We define FKK as a network set of interacting concept processes, the functioning of which is determined both by the influence of related concepts and formal procedures describing the real parameters of an object.

The topological model (TM) is described on the basis of graph structures traditional for the representation of cognitive models:

$$TM = \{K, D\}$$

corresponds to the meaning of the concept. $D = \{d_{j1, j2}\}$—a set of graph arcs reflecting causal relationships. In this connection, j_1 corresponds to the final node of the connection, j_2 is the outgoing node.

Here, the main provisions of cognitive modeling are presented and its basic ontologies—cognitive maps and methods of analyzing cognitive maps—are considered. Cognitive tools for dynamic analysis of strategies are presented, taking into account the peculiarities of enterprises in today's difficult conditions. The tools can be used to find effective (in one sense or another) strategies in an ever-changing business environment. An example of the use of these tools is given, and their applied capabilities are discussed. Cognitive modeling is a method of analyzing and managing complex (semi-structured [5]) systems and problem situations, carried out through:

(a)   building a model of a problem situation in the form of a cognitive map;
(b)   conducting model experiments in order to find effective strategies for managing a problem situation, carried out using the methods of analyzing the cognitive map.

1. Determination of the formal potential $(P_i^f)_{\Phi a3}$ _ phases.

At this stage, the indicator-indicator is calculated from the models that describe the real behavior of the object and take into account the real characteristics. The indicator-indicator calculated at this stage should be close in meaning to the concept under consideration and in many respects is a certain numerical assessment of the concept. In this case, applied specialized methods and variables are used.

2. Determination of the potential of the concept.

Within the framework of this stage, two tasks are solved:

(a) the actual formation of the meaning of the concept:

$$ki = f((Pik), (Pif)_{\Phi a3}).$$

In its simplest form, this dependence can be expressed in polynomial form.

$$ki = b_1 * (P_i^k), +b_2((P_i^f))_{\Phi a3}).$$

Here b_1, b_2 are empirical coefficients reflecting the degree of influence on the overall value. It can be noted that when $b_1 = 1, b_2 = 0$, expression (7) corresponds to a pure concept.

(b) Correcting the model
In accordance with the preferences of the designer, the formal component can be omitted in the modeling process. However, the formal component can be used to adjust the network model by identifying the concept residual:

$$e_i = e((P_i^k), (P_i^f)_{\Phi a3}),$$

where e_i—is the residual determination function.

In a simple case, one can define

$$e_i = \left(P_i^k\right) - \left(P_i^f\right)_{\Phi a3,}$$

Moreover, this refers to the implementation of operations in a fuzzy area. The ei value serves to determine the need to adjust the cognitive model and serves as certain parameters in the adjustment models.

3. Determination of the value of the potential output

In the general case, the value of the potential can be taken as the concept value, but the use of special functions allows you to introduce nonlinearity in the model. Activation models of neural systems can be used as such functions. This issue requires separate consideration and is beyond the scope of this article. Methods for analyzing a cognitive map make it possible to carry out model experiments on a cognitive map by changing the composition and values of the basic factors and the nature of the cause-and-effect relationships between them. Model experiments of this kind make it possible to study the propagation of external and control influences along the cognitive map and to solve a wide range of problems related to determining priority management decisions, assessing the attainability of management goals, developing alternative management strategies, finding effective (in one sense or another) management strategies, etc.

Existing methods for the analysis of cognitive maps are focused on analysis tasks of two types: static and dynamic. Static analysis, or influence analysis, is a range of tasks aimed at studying the structure of the mutual influences of the factors of the cognitive map. Dynamic analysis underlies the generation and analysis of possible scenarios for the development of the situation in time in the modes of "self-development" and "controlled development".

The theory of dynamic analysis of cognitive maps is based on the apparatus of linear dynamic systems [5]. The dynamics are simulated by setting in discrete moments of time $t = 0,1,2, \ldots$ of sequential impulse influences on controlled factors (factors-causes) and modeling of a wave of influences of these influences on target factors (factors-consequences).

The control action on the situation is set by impulse change in the value of the controlled factor $x\_in \in X\_ (i \wedge \{\cdot\})$.

The change in the values of the effect factors is determined using the "impulse process rule":

$$x_i(t_0 + 1) = x_i(t_0) + \sum_{i \in I} \mathrm{sgn}(x_j, x_i) w_{ij} \Delta x_j(t)$$

where $x_i(t_0)$ is the value of the i-th factor at time t_0 before the control action; $x_i(t_0 + 1)$ is the value of the i-th factor after the control action at the moment of time $(t_0 + 1)$

$$\mathrm{sgn}(x_j, x_i) = \begin{cases} +w(x_j, x_i), & (x_j, x_i) \geq 0 \\ -w(x_j, x_i), & (x_j, x_i) \leq 0 \end{cases}$$

Thus, the value of the factor-effect at each moment is estimated as the sum of the values of the factor at the previous moment and all influences that came from neighboring (associated) factors. When assessing the resulting value of a factor, both the actual values of the influencing factors and the degree of their influence are taken into account.

When modeling the dynamics, along with the values of factors, tendencies of changes in factors can also be used. The forecast of the development of the situation is obtained in the form of vectors of the state of the situation at successive discrete moments of time $t, t + 1, \ldots, t + n \leq$, were t is the number of the step (cycle) of modeling.

The task of managing the situation is to transfer the situation from the initial state to some target state corresponding to the target image of the problem situation.

The target image defines the desired changes in the state of the problem situation from the standpoint of the subjects of management and is formally presented as

$$C = (X^C, R(X^C)),$$

Cognitive tools provide ample opportunities for solving critical tasks for modern management. First of all, these are issues related to the generation of alternative strategies and the assessment of their effectiveness in the context of the multifactor dynamics of the internal and external environment of the region's ecology.

$\Delta x_j (t_0)$—pulse increment of the j-th control factor at time t_0; w_{ij} is the weight (strength) of the influence of the factor x_j on the factor x_i;

I_i—the number of factors directly affecting the factor xi.

The technique of constructing cognitive models (like all other knowledge-based models) is specific and involves a gradual and multi-stage design: the development of a demonstration prototype, the development of a research prototype, then a working prototype and, finally, the creation of an industrial prototype suitable for solving real problems. We consider it necessary to emphasize once again that the article presents just a demonstration prototype of a cognitive model, which, in our opinion, is quite convincing evidence of the high potential of the applied capabilities of this class of models. The creation of industrial prototypes, taking into account the numerous features of a particular enterprise and its external environment, is usually laborious and, in terms of costs, can significantly exceed the costs of developing a demonstration prototype.

In real projects, cognitive maps can have a more complex structural and functional organization. The question of the reliability of cognitive maps is of great importance in the cognitive modeling of strategies. The solution to this issue largely depends on the correct choice of basic factors and cause-and-effect relationships of the constructed map. To date, in the theory of cognitive modeling, this question remains open.

2 Conclusion

In conclusion, we note that the models introduced in this article are aimed at providing the following capabilities that expand the scope of the use of cognitive modeling:

(a) the ability to take into account the real characteristics of the modeled object in the process of cognitive analysis;

(b) the possibility of taking into account the adjustment of cognitive structures due to the connection with real indicators of functioning. An important advantage (not excluded—unique) of cognitive tools is the opportunity they open to explore the fine structure of management strategies (the necessary sequence of switching on management influences, the required degree of activity of these influences, the study of the dynamic stability of strategies, etc.). None of the known strategic management support tools have such capabilities. Another important advantage of cognitive models is the fact that they allow one to study the dynamics of strategies at a qualitative level, without using for this purpose difficult to access and not always reliable quantitative statistics. This is extremely important in today's rapidly changing business environment and the growing pace of technological innovation. Cognitive dynamic analysis significantly expands the instrumental base of strategic management, which is based today mainly on the means of static situational analysis and recipe decision-making schemes.

References

1. Ginis, A., Kolodenkova, A.E.: Fuzzy cognitive modeling to prevent risk situations at critical infrastructure facilities. Bull. UTATU. 2017 **21**, 4(78), 113–120 (2017).
2. Clerk Maxwell, J.: A Treatise on Electricity and Magnetism, 3rd ed., vol. 2. Oxford, Clarendon, pp. 68–73 (1892)
3. Kolesnikov, V.L., Brakovich, A.I., Zhuk, Y.A.: Phasing and defusing data for solving multicriteria problems. Physics and mathematics and informatics. In: Works of BSTU. 2013 vol. 6, pp. 125–127. K. Elissa, "Title of Paper If Known," unpublished
4. Pervozvanskiy, A.A.: Course of the Theory of Automatic Control. Moscow, Nauka (1986)
5. Bozhenyuk, A.V., Ginis, L.A.: Application of fuzzy models for the analysis of complex systems. Control Syst. Inf. Technol. **51**(1.1), 122–126 (2013)

Soft Computing and Logic Aggregation

Which Are the Correct Membership Functions? Correct "And"- and "Or"-Operations? Correct Defuzzification Procedure?

Olga Kosheleva, Vladik Kreinovich, and Shahnaz N. Shahbazova

Abstract Even in the 1990s, when many successful examples of fuzzy control appeared all the time, many users were somewhat reluctant to use fuzzy control. One of the main reasons for this reluctance was the perceived subjective character of fuzzy techniques—for the same natural-language rules, different experts may select somewhat different membership functions and thus get somewhat different control/recommendation strategies. In this paper, we promote the idea that this selection does not have to be subjective. We can always select the "correct" membership functions, i.e., functions for which, on previously tested case, we got the best possible control. Similarly, we can select the "correct" and- and or-operations, the correct defuzzification procedure, etc.

Keywords Fuzzy logic · Membership function · "And"-operation · "Or"-operation · Defuzzification

1 Formulation of the Problem: Is There Such a Thing as the Correct Membership Function?

Need to translate imprecise (fuzzy) expert rules into a precise control strategy. In many application areas, be it medicine or engineering, some professionals are

O. Kosheleva · V. Kreinovich (✉)
University of Texas at El Paso, El Paso, TX 79968, USA
e-mail: vladik@utep.edu

O. Kosheleva
e-mail: olgak@utep.edu

S. N. Shahbazova
Azerbaijan State University of Economics, UNEC, 6, Istiqlaliyyet str., Baku, Azerbaijan
e-mail: shahbazova@gmail.com; shahnaz_shahbazova@unec.edeu.az

© The Author(s), under exclusive license to Springer Nature Switzerland AG 2023
Sh. N. Shahbazova et al. (eds.), *Recent Developments and the New Directions of Research, Foundations, and Applications*, Studies in Fuzziness and Soft Computing 422, https://doi.org/10.1007/978-3-031-20153-0_15

more successful and more skilled than others. It is therefore desirable to somehow incorporate the knowledge and skills of the top experts into a computer-based system that would help other experts in their decisions.

Top experts are usually willing (and even eager) to share their knowledge. Some part of this knowledge they describe in precise numerical terms, such as "if the body temperature is above 100 F, take aspirin". The problem is that a large part of the experts' knowledge is described not in such precise numerical terms, but in terms of imprecise ("fuzzy") words of natural language like "high" or "small". To incorporate this knowledge inside a computer-based system, it is important to translate such knowledge into a precise form.

This was the main motivation of Lofti Zadeh when he invented fuzzy techniques; see, e.g., [1, 4, 9, 13, 14, 18].

Fuzzy techniques: general idea. To explain the problem with which we deal win this paper, let us briefly refresh the main ideas of fuzzy techniques. We will explain these ideas on the example of the most frequent application of fuzzy technique—fuzzy control. Specifically, we will describe the main ideas behind the simplest (and most widely used) approach—Mamdani's approach.

We consider a system whose state, at any given moment of time, is characterized by n numerical values $x_1, \ldots, x_n$. For example:

- For a simple moving point-size object, the first three parameters are coordinates and the second three parameters are three components of the velocity.
- For a car, we have two spatial coordinates, two components of velocity, and parameters that describe the orientation of the wheels, the engine's regime, etc.

We want, based on the values x_i describing the system's state, to recommend an appropriate control. The control may also be described by several parameters. Let us concentrate on recommending one of these parameters; we will denote this parameter by u. In these terms, we want, for each tuple $x = (x_1, \ldots, x_n)$, to find an appropriate control value u.

To find this value u, we have several (K) if-then rules provided by the experts, i.e., rules of the type "if $A_{k1}(x_1)$ and ... and $A_{kn}(x_n)$, then $B_k(u)$", where $k = 1, \ldots, K$ is the number of the rule, and A_{ki} and B_k are imprecise properties like "small".

In general, once we have such rules, then the control value u is appropriate for the state x if:

- either the first rule is applicable—i.e., its conditions $A_{1i}(x_i)$ are all satisfied and its conclusion $B_1(u)$ is also satisfied,
- or the second rule is applicable—i.e., its conditions $A_{2i}(x_i)$ are all satisfied and its conclusion $B_2(u)$ is also satisfied, etc.

If we describe the property "u is a reasonable control for the state x" by $R(x, u)$, then the above statement can be described in the following symbolic form:

$$R(x, u) \Leftrightarrow$$

$$(A_{11}(x_1) \& A_{12}(x_2) \& \ldots \& B_1(u)) \vee (A_{21}(x_1) \& A_{22}(x_2) \& \ldots \& B_2(u)) \vee \ldots$$
(1)

First step: membership functions. The formula (1) may looks like a normal logical formula, but it is not:

- in a usual logical formula, every statement is either true or false, but
- here, a statement like "98° is a high fever" is only true to some extent.

In the computer, true is represented as 1, and false as 0. Thus, to represent statements which are not exactly true and not exactly fully false, Zadeh proposed to use real numbers intermediate between 0 and 1.

Eliciting such numbers for each statement is a usual thing in polls, when we are asked, e.g., to estimate, on a scale from 0 to 10, how satisfied we are with the given service. People usually have no problem assigning some numerical value to such statements. The only thing we need to do to get a value from 0 to 1 is to normalize this number—i.e., in our example, to divide it by 10.

For each of the quantities x_i, we can select several possible values $x_{i1}, \ldots, x_{ip}$. Then, for each of these values x_{iv} and for each of the corresponding properties A_{ki}, we can ask the expert to estimate the degree to which this property is satisfied by the given value. This degree is usually denoted by $\mu_{ki}(x_{iv})$.

Of course, there are infinitely many possible real numbers x_i, and it is not possible to ask the expert about each of them. So, to find the degree $\mu_{ki}(x_i)$ for all other possible values x_i, we need or perform some interpolation or extrapolation. As a result, we get a function $\mu_{ki}(x_i)$ that assigns, to each possible value x_i, the degree to which this value satisfies the property A_{ki}. This function is called a *membership function* corresponding to the property A_{ki}.

Similarly, we design a membership function $\mu_k(u)$ describing to what extent the value u satisfies the property B_k.

Second step: "and"- and "or"-operations. Once we know the expert's degrees of confidence $\mu_{ki}(x_i)$ and $\mu_k(u)$ in statement $A_{ki}(x_i)$ and $B_k(u)$, we need to estimate the degree to which the statement $R(x, u)$ is true. For this purpose, fuzzy technique uses special functions $f_\&(a, b)$ and $f_\vee(a, b)$ known as "and"- and "or"-operations (or, for historical reasons, t-norms and t-conorms):

- the function $f_\&(a, b)$ transforms the expert's degree of confidence a and b in statement A and B into an estimate for the expert's degree of confidence in the statement $A \& B$; and
- the function $f_\vee(a, b)$ transforms the expert's degree of confidence a and b in statement A and B into an estimate for the expert's degree of confidence in the statement $A \vee B$.

By using these operation, we can find, for each u, the degree $\mu_r(x, u)$ to which the control u is reasonable for the input x as

$$\mu_r(x, u) = f_\vee(f_\&(\mu_{11}(x_1), \mu_{12}(x_2), \ldots, \mu_1(u)),$$

$$f_\&(\mu_{21}(x_1), \mu_{22}(x_2), \dots, \mu_2(u)), \dots). \qquad (2)$$

Final step: defuzzification. If we are designing a recommender system, then the degrees (2) is all we want.

However, if we are designing a system for automatic control, then we need to transform the membership function (2) into an exact value $\bar{u}$. This process is known as *defuzzification*.

The most frequently used defuzzification is based on minimizing the mean squared difference $(\bar{u} - u)^2$ between the estimate and the actual (unknown) value of the optimal control—so that the term corresponding to each u is weighted by the degree $\mu(x, u)$ to which this value u is reasonable. In precise terms, we minimize the expression

$$\int \mu(x, u) \cdot (u - \bar{u})^2 \, du.$$

Differentiating this expression with respect to $\bar{u}$ and equating the resulting derivative to 0, we get the following formula

$$\bar{u} = \frac{\int u \cdot \mu(x, u) \, du}{\int \mu(x, u) \, du}.$$

This formula is known as *centroid defuzzification*.

Problem: this is all very subjective. One of the main problems is that all this sounds very subjective. One expert may draw one membership function, another expert may draw a somewhat different one. Because of this subjectivity, many users are reluctant to use fuzzy techniques.

A natural question may be: which membership function is a "correct" one? At present, the usual answer is: there is no correct, objective membership function, it is all subjective.

Our opinion. What we argue in this paper is that while, yes, the existing membership functions are subjective, there *is* such a thing as a correct membership function—as well the correct "and"- and "or"-operations, etc.

Actually, our arguments will be very straightforward and simple. However, they seem to be novel and, in our opinion, are important, so we decided to write them down and to promote them as much as possible.

2 In Each Application Area, There Are the Correct Membership Functions: Our Arguments

Towards an idea. In the control case—where fuzzy techniques are most frequently used—the main objective of the fuzzy technique is to transform the natural-language-based fuzzy rules into a precise control strategy.

If we use different membership functions, we get different control strategies. For each specific system, some strategies are better, some are somewhat worse.

There is usually an objective function that describes what exactly we want. For example, when planning a route for a self-driving car:

- in some situations, e.g., in emergency, we want to find the fastest way from point A to point B;
- in other cases, we want to find a way that—within a given time limit—will lead to the smallest amount of pollution;
- for sending delicate fragile objects, the objective may be to make sure that the ride is as smooth as possible—in some precise sense.

Thus, there exist membership functions for which this objective function attains the largest possible value—these membership functions are the best of this particular situation and are, thus, the *correct* ones.

So what? At first glance, what we just wrote may sound reasonable, but on second thought, this does not sound convincing at all. The only way to find the correct membership function is to know the exact system, and if we know the exact system, then we do not really need the fuzzy techniques: instead of optimizing membership functions, we can simply optimize the control itself.

Still, there are correct membership functions. If we only had one system to control, then, true, the above idea would make no sense at all. In practice, however, we have many different systems to control—and for each of these systems, we may have many different situations to control, with many different objective functions (as we have mentioned above).

In all these situations, the same users provide rules by using largely the same natural-language words. So, a natural idea is as follows.

We consider all the situations $s_1, \ldots, s_a$ for which the users provided the rules *and* for which the system is actually known. For each of these situations, we know the corresponding objective function. Let us denote by $F_i(\mu)$ the value of the objective function which is obtained in the i-th situation if we use a combination

$$\mu = (\mu_{11}, \mu_{12}, \ldots, \mu_1, \mu_{21}, \mu_{22}, \ldots, \mu_2, \ldots)$$

of the membership functions. To each of the situations, we assign a weight w_i describing this situation's importance. In this sense, the overall objective function has the form

$$F(\mu) = w_1 \cdot F_1(\mu) + w_2 \cdot F_2(\mu) + \ldots + w_a \cdot F_a(\mu).$$

Then, we find the combination of membership functions μ for which this overall objective function attains its largest possible value. The membership functions from this combination are thus the best (most effective), i.e., *correct* membership functions.

How are these correct membership functions useful? Why do we need to know which membership functions are the best for known situations? Because usually, there are many *unknown* situations, i.e., situations:

- in which we do not have exact expressions for the system, and
- for which the only thing we know are expert rules.

These are exactly the types of situations for which fuzzy techniques were invented in the first place. Since the "correct" membership functions worked well on previous situations, it is highly possible that they will work well in the new situation as well.

Finding the correct membership functions will make applications of fuzzy techniques easier. If we use the above idea, then, in each application area, after we have found the correct membership function, there is no longer any need to elicit membership degrees etc., no need for any subjectivity at all—all this will be now done "under the hood", inside the system.

All the expert has to do is supply rules, and the system will automatically translate it into a control strategy—by using the stored "correct" membership functions.

Of course, with new applications, we may need to update what we mean by correct membership functions. Of course, for new systems, we may also gain some information about how they actually work, and thus, be able to estimate possible numerical consequences of different combinations of membership functions.

We will then be able to add this new system to the list of situations based on which we search for the correct membership function, and thus, maybe come up with new, slightly different "correct" membership functions.

Caution. We are *not* yet describing *how* to find the "correct" membership functions, we are just explaining why finding such correct membership functions may be very useful. There are many different optimization techniques, so we should not worry about how to optimize—although, of course, it would be nice to see which optimization techniques are more efficient for finding such "correct" membership functions (see our discussion in the next section).

3 Not Only Membership Functions

Not only membership functions. Why limit ourselves to selecting membership functions? We can similarly select the best "and"- and "or"-operations, the best defuzzification procedure, etc.:

- we find the "and"- and "or"-operations and/or a defuzzification procedure that work the best on known cases, and
- we can then use these optimal ("correct") "and"- and "or"-operations and defuzzification in new situations.

This has been partly done. For a medicine-related recommendation system, this was done by the designers of the world's first expert system MYCIN, an expert system for diagnosing rare blood diseases; see, e.g., [2]. To find the best "and"-operation, the designers of MYCIN selected numerous pairs of related statements (A_i, B_i), $i = 1, 2, \ldots$, and asked the experts to estimate their degree of confidence a_i, b_i, and c_i in the statements A_i, B_i, and $A_i \& B_i$. Then, they found an "and"-operation $f_\&(a, b)$ that provided the best match for all these cases, i.e., for which $f_\&(a_i, b_i) \approx c_i$ for all i. This can be done, e.g., by the usual Least Squares technique, when we minimize the sum

$$(f_\&(a_1, b_1) - c_1)^2 + (f_\&(a_2, b_2) - c_2)^2 + \ldots$$

This selection—as well as the similar selection of the most efficient "or"-operation—led to a very successful expert system.

Interestingly, at first, the MYCIN designers thought that they found general laws of human reasoning. However, when they tried to apply their "and"- and "or"-operations to a different application domain—geosciences—the results were not good. It turned out that in different application domains, different "and"- and "or"-operations work best. This makes perfect sense; for example:

- in medicine, unless we have a real emergency, it is important to be very cautious, so as not to harm the patient; so, the doctors do not recommend a serious procedure like surgery unless they are reasonably sure that it will be successful—and they perform a lot of tests if they are not sure;
- on the other hand, if an oil company is too cautious, its competitors will be the first to exploit the new possibility, so actions need to be bold; empty wells and resulting losses are OK (and inevitable) as long as overall, the company is profitable.

For control, also, depending on different objective functions—whether we want stability or smoothness—different "and"- and "or"-operations turn out to be better; see, e.g., see, e.g., [6–8, 10–13, 16, 17].

An interesting possibility. It may turns out that several different combinations of membership functions, "and"- and "or"-operations, and defuzzification lead to exactly the same control and are, thus, equally good.

This happened in physics, where, as the famous mathematician and physicist Henri Poincaré noticed in the early 20 century, we can describe the same gravity phenomena is different geometries—we just need to modify physical equations:

- we can have Einstein's curved space, then particles move along inertial lines, or
- we can have the flat Minkowski space-time in which gravity is an extra force; see, e.g., [15] (see also [3, 5]).

If such a situation occurs in fuzzy—that several options lead to equally good control (or equally good recommendations)—then out of several equally good options we can select, e.g., the one which is the easiest to compute.

We can go even further. Why limit ourselves to the usual [0, 1]-based fuzzy sets? We can alternatively consider interval-valued (or, more generally, type-2) interpretations of natural-language words, in which, for each possible value x of the corresponding quantity, instead of a single degree $\mu(x)$, we have an interval $\left[\underline{\mu}(x), \overline{\mu}(x)\right]$ of possible degree values.

It would be interesting to know which technique leads to the best control and/or to the best recommendations. This will convert current philosophical and case-by-case discussions into a practical problem: as the proverb says, the proof of the pudding is its eating. Maybe in some application areas, type-1 will be better, in others type-2?

We can similarly compare Mamdani's technique with a sometimes used fuzzy logic technique, in which we use fuzzy implication $f_\rightarrow(a, b)$ to get a more direct description of the expert's rules:

$$\mu_r(x, u) = f_\&(f_\rightarrow(f_\&(\mu_{11}(x_1), \mu_{12}(x_2), \ldots), \mu_1(u)),$$

$$f_\rightarrow(f_\&(\mu_{21}(x_1), \mu_{22}(x_2), \ldots), \mu_2(u)), \ldots).$$

It would be interesting to know which one leads to better control or better recommendations.

Similarly, we can compare the fuzzy approach with probabilistic approaches proposed for describing natural-language rules.

Opportunities are endless, let us follow them, there is a lot of work ahead!

Acknowledgements This work was supported in part by the US National Science Foundation grants 1623190 (A Model of Change for Preparing a New Generation for Professional Practice in Computer Science) and HRD-1242122 (Cyber-ShARE Center of Excellence).

References

1. Belohlavek, R., Dauben, J.W., Klir, G.J.: Fuzzy Logic and Mathematics: A Historical Perspective. Oxford University Press, New York (2017)
2. Buchanan, B.G., Shortliffe, E.H.: Rule Based Expert Systems: The MYCIN Experiments of the Stanford Heuristic Programming Project. Addison-Wesley, Reading (1984)
3. Gupta, S.N.: Einstein's and other theories of gravitation. Rev. Modern Phys. **29**, 334–336 (1957)
4. Klir, G., Yuan, B.: Fuzzy Sets and Fuzzy Logic. Prentice Hall, Upper Saddle River (1995)
5. Kreinovich, V.: Gupta's derivation of Einstein equations. Soviet Acad. Sci. Doklady **222**(2) 319–321 (1975). (in Russian); translated into English in Soviet Phys. Doklady **20**(5) 341–342 (1975)
6. Kreinovich, V.: From semi-heuristic fuzzy techniques to optimal fuzzy methods: mathematical foundations and applications. In: Papadopoulos, B., Syropoulos, A. (eds.) Current Trends and Developments in Fuzzy Logic, Proceedings of the First International Workshop, Thessaloniki, Greece, October 16–20, pp. 1–62 (1998)
7. Kreinovich, V., Kumar, S.: Optimal choice of &- and ∨-operations for expert values. In: Proceedings of the 3rd University of New Brunswick Artificial Intelligence Workshop, Fredericton, New Brunswick, pp. 169–178 (1990)
8. Kreinovich, V., Quintana, C., Lea, R., Fuentes, O., Lokshin, A., Kumar, S., Boricheva, I., Reznik, L.: What non-linearity to choose? Mathematical foundations of fuzzy control. In: Proceedings of the 1992 International Conference on Fuzzy Systems and Intelligent Control, Louisville, Kentucky, pp. 349–412 (1992)

9. Mendel, J.M.: Uncertain Rule-Based Fuzzy Systems: Introduction and New Directions. Springer, Cham (2017)
10. Nguyen, H.T., Kreinovich, V.: Methodology of fuzzy control: an introduction. In: Nguyen, H.T., Sugeno, M. (eds.) Fuzzy Systems: Modeling and Control, pp. 19–62. Kluwer, Boston (1998)
11. Nguyen, H.T., Kreinovich, V., Tolbert, D.: On robustness of fuzzy logics. In: Proceedings of the 1993 IEEE International Conference on Fuzzy Systems FUZZ-IEEE'93, San Francisco, California, March 1993, vol. 1, pp. 543–547 (1993)
12. Nguyen, H.T., Kreinovich, V., Tolbert, D.: A measure of average sensitivity for fuzzy logics. Int. J. Uncertainty, Fuzziness, Knowl.-Based Syst. **2**(4), 361–375 (1994)
13. Nguyen, H.T., Walker, C.L., Walker, E.A.: A First Course in Fuzzy Logic. Chapman and Hall/CRC, Boca Raton (2019)
14. Novák, V., Perfilieva, I., Močkoř, J.: Mathematical Principles of Fuzzy Logic. Kluwer, Boston (1999)
15. Poincaré, H.: Science and Hypothesis. Bloomsbury Academic, London (2018)
16. Smith, M.H., Kreinovich, V.: Optimal strategy of switching reasoning methods in fuzzy control. In: Nguyen, H.T., Sugeno, M., Tong, R., Yager, R. (eds.) Chapter 6 Theoretical Aspects of Fuzzy Control, pp. 117–146. Wiley, New York (1995)
17. Tolbert, D.: Finding "and" and "or" operations that are least sensitive to change in intelligent control, University of Texas at El Paso, Department of Computer Science, Master's Thesis (1994)
18. Zadeh, L.A.: Fuzzy sets. Inf. Control **8**, 338–353 (1965)

Need for Simplicity and Everything Is a Matter of Degree: How Zadeh's Philosophy is Related to Kolmogorov Complexity, Quantum Physics, and Deep Learning

Vladik Kreinovich, Olga Kosheleva, and Andres Ortiz-Muñoz

Abstract Many people remember Lofti Zadeh's mantra—that everything is a matter of degree. This was one of the main principles behind fuzzy logic. What is somewhat less remembered is that Zadeh also used another important principle—that there is a need for simplicity. In this paper, we show that together, these two principles can generate the main ideas behind such various subjects as Kolmogorov complexity, quantum physics, and deep learning. We also show that these principles can help provide a better understanding of an important notion of space-time causality.

Keywords Zadeh · Fuzzy logic · Simplicity · Deep learning · Kolmogorov complexity · Quantum physics

1 Need for Simplicity and Everything Is a Matter of Degree: Two Main Principles of Zadeh's Philosophy

Two main ideas of Zadeh's philosophy: in brief. In a nutshell, the main principle behind Lotfi Zadeh's research is a phrase that he himself repeated many times: Everything is a matter of degree. However, it is easy to forget—especially for mathematicians by training (like us)—that he also actively pursued another principle: the idea that there is a need for simplicity.

V. Kreinovich (✉) · O. Kosheleva
University of Texas at El Paso, El Paso, Texas 79968, USA
e-mail: vladik@utep.edu

O. Kosheleva
e-mail: olgak@utep.edu

A. Ortiz-Muñoz
California Institute of Technology, Pasadena, CA 91125, USA
e-mail: aoortiz@caltech.edu

© The Author(s), under exclusive license to Springer Nature Switzerland AG 2023
Sh. N. Shahbazova et al. (eds.), *Recent Developments and the New Directions of Research, Foundations, and Applications*, Studies in Fuzziness and Soft Computing 422,
https://doi.org/10.1007/978-3-031-20153-0_16

These two principles underlie his ideas of fuzzy logic: yes, the first principle is that everything is a matter of degree, so instead of 0 (false) and 1 (true) we have all possible numbers from the interval [0, 1]. However, to flesh out this argument, Zadeh used simplicity as a guiding principle—thus, he selected the simplest "and"- and "or"-operations (i.e., t-norms and t-conorms), he selected the simplest membership functions, etc.; see, e.g., [13].

Many of us remember that every time we came to him with some complex developments, his first reaction always was: can we make it simpler?

What we do in this paper. At first glance, it may seem that these two principles—the need for simplicity and everything is a matter of degree—are specific for fuzzy logic and related areas. However, what we show in this paper is that these principles are valid way beyond fuzzy logic—i.e., beyond the description of imprecise natural language statements.

We show how these principles can be naturally applied:

- first, to the principles themselves,
- then to the description of the physical world, and
- finally, to the way we gain knowledge about the world.

As a result, we naturally arrive at the main ideas behind, correspondingly, Kolmogorov complexity, quantum physics, and deep learning.

This shows that there probably is a lot of remaining potential in these principles—which will enable us to solve many useful problems in the future as well.

2 Simplicity and Everything-Is-a-Matter-of-Degree Principles Naturally Lead to Kolmogorov Complexity

Let us apply the principles to themselves. What are the possible applications of the above two principle? The first natural idea is to apply these principles to the principles themselves: namely, to the description of what is simple and what is complex.

Need for degree of complexity. In accordance with the everything-is-a-matter-of-degree principle:

- we should not naively divide the constructions into simple and complex;
- instead, we should assign, to each construction, a degree of complexity.

What is the simplest degree of complexity? The complexity of an object is naturally described by how complex it is to design it. In particular, for objects generated by a computer, a natural measure of complexity is the complexity of a program that generated this object.

What is the simplest possible measure of complexity? The simplest possible idea is just to count the number of symbols in a program: if the program is short, it is probably rather simple, and, vice versa, if the program is long, it is probably complicated. So, a natural measure of a program complexity is its length.

For an object, the natural measure of its complexity is thus the length of the program that generates this object. Of course, we can have different programs for generating the same object:

- we can have reasonably short programs that compute this object directly, and
- we can have over-complicated programs that beginning students often write.

In the US, such over-complicated constructions are known as Rube Goldberg machines—after a cartoonist who excelled in describing unnecessarily complex ways of performing simple tasks.

From this viewpoint, to gauge the complexity of an object, we should take not just *any* program that generates this object, but the *simplest* such program—and we decided to measure the program's complexity by its length, the shortest such program. Thus, we arrive at the following definition.

We naturally arrive at Kolmogorov complexity. As we mentioned, the complexity of an object x can be measured by the length of the shortest program that generates this object. This definition is known as *Kolmorogov complexity* (see, e.g., [5]) and is usually denoted by $K(x)$.

Comments. Simultaneously with Kolmogorov, similar ideas were proposed by Solomonoff and Chaitin [5].

The length of a program depends on the programming language. However, as Kolmogorov has shown, the definitions $K_1(x)$ and $K_2(x)$ of Kolmogorov complexity based on two different languages differ by a constant:

$$|K_1(x) - K_2(x)| \le C_{12}$$

for all x. Thus, in effect, different programming languages lead to a similar notion of Kolmogorov complexity.

Why was Kolmogorov complexity invented—and how is it useful? While we showed that Kolmogorov complexity naturally follows from the two main principles of Zadeh's philosophy, this is not how this notion was originally invented. This notion was invented by Andrei Kolmogorov, a 20 century Russian mathematician who was one of the pioneers of modern (mathematically precise) probability theory. In addition to doing pure mathematics, Kolmogorov spent a lot of time and effort on applications. Because of this, he noticed a discrepancy between the traditional mathematical description of randomness and the physicists' and engineers' intuitive ideas of randomness.

Indeed, from the purely mathematical viewpoint, when you flip a fair coin n times, all resulting binary sequences—describing the n results—have the exact same probability 2^{-n} and are, in this sense, completely equal. However, from the practitioner's viewpoint:

- some of these sequences are random—in the sense that one can expect to get them when flipping an actual coin, and

- some sequences like 000...0 or 010101...01 are *not* random and thus, not expected.

From the purely mathematical viewpoint, if the same person wins the main multi-million prize of a super-lottery several years in a row, it is possible. However, if this would happen in real life, it would be clear that the lottery is rigged.

   To eliminate this discrepancy, Kolmogorov decided to formalize the intuitive notion of randomness of a finite sequence. His idea was very straightforward:

- Why is a sequence 000...0 not possible? Because this sequence is very simple, it can be obtained by running a very simple program.
- Similarly, a sequence 010101...01 can be computed by a very simple program: just continue printing 01 in a loop.
- On the other hand, if we have a sequence of actual coin-flipping results, there will be hardly any dependencies, so the only way to generate this sequence 011... is to write something like `print(011...)`. The length of such a program is about the same as the length $n = \text{len}(x)$ of the original binary sequence.

In other words:

- for a truly random sequence x, the length $K(x)$ of the shortest program that generates this sequence is close to len(x): $K(x) \approx \text{len}(x)$, while
- for a sequence which is not random, such length is much smaller:

$$K(x) \ll \text{len}(x).$$

This led Kolmogorov and other researchers to define a random sequence as a sequence for which $K(x) \geq \text{len}(x) - C$, for some appropriate constant C.

   This definition depends on C.

- If we select C to be too small, we may miss a sequence which is actually random and which accidentally has some dependencies.
- On the other hand, if we select C to be too large, we risk naming not-very-random sequences as random.

So, instead of a single definition of what is random and what is not random, we get, in effect, a *degree* of randomness of a sequence—gauged by the smallest value C for which the above inequality holds. This smallest value, as one can easily check, is equal to the difference $C = \text{len}(x) - K(x)$.

- When this difference is small, the sequence is truly random.
- When this difference is large, the sequence is not so random.

   Kolmogorov complexity and the related notion of randomness have indeed been very useful, both in physics applications and in the analysis of algorithms; see, e.g., [5]. And if this notion was not invented by Kolmogorov—it was invented in the 1960s, at the same time as fuzzy logic—Zadeh's principles could have helped invent it.

3 Simplicity and Everything-Is-a-Matter-of-Degree Principles Naturally Lead to the Main Ideas Behind Quantum Physics

Let us apply these principles to the description of the physical world. Fuzzy logic originated by the observation that:

- in most cases, we cannot say that someone is absolutely young or absolutely old;
- for most people, "young" is a matter of degree.

If we apply the same everything-is-a-matter-of-degree principle to the physical world, we conclude that for every two possible states:

- we should not expect the system to be definitely in one state or definitely in another state;
- in many cases, we should have a system which is, with some degree, in one state, and with some degree, in another state.

How can we describe this in precise mathematical form? We have two original states s_1 and s_2. Each state can be described by, e.g., the values of the corresponding quantities, i.e., by a tuple $s_i = (s_{i1}, \ldots, s_{in})$ of real numbers. We need a function that transforms the two states into a new combined state.

From the mathematical viewpoint, the simplest functions are linear functions. Thus, the simplest possible way to combine the two states is to have a linear combination $a_1 \cdot s_1 + a_2 \cdot s_2$. This is exactly what in quantum physics (see, e.g., [2, 12]) is called *superposition* of the two states—with the only difference that in quantum physics, the coefficients a_i can be complex numbers.

The existence of such superposition is one of the main ideas behind quantum physics—and the main idea behind *quantum computing* and its successes; see, e.g., [8].

What is the dynamics here? To fully describe a physical phenomenon, it is not sufficient to describe the current state, we need also to describe how this state changes in time—i.e., what is the corresponding dynamics.

We thus need to describe how the state s of a physical system changes with time, i.e., how the rate of change $\dfrac{ds}{dt}$ depends on the state: $\dfrac{ds}{dt} = F(s)$, for some function $F(s)$.

Which function $F(s)$ should we select? For this selection, it is reasonable to use the simplicity principle. As we have mentioned earlier, the simplest functions are linear functions, so it is reasonable to use linear function $F(s)$—and this is exactly what is happening in quantum physics, where the Schroedigner's equation describes the rate of change in a state as a linear function of this state [2, 12].

4   Simplicity and Everything-Is-a-Matter-of-Degree Principles Naturally Lead to the Main Ideas Behind Deep Learning

How do we learn? Let us now go from the description of a physical world to the description of how we learn things about this world. First, according to the everything-is-a-matter-of-degree principle, at each given moment of time, about each statement about the world, we do not necessarily have a precise opinion. Mostly, we have a degree of confidence about the given statement.

How do these degrees of confidence change? The simplest possible functions, as we have mentioned several times, are linear functions. So, a natural transformation of degrees is linear.

The problem with using only linear transformations is that by combining several linear transformations, we can only have linear functions—and some real-world processes that we want to describe are not linear. So, we need some non-linear transformations as well.

What are the simplest nonlinear transformations? First, in general, the more inputs, the more complex the transformation. From this viewpoint, the easiest are the functions of one variable.

What are the simplest functions of one variable which are not linear? A natural idea is to use functions which are *piece-wise linear*, i.e., for example, consist of two parts on each of which they are linear. The simplest value separating the two parts is 0, so we end up with a function $s(x)$ which is equal to $c_- \cdot x$ for $x \leq 0$ and to $c_+ \cdot x$ for $x \geq 0$, for some appropriate values c_- and c_+.

What are the simplest real numbers? Clearly, 0 and 1, so we end up with a function which is equal to $0 \cdot x = 0$ for $x \leq 0$ and to $1 \cdot x = x$ for $x \geq 0$. This function can be equivalently described as $s(x) = \max(0, x)$. So, we arrive at the following description of the corresponding computational schemes.

We naturally arrive at deep learning. The input to the non-linear transformer comes from a linear combination of inputs. So, we have transformations of the type

$$x_1, \ldots, x_n \to \max\left(0, \sum_{i=1}^{n} w_i \cdot x_i - w_0\right)$$

for appropriate wights w_i. This is exactly the transformation performed by each neuron in a deep neural network; see, e.g., [3]. Thus, by combining such transformations, we get the main structure of deep neural networks.

If instead of nonlinear functions of one variable, we allow nonlinear functions of several variables, then the simplest are—like in fuzzy logic—min and max. Such transformations form the basis of convolutional neural networks—another component of (and type of) deep learning [3].

So, Zadeh's ideas are in good accordance with deep learning as well.

5 What Next?

What next? There are two natural ideas.

First is to continue applying Zadeh's ideas to other areas—and we will provide an example of such application in the next section.

Another idea is to take into account that a degree does not have to be a single number. We have already seen this in quantum physics—where the coefficients a_i are, in, general complex numbers $a + b \cdot i$, i.e., in effect, pairs of real numbers. Similar ideas of using interval-valued and more general fuzzy degrees—i.e., degrees described by several numbers—have been successfully in fuzzy applications as well; see, e.g., [6]. And even in applications like education, where traditionally the degree of student's knowledge was described by a single number, the new tendency is to use several numbers, e.g., describing the degree of student's knowledge of different topic—and thus providing a more adequate description of the student's knowledge; see, e.g., [1].

6 Causality as a Matter of Degree

6.1 Motivations

Space-time causality is one of the fundamental notions of modern physics; however, it is difficult to define in observational physical terms. Intuitively, the fact that a space-time event $e = (t, x)$ can causally influence an event $e' = (t', x')$ means that what we do in the vicinity of e changes what we observe at e'. If we had two copies of the Universe, we could perform some action at e in one copy but not in another copy; if we then observe the difference at e', this would be an indication of causality. However, we only observe one Universe, in which we either perform the action or we do not. At first glance, it may seem that in this case, there is no meaningful way to provide an operational definition of causality. In this section, we show that such a definition is possible if we use the notions of algorithmic randomness and Kolmogorov complexity. The resulting definition leads to a conclusion that space-time causality is a matter of degree.

Comment. The main idea of this section first appeared in [4].

6.2 Defining Causality Is Important

Space-time causality is important. Causal relation between space-time events (i.e., points in space-time) is one of the fundamental notions of physics; see, e.g., [2, 7,

12]). Because of this, many fundamental physical theories describe, among other things, the causal relation between space-time events.

According to modern physics, space-time causal relation is non-trivial. In Newton's physics, it was assumed that influences can propagate with an arbitrary speed, constituting, in effect, immediate action-at-a-distance. Under this assumption, an event $e = (t, x)$ occurring at moment t at location x can influence an event $e' = (t', x')$ occurring at moment t' at location x' if and only if the second event occurs later than the first one, i.e., if and only if $t < t'$.

In special relativity, the speeds of all the processes are limited by the speed of light c. In this theory, an event $e = (t, x)$ can influence an event $e' = (t', x')$ if during the time $t' - t$, the faster possible process—light—can cover the distance $d(x, x')$ between locations x and x', i.e., if

$$c \cdot (t' - t) \geq d(x, x').$$

In the general relativity theory, the space-time is curved, so the corresponding causal relation is even more complex. This relation is also complex in alternative gravitation theories; see, e.g., [7].

Need for experimental verification of space-time casuality. Different theories, in general, make different predictions about the causality. So, to experimentally verify fundamental physical theories, we need to be able to experimentally verify the corresponding space-time causality. In other words, we must be able to experimentally check, for every two space-time events a and b, whether the event a can causally influence the event b.

Need for a theory-free verification of space-time causality. Since the space-time causality is fundamental, more fundamental than specific partial differential equations that describe the physical fields and/or their relation with space-time, it is desirable to be able to experimentally check this causality in a theory-free way, without invoking other fields and corresponding differential equations.

In this section, we describe a possible way of such theory-free experimental validation of space-time causality.

6.3 Defining Causality: Challenge

Intuitive meaning of space-time casuality. Intuitively, the fact that a space-time event e can causally influence an event e' means that:

- what we do in the vicinity of e
- changes what we observe at e'.

How to transform this meaning into a definition: a hypothetical idea. The above intuitive meaning of space-time causality can easily lead to a observational definition *if* we had two (or more) copies of the Universe. In this case, to check that e can causally influence e', we could do the following (see, e.g., [9]):

- in one copy of the Universe, we perform some action at e, and
- we do not perform this action in the second copy of the Universe.

If the resulting states at e' are different in the two copies of the Universe, this would be an indication of causal relation between e and e'.

Comment. This interpretation of causality is known as a *counterfactual* interpretation; see, e.g., [10]. This name comes from the usual interpretation of *counterfactual* statements, i.e., statements of the type "If we were born in Sahara, we would have been better adjusted for the warm climate." These statements are called counterfactual because the premise (we are born in Sahara) contradicts to the facts. The usual interpretation of such statements is to consider not just our world, but also the whole set of possible worlds. To check whether a counterfactual statement is true we select, among all possible worlds in which the premise is satisfied, the one which is the closest to our own world. The statement is considered true if the conclusion holds in this selected world.

Similarly, in the counterfactual interpretation of causality, instead of considering only one world, we consider all possible worlds. We then say that e casually influences e' if in every world in which e occurs, this occurrence affects e'. For example, we want to check whether a rain dance (e) causes rain (e'). In our world, we observe a rain dance, and we observe rain, but we cannot tell whether the rain was caused by the rain dance or not. Intuitively, the way to check is to see if rain dances leads to rain. So, in one possible world, we perform a rain dance, in another possible world, we do not perform it. If, as a result, we see rain in the first world but not in the second one, this is good indication that the rain dance indeed causes rain.

Can we make this idea practical? In reality, we only observe one Universe, in which we either perform the action or we do not.

At first glance, it may seem that in this case, there is no meaningful way to provide an operational definition of space-time causality.

Our idea. In this section, we show that a meaningful operational definition of space-time causality is possible if we use the notions of algorithmic randomness and Kolmogorov complexity. Before we explain our idea, let us briefly recall the corresponding notions.

6.4 Algorithmic Randomness and Kolmogorov Complexity: A Brief Reminder

The corresponding notion of independence. In probability theory, in addition to analyzing what is *random* and what is not, it is also important to decide when the two events are *independent* and when they are not. Once we have two finite binary sequences x and y, the idea that y is independent on x can be described in a similar way:

- if y is independent on x, then knowing x does not help us generate y;
- in contrast, if y depends on x, then knowing x can help us compute y.

For example, if we know the locations and velocities x of a mechanical system at some moment of time t, we can use this information to easily compute the locations and velocities y at the next moment of time $t + \Delta t$. In contrast, an irrelevant information x (e.g., locations and velocities of particles on another planet) does not help in computing y.

To formalize this intuition, we should consider programs that use x as an input to generate y.

Definition 1 Let a programming language be fixed. By a *relative Kolmogorov complexity $K(y \mid x)$* of a finite binary string y in relation to a binary string x, we mean the shortest length of a program that, when using x as an input, generates y:

$$K(y \mid x) \stackrel{\text{def}}{=} \min\{\text{len}(p) : p(x) \text{ generates } y\}.$$

Comment. Intuitively, if using x helps to compute y, i.e., if $K(y \mid x) \ll K(y)$, this means that y depends on x. Vice versa, if using x does not help to compute y, i.e., if $K(y \mid x) \approx K(y)$, this means that x and y are independent. We can describe this in a way similar to the above definition of randomness:

Definition 2 Let an integer $C > 0$ be fixed.

- We say that a string y is *independent* of the string x if

$$K(y \mid x) \geq K(y) - C.$$

- We say that a string y is *dependent* on the string x if

$$K(y \mid x) < K(y) - C.$$

6.5 How to Define Space-Time Causality: Analysis of the Problem and the Resulting Definition

First seeming reasonable idea. At first glance, the above notion of dependence can already lead to a natural definition of space-time causality:

- First, we perform some measurements and observations in the vicinity of the event e. Since most nowadays measuring instruments are computer-connected, each such measurement produces a computer-readable output. In the computer, everything is represented as a sequence of 0s and 1s, so the results of all the measurements and observations will also be represented as a sequence x of 0s and 1s.

- We also perform measurements and observations in the vicinity of the event e', and also produce a sequence x' of 0s and 1s.
- If x' depends on x, i.e., if $K(x' \mid x) \ll K(x')$, then we claim that e can casually influence e'.

Unfortunately, this idea does not always work. Yes, if e can casually influence e', then we indeed expect that knowing what happened at e can help us predict what is happening at e'. However, the inverse is not necessarily true: we may have identical observations $x = x'$ at events e and e' simply because they are both caused by the same event e'' from the joint past of events e and e'.

For example, if two people at different locations are watching the same movie, then their observations are identical, but not because they causally influence each other, but because they are both influenced by a past event e'' (of making this movie).

How to transform the above idea into a working definition. According to modern physics, the Universe is quantum in nature. For many measurements involving microscopic objects, we cannot predict the exact measurement results, we can only predict probabilities of different outcomes. The actual observations are truly random.

Moreover, for each space-time event e, we can always set up such random-producing experiments in the small vicinity of e, and generate a random sequence r_e. For example, we can locally set up a Stern-Gerlach experiment (see, e.g., [2, 12]), a quantum experiment that generates a truly random sequence.

This random sequence can affect *future* results, so if we know this random sequence, it may help us predict future observations. So, if e can causally influence e', then for some observations x' performed in the small vicinity of e', we have $K(x' \mid r_e) \ll K(x')$.

However, it is clear that this sequence cannot affect the measurement results which are in the *past* (or, more generally, not in the future) of the event e. So, if e cannot causally influence e', then observations x' made in the vicinity of e' are independent on r_e: $K(x' \mid r_e) \approx K(x')$. So, we arrive at the following semi-formal definition.

Definition 3 For each space-time event e, let r_e denote a random sequence that is generated by an experiment performed in the small vicinity of e. We say that the event e can *causally influence* the event e' if for some observations x' performed in the small vicinity of e', we have

$$K(x' \mid r_e) \ll K(x').$$

Historical comment. Our definition follows the ideas of casuality as mark transmission [10, 11], with the random sequence as a mark.

Discussion. We have argued that if e *does not* causally influence e', then, no matter what we measure in the vicinity of the event e', we get $K(x' \mid r_e) \approx K(x')$; so, in these cases, the above definition is in accordance with the physical intuition.

On the other hand, if e can causally influence e', this means that we can send a signal from e to e', and as this signal, we send all the bits forming the random sequence r_e. The signal x' received in the vicinity of e' will thus be identical to r_e, so generating x' based on r_e does not require any computations at all: $K(x' \mid r_e) = 0$. Since the sequence $x' = r_e$ is random, we have

$$K(x') \geq \text{len}(x') - C.$$

For a sufficiently long random sequence $r_e = x'$, namely for a sequence for which $\text{len}(x') > 2C$, we have

$$K(x') \geq \text{len}(x') - C > 2C - C = C,$$

so

$$0 = K(x' \mid r_e) < K(x') - C$$

and thus,

$$K(x' \mid r_e) \ll K(x').$$

So, in these cases, the above definition is also in accordance with the physical intuition.

6.6   Corollary of Our Definition: Space-Time Causality is a Matter of Degree

Our definition of causality uses the notion of randomness: namely, we say that there is a causal re/lation between e and e' if for some random sequence r_e generated in the vicinity of the event e and for measurement results x' produced in the vicinity of e', we have $K(x' \mid r_e) < K(x') - C$ for some large integer C.

The larger the integer C, the more confident we are that an event e can causally influence e'. It is therefore reasonable, for each pair of events e and e', to define a *degree of causality* c as the largest integer C for which $K(x' \mid r_e) < K(x') - C$. One can check that this largest integer is equal to the difference $c = K(x') - K(x' \mid r_e) - 1$. The largest this difference c, the more confident we are that e can influence e'. Thus, this difference can serve as *degree* with which e can influence e'.

In other words, just like randomness turns out to be a matter of degree, causality is also a matter of degree.

Corresponding open problems. It is desirable to explore possible physical meaning of such "degrees of causality": instead of describing the space-time causality, we now have a function $d(e, e')$ that:

- for each pair of events for which e causally precedes e',
- describes to what extent e can influence e'.

Maybe this function $d(e, e')$ is related to relativistic metric—the amount of proper time between e and e'?

Another open problem is related to the fact that he above definition works for localized objects, objects which are located in a small vicinity of one spatial location.

In quantum physics, not all objects are localize in space-time. We can have situations when the states of two spatially separated particles are entangled. It is desirable to extend our definition to such objects as well.

6.7 Conclusions

In this section, we propose a new operationalist definition of causality between space-time events. Namely, to check whether an event e can casually influence an event e', we:

- generate a truly random sequence r_e in the small vicinity of the event e, and
- perform observations in the small vicinity of the event e'.

If some observation results x' (obtained near e') *depend* on the sequence r_e (in the precise sense of dependence described in the section), then we claim that e *can* casually influence e'. On the other hand, if all observation results x' are independent on r_e, then we claim that e *cannot* casually influence e'.

This new definition naturally leads to a conclusion that space-time causality is a matter of degree, a conclusion that is worth physical analysis.

Acknowledgements This work was supported in part by the US National Science Foundation grants 1623190 (A Model of Change for Preparing a New Generation for Professional Practice in Computer Science) and HRD-1242122 (Cyber-ShARE Center of Excellence).

References

1. Astin, A.: Are You Smart Enough: How Colleges' Obsession with Smartness Shortchanges Students. Stylus Publishing, Sterling (2016)
2. Feynman, R., Leighton, R., Sands, M.: The Feynman Lectures on Physics. Addison Wesley, Boston (2005)
3. Goodfellow, I., Bengio, Y., Courville, A.: Deep Leaning. MIT Press, Cambridge (2016)
4. Kreinovich, V., Ortiz, A.: Towards a better understanding of space-time causality: Kolmogorov complexity and causality as a matter of degree. In: Proceedings of the Joint World Congress of the International Fuzzy Systems Association and Annual Conference of the North American Fuzzy Information Processing Society IFSA/NAFIPS'2013, Edmonton, Canada, June 24–28, 2013, pp. 1349–1353
5. Li, M., Vitanyi, P.: An Introduction to Kolmogorov Complexity and Its Applications. Springer, Berlin (2008)
6. Mendel, J.M.: Uncertain Rule-Based Fuzzy Systems: Introduction and New Directions. Springer, Cham (2017)
7. Misner, C.W., Thorne, K.S., Wheeler, J.A.: Gravitation. W. H. Freeman, New York (1973)

8. Nielsen, M., Chuang, I.: Quantum Computation and Quantum Information. Cambridge University Press, Cambridge (2000)
9. Pearl, J.: Causality: Models, Reasoning, and Inference. Cambridge University Press, Cambridge (2000)
10. Psillos, S.: Causality and Explanation, McGill-Queen's University Press. Kingston, and Ithaca, Montreal (2002)
11. Salmon, W.: Causality and Explanation. Oxford University Press, Oxford (1997)
12. Thorne, K.S., Blandford, R.D.: Modern Classical Physics: Optics, Fluids, Plasmas, Elasticity, Relativity, and Statistical Physics. Princeton University Press, Princeton (2017)
13. Zadeh, L.A.: Fuzzy sets. Inf. Control **8**, 338–353 (1965)
14. Zadeh, L.A.: Causality is undefinable: toward a theory of hierarchical definability. In: Proceedings of the IEEE International Conference on Fuzzy Systems FUZZ-IEEE'2001, Melbourne, Australia, December 2–5, 2001, pp. 67–68

LSPeval—An Educational Decision Support Tool Based on Soft Computing Logic Aggregation

Jozo Dujmović🅓 and Ketan Kapre

Abstract LSPeval is a free Internet-based soft computing educational software tool. It is successfully tested in project-oriented graduate and advanced undergraduate courses that introduce students to the design and use of decision support systems. Instructors can use LSPeval to expose students to mathematical models for decision making, based on soft computing, fuzzy reasoning, and graded logic implemented in the LSP method. Students use LSPeval to solve professional decision problems of medium complexity. Such problems include both functional and financial inputs. Students are also exposed to graded logic and become familiar with hard and soft models of graded simultaneity and substitutability, conjunctive mandatory-optional criteria, financial aspects of decision making, and interpretability and explainability of proposed decisions. LSPeval is designed to provide professional evaluation experiences with minimum mathematical prerequisites. Consequently, it can be used in initial segments of soft computing and decision support courses. This paper includes all information necessary for instructors and students to use the LSP evaluation methodology and to experience realistic evaluation projects. An educational example and a sample project are presented in detail. The contributions of this paper include the structure and functionality of the LSPeval tool, the project-oriented stepwise refinement approach to presenting soft computing logic and decision problems, and the pedagogical methodology and software for including evaluation projects in computational intelligence courses.

Keywords LSP method · Evaluation · Educational tool · Decision support

1 Introduction

Human cognitive processes are observable and in the case of decision making they can be described using mathematical models based on concepts of fuzziness [16], fuzzy

J. Dujmović (✉) · K. Kapre
San Francisco State University, San Francisco, CA 94132, USA
e-mail: jozo@sfsu.edu

© The Author(s), under exclusive license to Springer Nature Switzerland AG 2023
Sh. N. Shahbazova et al. (eds.), *Recent Developments and the New Directions of Research, Foundations, and Applications*, Studies in Fuzziness and Soft Computing 422, https://doi.org/10.1007/978-3-031-20153-0_17

logic, and soft computing [17, 18]. AI systems frequently include decision support applications that use a variety of mathematical models [10, 15]. Unsurprisingly, the credibility of decision models depends on their consistency with observable properties of human reasoning and decision making [7].

Human decision making is a mental logic process based on evaluation and comparison of alternatives. Decision makers first identify alternatives (or competitive objects) and then evaluate the suitability of each of them. In the special case of a single object, the object must be individually evaluated, and the result of evaluation is an overall suitability degree $X \in [0, 1]$, where $X = 1$ denotes complete satisfaction of decision maker's requirements, and $X = 0$ denotes no satisfaction (i.e. the rejection of evaluated object/alternative). All values between 0 and 1 denote partial satisfaction, or the degree of membership in a fuzzy set. The overall suitability degree X is computed as a function of n suitability attributes, $X = E(a_1, \ldots, a_n)$. The suitability attributes $a_1, \ldots, a_n$ include all inputs that affect the overall suitability X, and E is the evaluation criterion function.

The overall suitability X is interpreted as the degree of truth of a value statement that claims the complete satisfaction of requirements that the decision maker specified in the multiattribute criterion function E. Alternatively, X can be the degree of membership in the fuzzy set of objects that completely satisfy all requirements. The method for evaluating a single object can also be used in the case of $k > 1$ objects or alternatives. The ranking of k competitive objects is based on the decreasing sequence of overall suitability degrees $X_1 \geq \ldots \geq X_k$ and the winner is the highest ranking (first) object.

The critical question is this: how to compute the overall suitability degree, based on a justifiable and explainable criterion function which is consistent with observable properties of human decision making? To achieve this goal, we use the Logic Scoring of Preference (LSP) method [7], which is based on soft computing graded logic [6, 8].

The study of graded logic requires an educational effort and the LSPeval tool is developed to help in that direction. The initial version of LSPeval [12] is developed with the goal to create and use LSP criteria without the prerequisite of knowing all details of graded logic. After the educational experience with LSP evaluation based on LSPeval, students are well prepared to use advanced professional tools, such as LSP.NT [14].

The paper has six sections. In Sect. 2 we present the structure of LSP criteria, and outline the process of evaluation based on mandatory and optional suitability attributes. Section 3 presents the structure and functionality of LSPeval. Sections 4 and 5 show examples of the use of LSPeval, and proposal for educational use of LSPeval in student projects. Section 6 contains conclusions.

2   The Structure of LSP Criteria

The development of LSP suitability evaluation criteria includes three main steps:

- Specification of suitability attributes
- Definition of suitability attribute criteria
- Logic aggregation of attribute suitability degrees.

The first step consists of specifying suitability attributes $a_1, \ldots, a_n$. Suitability attributes are all those attributes that justifiably contribute to the satisfaction of decision maker's requirements. A general method for development of a suitability attribute tree is illustrated in Fig. 1. The root of the suitability attribute tree is the overall suitability. If the overall suitability can be decomposed into simpler subsystems, that process is systematically repeated, and at the end generates simple attributes that cannot be further decomposed and can be directly evaluated. For example, a homebuyer can evaluate a home location using the following tree:

1. Suitability of home location
    1.1 Suitability for working parents
        1.1.1 Job distance for mother
        1.1.2 Job distance for father
    1.2 Suitability for school children
        1.2.1 School distance for son
        1.2.2 School distance for daughter
    1.3 Proximity to recreation areas
        1.3.1 Proximity to parks and lakes
        1.3.2 Proximity to sport centers

Job and school distances cannot be further decomposed; they are suitability attributes directly measured as travel times. Each of these attributes can be classified in one of two basic categories: (a) mandatory, or (b) optional. For example, the parents can consider that school access time for children must be less than 20 min, and locations where the school is not accessible in less than 20 min are going to be rejected. In other words, it is mandatory to satisfy the school distance requirements. On the other hand, the proximity to parks, lakes, and sport centers can be desired but not mandatory, and such attributes are considered optional. Even if there are no close parks and lakes, a home location can be acceptable if other attributes are sufficiently satisfied.

Fig. 1 General suitability attribute tree

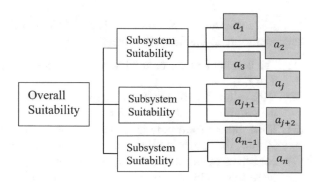

Fig. 2 Mandatory/optional
suitability attribute tree

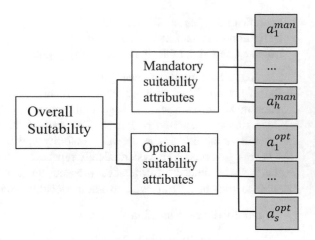

To provide simplicity that is desirable for beginners, the set of n suitability attributes can be partitioned as h mandatory and s optional attributes, as shown in Fig. 2. Of course, that is not the most general solution for all criteria, but it is convenient for educational purposes.

In the special case of a very demanding criterion function, all suitability attributes can be mandatory ($s = 0$). Similarly, in the special case of a very relaxed evaluation criterion, all attributes can be optional ($h = 0$). In most frequent cases, however, we assume $h > 0$, $s > 0$, and $h + s = n$.

All suitability attributes must be separately evaluated using attribute criteria. An attribute criterion is a mapping of an attribute value a_i (which is usually a real number) to the corresponding attribute suitability degree $x_i = g_i(a_i) \in [0, 1]$. Suppose that a_i denotes the job distance, measured in minutes, and let the decision maker be completely satisfied ($x_i = 1$) if the time for traveling from home to work is $a_i \leq 15$ minutes. If the time is 30 min the satisfaction is 80%, and if the time is 60 min or more, that is not acceptable and such location will be rejected. Such a criterion can be denoted using the following vertex notation, defined as the set of breakpoints:

$$Crit(job\,distance\,[min]) = \{(15, 100), (30, 80), (60, 0)\}.$$

Each breakpoint is a parenthesized pair (attribute value, attribute suitability [%]), used according to the following rules:

(1) The breakpoints are ordered according to increasing values of attribute ($15 < 30 < 60$).
(2) If the value of attribute is less than the value of the initial breakpoint (e.g. 10 min) the suitability is the same as for the initial point (i.e. 100%).
(3) If the value of attribute is greater than the value of the final breakpoint (e.g. 90 min) the suitability is the same as for the final point (i.e. 0).
(4) For values of attribute between breakpoints we use the linear interpolation (e.g. for $a_i = 45$ min, the corresponding attribute suitability is 40%).

Using attribute criterion functions $g_1, \ldots, g_n$, for each evaluated object we create n attribute suitability degrees $x_i = g_i(a_i), i = 1, \ldots, n$. The attribute suitability degrees are then aggregated to create the overall suitability degree $X = L(x_1, \ldots, x_n)$, as shown in Fig. 3.

The structure of function $L : [0, 1]^n \rightarrow [0, 1]$ is a delicate problem. In the simplest special case L can be selected as one of means [2, 3]. The theory of aggregation functions [1, 11] offers a wide variety of functions that can be used for aggregation of attribute suitability degrees and the computation of overall suitability. However, it is easy to see that L is a logic function, because all arguments are the degrees of truth and the resulting overall suitability is also the degree of truth. Such functions are called graded logic aggregators [6, 8] and L is a formula of graded propositional calculus [7].

The study of graded logic starts with the concepts of conjunction degree (andness) and disjunction degree (orness) [4] and their use as adjustable parameters of the graded conjunction/disjunction (GCD) aggregator that is used to model simultaneity and substitutability [5]. According to graded logic conjecture [7], all graded logic functions can be developed using the superposition of ten fundamental functions: nine characteristic versions of GCD and negation. The study of these aggregators and their use is not trivial and requires appropriate preparation. LSPeval is developed with the main goal to help in such preparation using a specific fixed structure of L function shown in Fig. 3. That is a frequently used canonical conjunctive structure that requires simultaneous satisfaction of n attribute criteria, where h criteria must be (to some extent) satisfied (mandatory attributes), and the satisfaction of s criteria is desired, but not mandatory (optional attributes).

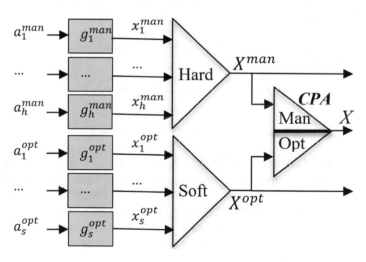

Fig. 3 The structure of the LSPeval criterion function

The aggregation function L consists of two conjunctive aggregators denoted "hard" and "soft", and an asymmetric conjunctive aggregator denoted CPA (conjunctive partial absorption). All mandatory attributes are separately evaluated, and the corresponding attribute suitability degrees are aggregated using a hard partial conjunction aggregator $X^{man} = H(x_1^{man}, \ldots, x_h^{man})$, which has the following properties:

$$\forall i \in \{1, \ldots, h\}, x_i^{man} = 0 \Rightarrow H(x_1^{man}, \ldots x_h^{man}) = 0;$$
$$\forall x \in [0, 1], H(x, \ldots, x) = x;$$
$$\min(x_1^{man}, \ldots x_h^{man}) \leq H(x_1^{man}, \ldots, x_h^{man});$$
$$< (x_1^{man} + \ldots + x_h^{man})/h.$$

These properties assume that all mandatory attributes are equally important. However, to support semantic aspects of aggregation, all LSP aggregators are weighted, and LSPeval offers degrees of importance that can be selected from the following verbalized rating scale: (1) *lowest*, (2) *very low*, (3) *low*, (4) *medium–low*, (5) *medium*, (6) *medium–high*, (7) *high*, (8) *very high*, (9) *highest*. E.g., in the case of 4 inputs, if the user selects verbalized degrees of importance low, medium–low, medium, and very high (3,4,5,8), then LSPeval will compute the corresponding normalized numerical weights that have the sum 1, as follows: $3/20 = 0.15$, $4/20 = 0.2$, $5/20 = 0.25$, $8/20 = 0.4$. Consequently, the weights satisfy $0.15 + 0.2 + 0.25 + 0.4 = 1$.

A conjunctive aggregator is *hard* if it supports the annihilator 0 and *soft* if it does not support the annihilator 0. The soft aggregator $X^{opt} = S(x_1^{opt}, \ldots, x_s^{opt})$, in Fig. 3 has the following properties:

$$\forall i \in \{1, \ldots, s\}, x_i^{opt} > 0 \Rightarrow S(x_1^{opt}, \ldots, x_s^{opt}) > 0;$$
$$\forall x \in [0, 1], S(x, \ldots, x) = x;$$
$$\min(x_1^{opt}, \ldots, x_s^{opt}) \leq H(x_1^{opt}, \ldots, x_s^{opt})$$
$$< S(x_1^{opt}, \ldots, x_s^{opt})$$
$$\leq (x_1^{opt} + \ldots + x_s^{opt})/s.$$

The optional attributes are also weighted, the same way as the mandatory attributes. According to Fig. 3, LSPeval supports a special case where all attributes are mandatory (there are no optional attributes, $s = 0$, and the overall output suitability is X^{man}). Similarly, LSPeval supports a special case where all attributes are optional (there are no mandatory attributes, $h = 0$, and the overall output suitability is X^{opt}). However, the general case of this type of LSP criterion is the combination of mandatory and optional attributes.

In a general case, the aggregated suitability of mandatory attributes X^{man} and the aggregated suitability of optional attributes X^{opt} are additionally aggregated using the asymmetric aggregator called the conjunctive partial absorption $X = CPA(X^{man}, X^{opt})$ which has the following properties [7]:

$$\forall x \in [0, 1], CPA(x, x) = x$$
$$X^{man} = 0, X^{opt} > 0 \Rightarrow CPA(X^{man}, X^{opt}) = 0$$
$$X^{man} > 0, X^{opt} = 0 \Rightarrow CPA(X^{man}, X^{opt}) = X^{man} - P$$
$$0 < X^{man} < 1, X^{opt} = 1 \Rightarrow CPA(X^{man}, X^{opt}) = X^{man} + R$$

Therefore, if we want a positive overall suitability X, all mandatory attribute criteria must be to some extent satisfied and X^{man} must be positive. If $0 < X^{man} < 1$ then the zero optional input causes an output penalty P, and the fully satisfied optional input causes a reward R. So, P is a decrement of output suitability and R is an increment of output suitability with respect to the achieved value of X^{man}.

Generally, the LSP method uses P and R as freely adjustable parameters of CPA, and advanced LSP tools support the precise adjustment of these parameters. A detailed study of CPA aggregator can be avoided by selecting three characteristic P,R pairs, denoted *relaxed* (low penalty, high reward), *standard* (medium penalty, medium reward), and *strict* (high penalty, low reward). Users can try these three options (see the bottom of Fig. 4) and select one that generates the most justifiable results.

The overall suitability X reflects the satisfaction of all mandatory and optional attribute criteria, but if evaluated objects have costs, then the LSP method assumes that cost is not one of suitability attributes but a separate input that is combined with the overall suitability X. Let C be the cost of an evaluated object and let the number of competitive objects be $m > 1$. Thus, the overall suitability degrees are $C_1, \ldots, C_m$ and the corresponding overall suitability degrees are $X_1, \ldots, X_m$. In most evaluation projects there is the maximum available/approved funding C^*, and all competitors that are more expensive than C^* are automatically rejected. Similarly, there is the minimum acceptable suitability X^* and all objects that are less suitable are automatically rejected. So, we assume that $\{C_i, X_i\} \in]0, C^*] \times [X^*, 1], i = 1, \ldots, m$. To complete the evaluation process, we need to compare the competitors and select the best C_i, X_i pair.

The ranking of m competitors is based on logic requirement to *simultaneously* achieve high suitability and low cost. The minimum cost is $C_{min} = \min(C_1, \ldots, C_m)$ and the cost normalized as a logic variable is $C_{min}/C_i \leq 1$. Similarly, the maximum suitability of competitors is $X_{max} = \max(X_1, \ldots, X_m)$ and the normalized suitability (another logic variable) is $X_i/X_{max} \leq 1$. As a model of simultaneity, it is convenient to use the weighted geometric mean, so that users can analyze cases where low cost and high performance are not equally important. The corresponding model of the overall relative value is the following:

$$V_i(w) = \left(\frac{C_{min}}{C_i}\right)^{1-w} \left(\frac{X_i}{X_{max}}\right)^w, 0 < w < 1, i = 1, \ldots, m.$$

The weight $1-w$ shows the relative importance of low cost, and the weight w is the relative importance of high suitability scores. The overall value $V_i(w)$ is a function

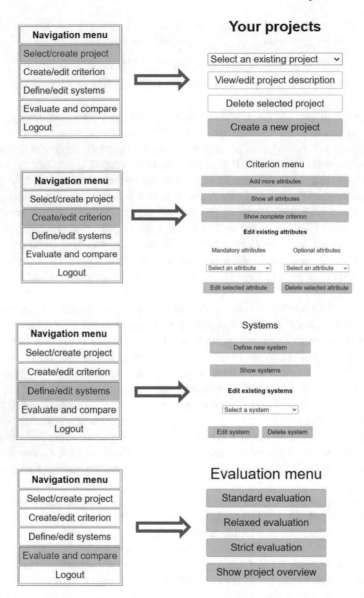

Fig. 4 Four main functional components of the LSPeval tool

of weight w and the evaluation tools (e.g., LSPeval) should produce the table of this function. After selecting the desired value of w, the winner of competition is the object with the highest value $V_i(w)$.

3 The Structure and Functionality of LSPeval

The main goal of the LSPeval tool is to minimize prerequisites for using LSP evaluation method in the context of decision support courses. LSPeval is developed to provide all important aspects of the LSP evaluation process without including graded logic components that require mathematical details and longer preparation of students. On the other hand, LSPeval contains all components that can be included with modest prerequisite effort: (1) students are exposed to real-life decision problems, (2) the process of selection of suitability attributes is fully supported, (3) attribute suitability criteria are developed by students, (4) evaluation process has visible logic components, and students select some of them, (5) justifiable numeric results based on functional and financial components are derived and applicable, (6) the LSP evaluation model is derived from observing properties of human decision making, (7) students can see the need and possibility for mathematical refinement and generalization of the LSPeval model, and (8) advanced topics of explainability can be introduced as next steps in development of decision support methodology [9]. In this way LSPeval can be used in early segments of decision support courses. It is a suitable educational tool in both graduate and advanced undergraduate courses. This educational experience shows important relationship between a natural decision process and its mathematical model, and the need for increasing precision and the expansion and improvement of the used model.

LSPeval supports multiple evaluation projects. It includes user login followed by four main phases, visible in the navigation menu and shown in each LSPeval screen (Fig. 4):

1. Creation or selection of an evaluation project
2. Creation or editing of the LSP criterion
3. Definition of competitive objects/alternatives
4. Evaluation and comparison of competitors

The user login process with information about the location of a working LSPeval system can be found in the LSPeval User Manual [13]. Initially the user must create her/his LSPeval account with a username and a password. The new user is given an initial complete educational project (evaluation of home location for a senior citizen [7]). This project can be used for learning the use of LSPeval, and modifying and investigating results. It should be presented by instructor during lectures.

The LSPeval user can use any of existing projects and/or create any number of new projects. The first phase, shown in Fig. 4, has the options of creating/viewing/editing/deleting a selected project. In the second phase, users create or edit LSP criteria: initially they add mandatory or optional attributes and define attribute criteria using tables supporting the vertex notation. In the third phase the user can define any number of competitive objects, and in the fourth phase it is possible to perform relaxed/standard/strict evaluation.

Each phase shown in Fig. 4 has submenus that provide additional commands and presentation of numeric results. All LSP criteria, competitive objects, and results are stored in a database that belongs to the user/owner and can be reused in next projects. That provides incremental development of educational projects and offers instructors the possibility to adjust the student projects according to the level of course and the available student population.

4 The Location Sample Project and Its Results

LSPeval offers to all users a predefined project called "Location" that illustrates the use of the LSP method and capabilities of the LSPeval tool. This project evaluates the suitability of a home location for a senior citizen who needs proximity to six selected points of interest. An overview of the project and its results, generated by LSPeval, is shown in Figs. 5, 6, 7 and 8. This project documentation shows all attributes, attribute criteria, the evaluation criterion, all competitive locations, and all evaluation results.

Assuming that the senior citizen does not have a car, the mandatory requirements include the proximity to bus stations, train stations, and food stores. The proximity to public library, parks, and restaurants are desired attributes but they are defined as optional. Each of these attributes has specific verbalized degree of importance specified in Fig. 5.

There are five competitive locations denoted Alpha, Beta, Gamma, Delta, and Zero. They have normalized prices where the highest price is 1. Each location is characterized by its price and six attribute values that are shown in Fig. 6. Detailed results of standard suitability evaluation are presented in Fig. 7. All suitability scores are multiplied by 100 and interpreted as percentages of satisfied requirements. The LSP criterion generates overall suitability scores, where the highest score for location Gamma is 86.41%. The minimum positive suitability of 58.04% corresponds

LSP criterion for project Location

1. Mandatory

#	Attribute name	Importance	Attribute criterion
1	Distance from bus station [m]	Medium	{ (50.0, 100.0), (700.0, 0.0) }
2	Distance from food store [m]	Very high	{ (100.0, 100.0), (400.0, 50.0), (600.0, 0.0) }
3	Distance from train station [m]	Medium-low	{ (400.0, 100.0), (1000.0, 0.0) }

2. Optional

#	Attribute name	Importance	Attribute criterion
1	Distance from library [m]	Low	{ (1000.0, 100.0), (3000.0, 0.0) }
2	Distance from park [m]	High	{ (0.0, 100.0), (3500.0, 0.0) }
3	Distance from restaurant [m]	Medium	{ (0.0, 40.0), (150.0, 100.0), (300.0, 100.0), (900.0, 0.0) }

Fig. 5 Evaluation criterion for the default project "Location"

to the location Beta. The location Zero is rejected (overall score 0) because it is not providing the necessary proximity to bus stations. It is intuitively clear that satisfying 58% requirements might be insufficient, while the satisfaction of 86% of requirements can be very satisfactory, provided that the price is not excessive. So, we need the results of the suitability/price analysis.

The results presented in Fig. 8 show the impact of prices. There are two tables: standard and normalized. They show corresponding overall values for all five locations and for all weights $w = 0, 0.1, \ldots, 1$. The user should select the most appropriate value of w, and the ranking of competitors is visible in the column under the selected value of w. For example, if the user is in situation where the suitability and price equally affect the final decision, then in the column $w = 0.5 (i.e. 50\%)$ of the upper table in Fig. 8 we see that the best location is Gamma, which attains the high relative value of 94.59%, considering simultaneously both suitability and price requirements.

The lower table in Fig. 8 shows normalized results where the results in each column are normalized so that the best competitor has the value 100%. These results show a more general aspect of comparison of the competitive locations. In the case where the importance of high suitability is less than or equal to 30% and the importance of low price is 70% or more, the winner is the location Delta. The location Gamma becomes the winner only in the range where the relative importance of high suitability is 40% or more. This table shows the region of dominance of leading competitors.

LSPeval includes three logic forms of CPA aggregation, denoted relaxed, standard, and strict. The default aggregation is the standard form. The strict form is more demanding, and the relaxed form is less demanding, resulting in overall suitability relation $strict < standard < relaxed$. This is visible in Figs. 8, 9, and 10.

LSPeval can be used for various evaluation experiments. E.g., let us ask the following question: if $w = 50\%$, what discount should offer the location Delta to win this competition? The result ($C_{Delta} \leq 0.787$) is shown in Fig. 11.

Attributes	Importance	Systems and their prices					Attribute criterion
		Alpha	Beta	Delta	Gamma	Zero	
		1.000	0.920	0.850	0.950	1.000	
Mandatory							
Distance from bus station [m]	Medium	120	200	250	100	777	{ (50, 100), (700, 0) }
Distance from food store [m]	Very high	350	410	80	200	111	{ (100, 100), (400, 50), (600, 0) }
Distance from train station [m]	Medium-low	600	800	750	300	444	{ (400, 100), (1000, 0) }
Optional							
Distance from library [m]	Low	500	1500	1900	2750	1111	{ (1000, 100), (3000, 0) }
Distance from park [m]	High	1100	900	200	250	111	{ (0, 100), (3500, 0) }
Distance from restaurant [m]	Medium	250	300	460	220	60	{ (0, 40), (150, 100), (300, 100), (900, 0) }
Final results							
Suitability score [%]		72.4	58.04	71.66	86.41	0.0	

Fig. 6 Suitability attributes and attribute criteria of five competitive locations

Attributes	Importance	Systems and their prices				
		Alpha	Beta	Delta	Gamma	Zero
		1.000	0.920	0.850	0.950	1.000
Total mandatory attributes score [%]		67.15	48.10	68.49	89.40	0.00
Distance from bus station [m]	Medium	89.23	76.92	69.23	92.31	0.00
Distance from food store [m]	Very high	58.33	47.50	100.00	83.33	98.17
Distance from train station [m]	Medium-low	66.67	33.33	41.67	100.00	92.67
Total optional attributes score [%]		85.33	83.00	79.44	79.17	85.41
Distance from library [m]	Low	100.00	75.00	55.00	12.50	94.45
Distance from park [m]	High	68.57	74.29	94.29	92.86	96.83
Distance from restaurant [m]	Medium	100.00	100.00	73.33	100.00	64.00
Final results						
Overall suitability score [%]		72.4	58.04	71.66	86.41	0.0

Fig. 7 The results of standard evaluation of five locations

Standard evaluation report for project Location

Overall value of system k: $100 * (\text{Score}_k / \text{Score}_{max})^w (\text{Cost}_{min} / \text{Cost}_k)^{1-w}$ [%]

System	Cost	Relative importance of high scores (w)											Overall score [%]
		0%	10%	20%	30%	40%	50%	60%	70%	80%	90%	100%	
Alpha	1.000	85.00	84.88	84.76	84.63	84.51	84.39	84.27	84.15	84.03	83.91	83.79	72.4
Beta	0.920	92.39	89.49	86.68	83.96	81.33	78.78	76.30	73.91	71.59	69.34	67.17	58.04
Delta	0.850	100.00	98.15	96.33	94.54	92.79	91.07	89.38	87.72	86.09	84.50	82.93	71.66
Gamma	0.950	89.47	90.47	91.49	92.51	93.54	94.59	95.65	96.72	97.80	98.89	100.00	86.41
Zero	1.000	85.00	0.00	0.00	0.00	0.00	0.00	0.00	0.00	0.00	0.00	0.00	0.0

Normalized report

System	Cost	Relative importance of high scores (w)											Overall score [%]
		0%	10%	20%	30%	40%	50%	60%	70%	80%	90%	100%	
Alpha	1.000	85.00	86.48	87.99	89.52	90.34	89.22	88.10	87.00	85.92	84.85	83.79	72.4
Beta	0.920	92.39	91.18	89.99	88.81	86.94	83.28	79.78	76.42	73.20	70.12	67.17	58.04
Delta	0.850	100.00	100.00	100.00	100.00	99.19	96.27	93.44	90.70	88.03	85.44	82.93	71.66
Gamma	0.950	89.47	92.18	94.98	97.85	100.00	100.00	100.00	100.00	100.00	100.00	100.00	86.41
Zero	1.000	85.00	0.00	0.00	0.00	0.00	0.00	0.00	0.00	0.00	0.00	0.00	0.0

Fig. 8 Overall values: the results of standard suitability/price analysis

Relaxed evaluation report for project Location

Overall value of system k: 100 * ($Score_k/ Score_{max})^W$ ($Cost_{min}/ Cost_k)^{1-W}$ [%]

System	Cost	Relative importance of high scores (w)											Overall score [%]
		0%	10%	20%	30%	40%	50%	60%	70%	80%	90%	100%	
Alpha	1.000	85.00	85.12	85.23	85.35	85.47	85.58	85.70	85.82	85.93	86.05	86.17	76.0
Beta	0.920	92.39	90.24	88.15	86.10	84.10	82.14	80.23	78.37	76.54	74.77	73.03	64.41
Delta	0.850	100.00	98.56	97.13	95.73	94.35	92.99	91.64	90.32	89.01	87.73	86.46	76.26
Gamma	0.950	89.47	90.47	91.49	92.51	93.54	94.59	95.65	96.72	97.80	98.89	100.00	88.2

Normalized report

System	Cost	Relative importance of high scores (w)											Overall score [%]
		0%	10%	20%	30%	40%	50%	60%	70%	80%	90%	100%	
Alpha	1.000	85.00	86.36	87.75	89.16	90.59	90.48	89.60	88.73	87.87	87.01	86.17	76.0
Beta	0.920	92.39	91.57	90.75	89.94	89.13	86.84	83.88	81.03	78.27	75.60	73.03	64.41
Delta	0.850	100.00	100.00	100.00	100.00	100.00	98.30	95.81	93.38	91.02	88.71	86.46	76.26
Gamma	0.950	89.47	91.80	94.19	96.64	99.15	100.00	100.00	100.00	100.00	100.00	100.00	88.2

Fig. 9 The results of evaluation using a relaxed CPA criterion

Strict evaluation report for project Location

Overall value of system k: 100 * ($Score_k/ Score_{max})^W$ ($Cost_{min}/ Cost_k)^{1-W}$ [%]

System	Cost	Relative importance of high scores (w)											Overall score [%]
		0%	10%	20%	30%	40%	50%	60%	70%	80%	90%	100%	
Alpha	1.000	85.00	85.19	85.38	85.57	85.76	85.95	86.14	86.33	86.52	86.72	86.91	71.03
Beta	0.920	92.39	89.42	86.55	83.76	81.07	78.46	75.94	73.50	71.13	68.85	66.63	54.46
Delta	0.850	100.00	97.67	95.39	93.17	90.99	88.87	86.80	84.77	82.80	80.87	78.98	64.55
Gamma	0.950	89.47	90.47	91.49	92.51	93.54	94.59	95.65	96.72	97.80	98.89	100.00	81.73

Normalized report

System	Cost	Relative importance of high scores (w)											Overall score [%]
		0%	10%	20%	30%	40%	50%	60%	70%	80%	90%	100%	
Alpha	1.000	85.00	87.22	89.50	91.85	91.68	90.86	90.06	89.26	88.47	87.69	86.91	71.03
Beta	0.920	92.39	91.56	90.73	89.91	86.66	82.95	79.39	75.99	72.74	69.62	66.63	54.46
Delta	0.850	100.00	100.00	100.00	100.00	97.27	93.95	90.75	87.65	84.66	81.77	78.98	64.55
Gamma	0.950	89.47	92.63	95.91	99.30	100.00	100.00	100.00	100.00	100.00	100.00	100.00	81.73

Fig. 10 The results of evaluation using a strict CPA criterion

Standard evaluation report for project Location

Overall value of system k: 100 * (Score$_k$/ Score$_{max}$)^W (Cost$_{min}$/ Cost$_k$)$^{1-w}$ [%]

System	Cost	Relative importance of high scores (w)											Overall score [%]
		0%	10%	20%	30%	40%	50%	60%	70%	80%	90%	100%	
Alpha	1.000	78.70	79.19	79.69	80.19	80.70	81.20	81.71	82.23	82.74	83.26	83.79	72.4
Beta	0.920	85.54	83.50	81.50	79.56	77.66	75.80	73.99	72.22	70.50	68.81	67.17	58.04
Delta	0.787	100.00	98.15	96.33	94.54	92.79	91.07	89.38	87.72	86.09	84.50	82.93	71.66
Gamma	0.950	82.84	84.42	86.02	87.65	89.32	91.02	92.75	94.51	96.31	98.14	100.00	86.41

Normalized report

System	Cost	Relative importance of high scores (w)											Overall score [%]
		0%	10%	20%	30%	40%	50%	60%	70%	80%	90%	100%	
Alpha	1.000	78.70	80.69	82.73	84.82	86.97	89.17	88.10	87.00	85.92	84.85	83.79	72.4
Beta	0.920	85.54	85.08	84.61	84.15	83.69	83.24	79.78	76.42	73.20	70.12	67.17	58.04
Delta	0.787	100.00	100.00	100.00	100.00	100.00	100.00	96.37	92.82	89.40	86.10	82.93	71.66
Gamma	0.950	82.84	86.01	89.30	92.72	96.26	99.95	100.00	100.00	100.00	100.00	100.00	86.41

Fig. 11 The results of evaluation after Delta reduces the price

5   A Sample Student Project Based on LSPeval

LSPeval can be used in various student projects. The project entitled "Evaluation using LSPeval" has the goal to teach students how to create a small LSP criterion with mandatory and optional input attributes. Students must provide domain expertise and show some creativity. The LSPeval project that can be prepared in two weeks is assigned as follows:

PART ONE

1. *Using access from seas.com/sw open an LSPeval account using your email as your username.*
2. *Download the LSPeval user manual V1.3. Read the manual.*
3. *Perform evaluation using the senior citizen home location project.*
4. *Modify the evaluation criterion according to your taste.*
5. *Add and/or modify competitive locations. Call new locations AAA, BBB, CCC, etc.*
6. *Perform the quantitative evaluation and prepare the evaluation report. Explain the advantages of the winning location (the reasons for the first place). What can be achieved by changing the price of locations? How much could you increase the price of the winner?*

PART TWO

1. *Go to* www.dell.com *and click on Products tab.*
2. *Select one of Dell products (e.g. laptops or desktops).*

3. *Create a small LSP criterion for selection of your computer (minimum 3 mandatory and 3 optional attributes).*
4. *Enter the LSP criterion for computer evaluation in LSPeval.*
5. *Select 2–3 Dell computers you want to evaluate. Enter their data and perform evaluation and comparison. Explain the results.*
6. *Go to www.hp.com.*
7. *Select 2–3 HP computers that are competitive to selected Dell computers (similar hardware, similar performance, and similar price range).*
8. *Perform evaluation of selected competitive Dell and HP computers using LSPeval, separately for standard, strict, and relaxed criteria.*
9. *Compare the results of evaluation obtained with the above three types of criteria.*
10. *Prepare the evaluation report and explain the reasons for the winning position of the best competitor. Submit a technical report that contains results of Part One and Part Two of this project. Try to write a conclusion that summarizes positive and negative experiences with this project. Critical remarks and improvement suggestions about the method and the LSPeval tool are very welcome (and can get extra credit!).*

The initial experiences obtained with this project are very positive. Frequently, students can also be offered the possibility to replace computer evaluation with other similar projects of their choice. That can create interesting evaluation projects for jobs, cars, homes, software products, web sites, universities, and other objects.

The presented student project comes before the classroom presentation of mathematical details related to fuzzy sets, soft computing, theory of aggregation, and graded logic. The suggested sequence of introductory topics presented in the classroom is the following:

1. Introduction to decision problems and decision support systems.
2. Basic properties of human evaluation reasoning.
3. Basic mathematical models of soft computing logic: partial conjunction, partial disjunction, and partial absorption.
4. Standard model of evaluation reasoning [7].
5. Classroom presentation of the LSPeval tool using the location comparison and selection problem.
6. Student project based on LSPeval.
7. Mathematical models used in fuzzy systems, soft computing, and graded logic.
8. Professional decision support problems, methods, and tools.

The study of these topics can be directly supported by [7] and subsequently expanded using [1–3, 10, 11, 15]. In this way, before the project in the step #6, students are gradually exposed to concepts of andness, orness, annihilators, soft and hard aggregation, criterion functions, semantic aspects of aggregation, the roles of stakeholders, evaluators, and domain experts, and basic computing with words. The goal is to avoid trivial toy problems and include small but realistic and useful multiattribute decision problems that use functional and financial inputs and show the need for logic aggregation. Using our stepwise refinement approach, after the proposed gentle introduction to decision engineering, students are well prepared to study advanced mathematical topics and then participate in research projects.

6 Conclusions

LSPeval is an educational tool developed with intention to expose students to evaluation decision models based on graded logic and the LSP method. It was satisfactorily tested and used in soft computing and decision support courses offered to graduate students and undergraduate seniors in a computer science program. The LSPeval tool and the supporting documentation are free and available over the Internet.

The main goal of LSPeval is to expose students to concepts of soft computing and fuzzy logic with minimum mathematical prerequisites. Thus, LSPeval can be used in initial segments of decision support, soft computing, fuzzy logic, and similar computational intelligence courses.

Educational benefits of LSPeval include the understanding of problems of modeling human mental activities, stimulation of creativity in problem solving, and critical thinking: understanding the limitations of mathematical models and software tools, and possibilities for their expansion and improvement. LSPeval is the first step towards building evaluation criteria of unrestricted complexity and using advanced professional tools such as LSP.NT. After the introductory experience with LSPeval, the study of mathematical aspects of graded logic and advanced topics of optimization, sensitivity, explainability, and reliability analyses become natural and easily understandable expansions.

References

1. Beliakov, G., Pradera, A., Calvo, T.: Aggregation Functions: A Guide for Practitioners. Springer, New York (2007)
2. Beliakov, G., Bustince Sola, H., Calvo Sanchez, T.: A Practical Guide to Averaging Functions. Studies in Fuzziness and Soft Computing, vol. 329. New York, Springer (2016)
3. Bullen, P.S.: Handbook of Means and Their Inequalities. Kluwer (2003) (2010)
4. Dujmović, J.: Weighted conjunctive and disjunctive means and their application in system evaluation. Univ. Beograd Publ. Elektrotehn. Fak., Ser. Mat. Fiz. 461(597), 147–158 (1974). (Available from jstor.org)
5. Dujmović, J., Larsen, H.L.: Generalized conjunction/disjunction. Int. J. Approx. Reason. **46**, 423–446 (2007)
6. Dujmović, J.: Weighted compensative logic with adjustable threshold andness and orness. IEEE Trans. Fuzzy Syst. **23**(2), 270–290 (2015)
7. Dujmović, J.: Soft Computing Evaluation Logic. Wiley, Hoboken (2018)
8. Dujmović, J.: Graded logic for decision support systems. Int. J. Intell. Syst. **34**, 2900–2919 (2019)

9. Dujmović, J.: Interpretability and explainability of LSP evaluation criteria. In: Proceedings of the 2020 IEEE International Conference on Fuzzy Systems (FUZZ-IEEE), pp. 1–8. Glasgow, UK (2020)
10. Fodor, J., Roubens, M.: Fuzzy preference modelling and multicriteria decision support. Kluwer Academic Publishers, Dordrecht, The Netherlands (1994)
11. Grabisch, M., Marichal, J.-L., Mesiar, R., Pap, E.: Aggregation Functions. Cambridge University Press, Cambridge (2009)
12. Kapre, K.: LSP evaluator – A suitability evaluation tool based on the LSP method. San Francisco State University, Computer Science Department, Report SFSU-CS-CE-17.18 (2017)
13. Kapre, K., Dujmović, J.: LSPeval – Evaluation using canonical mandatory/optional LSP criteria. User manual V 1.3 (2018) http://www.systemevaluationandselection.com/documenta tation/LSPeval_V1.3_User_Manual.pdf , Accessed: 2021–08–30.
14. SEAS Co. LSP.NT – LSP method for evaluation over the Internet. Short LSP.NT User Manual for Demo Users, V 1.1 (2017). http://www.systemevaluationandselection.com/documentation/ LSP.NT_V1.1_Demo_User_Manual.pdf. Accessed: 2021–08–30.
15. Torra, V., Narukawa, Y.: Modeling Decisions. Springer, Berlin (2007)
16. Zadeh, L.A.: Fuzzy sets. Inf. Control **8**(3), 338–353 (1965). https://doi.org/10.1016/s0019-995 8(65)90241-x
17. Zadeh, L.A.: Fuzzy logic, neural networks, and soft computing. Commun. ACM **37**(3), 77–84 (1994). https://doi.org/10.1145/175247.175255
18. Zimmermann, H.-J.: Fuzzy Set Theory and Its Applications. Springer Science + Business Media, New York (1996)

Job Offers Classifier Using Neural Networks and Oversampling Methods

Germán Ortiz[ID]**, Gemma Bel Enguix**[ID]**, Helena Gómez-Adorno**[ID]**, Iqra Ameer**[ID]**, and Grigori Sidorov**[ID]

Abstract Both policy and research benefit from a better understanding of individuals' jobs. However, as large-scale administrative records are increasingly employed to represent labor market activity, new automatic methods to classify jobs will become necessary. We developed an automatic job offers classifier using a dataset collected from the largest job bank of Mexico known as *Bumeran*. We applied machine learning algorithms such as Support Vector Machines, Naive-Bayes, Logistic Regression, Random Forest, and deep learning Long-Short Term Memory (LSTM). Using these algorithms, we trained multi-class models to classify job offers in one of the 23 classes (not uniformly distributed): Sales, Administration, Call Center, Technology, Trades, Human Resources, Logistics, Marketing, Health, Gastronomy, Financing, Secretary, Production, Engineering, Education, Design, Legal, Construction, Insurance, Communication, Management, Foreign Trade, and Mining. We used the SMOTE, Geometric-SMOTE, and ADASYN synthetic oversampling algorithms to

G. Ortiz
Posgrado en Ciencia e Ingeniería de la Computación Universidad Nacional Autónoma de México, Mexico City, Mexico
e-mail: jortizb@iingen.unam.mx

G. B. Enguix
Instituto de Ingeniería, Universidad Nacional Autónoma de México, Mexico City, Mexico
e-mail: gbele@iingen.unam.mx

H. Gómez-Adorno
Instituto de Investigaciones en Matemáticas Aplicadas y en Sistemas, Universidad Nacional Autónoma de México, Mexico City, Mexico
e-mail: helena.gomez@iimas.unam.mx

I. Ameer · G. Sidorov (✉)
Instituto Politécnico Nacional, Centro de Investigación en Computación, Mexico City, Mexico
e-mail: sidorov@cic.ipn.mx

I. Ameer
e-mail: iqra@nlp.cic.ipn.mx

© The Author(s), under exclusive license to Springer Nature Switzerland AG 2023
Sh. N. Shahbazova et al. (eds.), *Recent Developments and the New Directions of Research, Foundations, and Applications*, Studies in Fuzziness and Soft Computing 422,
https://doi.org/10.1007/978-3-031-20153-0_18

handle imbalanced classes. The proposed convolutional neural network architecture achieved the best results when applied the Geometric-SMOTE algorithm.

Keywords Jobs offer classification · Unbalanced dataset · Oversampling · Neural networks

1 Introduction

Mexico City has around 9 million[1] inhabitants and is the state with the highest population density in the entire country. During the last quarter of 2018, there were 4.5 million economically active inhabitants, of which 230 thousand were unemployed.[2]

One of the biggest problems between supply and demand in the labor market is the asymmetry of information, which decreases the efficiency in the search for job opportunities [14].

In this context, the Secretary of Labor and Employment Promotion (STyFE, *Secretaria de Trabajo y Fomento al Empleo*), developed the project Diagnosis of Skills in Demand (DiCoDE, *Diagnóstico de Competencias Demandadas*), a system that monitors the job offers displayed on job portals in Mexico City to perform analysis on them. Our work focuses on improving the functioning of the described system through the automatic classification of published offers in order to identify job profiles, skills and competencies demanded by employers and thus reduced the asymmetry mentioned above.

In this article, we develop two proposals for automatic classifiers of job offers for DiCoDE using two neural network architectures, a recurrent neural network and a convolutional neural network, in which we face the problem of unbalanced data set. A dataset is unbalanced if the number of instances of one or more classes is significantly higher or lower relative to the other [25] classes. The problem of class imbalance has gained more interest in recent years [21] due to the growth of the field of machine learning applications, such as sentiment analysis [3], disease detection, and fraudulent phone call detection, among others. Since these data sets represent real-world problems, it is impossible to have the same number of examples of each of the classes involved. One of the ways to solve this problem is oversampling, which consists of increasing the number of examples of the minority classes, either by replicating some instances of it or by generating new examples using some criterion, such as the k- nearest neighbors algorithm [4].

In this paper, we present a series of experiments using various oversampling techniques to improve the performance of classifiers using [17] Accuracy, Recall, Precision, and F1 as evaluation metrics. To balance the class distribution of our

---

[1] https://www.bumeran.com.mx/ Last visited: 19-01-2022.

[2] INEGI. Encuesta Nacional de Ocupación y Empleo (ENOE) https://www.inegi.org.mx/programas/enoe/15ymas/default.html#Tabulados Last visited: 19-02-2022.

dataset, we use the SMOTE [6] technique, the Geometric-SMOTE [9] variant, and the ADASYN [16] algorithm.

The content of the article is organized as follows; in Sect. 2 we describe related work to the oversampling methods used in our classifier proposals, as well as the use of neural networks in multiclass classification tasks. In Sect. 3, we detail the characteristics of the corpus used for training and testing the neural networks. Subsequently, we address the methodology of the experiments carried out, in which the operation of the SMOTE, Geometric-SMOTE, and ADASYN algorithms is explained. In addition, we describe the pre-processing carried out on the text, the proposed neural network architectures, the training parameters used, and additional experiments using traditional machine learning algorithms to compare their performance with our proposals. In Sect. 5, we show the results obtained and their comparison; Finally, in Sect. 6, we offer the conclusions of the work carried out and the direction of future work.

2 Related Work

In recent years, the problem of dealing with unbalanced data sets has gained more attention from the scientific community.

Regarding the generation of new examples of minority classes, *Mohasseb et al.* [24], propose the hierarchical-SMOTE algorithm to deal with imbalanced data in the Questions Classification task. This algorithm creates a grammar pattern to analyze each class in the dataset, then SMOTE algorithm is applied over all minority classes. The authors use a dataset that contains 1160 questions distributed in 6 classes and use a Naive Bayes classifier. The evaluation metrics they use are Precision, Recall, and F1, whose best results are 0.851, 0.865, and 0.847, respectively.

On the other hand, *Douzas et al.* [11] face the unbalanced learning problem in the land cover classification task, in which a data set composed of 1694 examples is distributed in 8 classes. The class with the largest number of examples has 761 instances, while the category with the fewest number of elements has only 4 elements. In a matter of experiments, the evaluation metrics Accuracy and F1 and geometric mean are used. The algorithms used as classifiers are Logistic Regression, K-nearest neighbors, decision trees, Gradient Boosting and Random Forest. The Geometric-SMOTE algorithm achieves the best results in terms of F1 and geometric mean in the 4 algorithms, with Random Forest being the highest values with 0.341 and 0.572, respectively.

In the field of the telecommunications industry, *Aditsania et al.* [1] deal with the problem of customer loss by developing a classifier that allows determining whether a customer is likely to stop using a service or not. To carry out this task, a data set belonging to an Indonesian telecommunications company was used, made up of 200,387 examples with 55 characteristics each, in which only 4% correspond to examples of clients labeled as prone to stop using the service. Counteracting the imbalance of the data set, the ADASYN algorithm was used. A multilayer neural network trained using the backpropagation algorithm was used as a classifier.

The results of the experiments show an interesting behavior since the highest Accuracy value is reached when the oversampling algorithm is not used. However, for this case, the F1 value is equal to zero, which implies that the system can only predict the majority class. With the implementation of the ADASYN algorithm, the values of Accuracy and Value-F are 0.93 and 0.46, respectively.

In the field of neural networks, *Nowak et al.* [26] present a short text classifier using recurrent neural networks, specifically of the Long Short Term Memory (LSTM) type. 3 data sets were used to carry out their experiments, the first of them on messages classified as spam and non-spam with a distribution of 13.4% and 86.6%, respectively. The second one is about advertising, made up of 53.3% of examples considered accepted and 46.7% considered messages rejected. The last data set used consisted of book reviews classified as negative, positive, and neutral. As part of the preprocessing of the text, the text was converted to lowercase, punctuation marks and special characters were removed, numbers were removed, and a dictionary of unique words was created for each corpus. The evaluation was made based on the Accuracy obtaining values of 99.798%, 94.497%, 84.415%, respectively.

In the same field, *Lai et al.* [19] present a classifier model based on a recurrent and convolutional neural network architecture to capture contextual information during the learning of word representations, as well as identify which words have a key role in the texts. To carry out their experiments, 4 data sets were used, the first of them: *20newsgroup*, of which only 4 classes of 20 available are considered; the second: *fudan*, composed of texts in Chinese distributed in 20 classes; third: *ACL Anthology network*, made up of scientific papers divided into 5 classes and finally, the fourth data set: *Stanford Sentiment Treebank*, made up of movie reviews in 5 classes. For performance evaluation, the first data set is evaluated using F1; the remaining sets use Accuracy as the evaluation metric. *word embeddings* were used as word representations by the Skip-gram algorithm using Wikipedia for both Chinese and English. Additionally, comparisons were made using traditional machine learning methods and various neural network architectures. The model performs best on 3 of the 4 data sets, with an F1 of 96.49 for the *20newsgroup* set, and Accuracy of 95.20 and 49.19 for the *fudan* and *ACL sets. Anthology network*, respectively.

3 Dataset

Our dataset consists of 979,956 job offers written in Spanish, each of them belonging to one of the following 23 categories: Sales, Administration, Call Center, Technology, Trades, Human Resources, Logistics, Marketing, Health, Gastronomy, Financing, Secretary, Production, Engineering, Education, Design, Legal, Construction, Insurance, Communication, Management, Foreign Trade, and Mining. The job offers, as well as the categories used, were collected from the website Bumeran. Using a web scraping program [8] written in the R programming language. After performing the preprocessing described in Sect. 4.1 the distribution of the dataset is shown in Table 1.

Table 1 Dataset statistics after eliminating equal job offers

Category	Instances	Percentage (%)
Sales	13,002	22.59
Administration	8,730	15.16
Call center	8,453	14.68
Technology	5,559	9.65
Trades	3,973	6.90
Human Resources	2,359	4.10
Logistics	2,206	3.83
Marketing	1,663	2.89
Health	1,610	2.80
Gastronomy	1,343	2.33
Financing	1,267	2.20
Secretary	1,236	2.15
Production	1,129	1.96
Engineering	881	1.53
Education	702	1.22
Design	661	1.15
Legal	645	1.12
Construction	622	1.08
Insurance	573	0.99
Communication	417	0.72
Management	272	0.47
Foreign Trade	228	0.40
Mining	41	0.07
Total	57,572	

4 Methodology

This section briefly describes the oversampling algorithms used in this work. We explain the pre-processing performed on the text, the neural networks architectures used to perform classification as well as the parameters used to carry out the experiments.

4.1 Pre-processing

Text Pre-processing was the following: first, the text was converted to lower-case, then punctuation marks were eliminated as well as special characters such as tab character and new-line symbol. Finally, we delete Spanish stopwords using the NLTK library [2, 22].

After performing text pre-processing, we searched to eliminate repeated job offers to avoid overfitting our neural network architecture. Table 1 shows the distribution of samples after the removal of equal job offers.

4.2 Oversampling Algorithms

Unlike random oversampling, in which examples belonging to the minority class are randomly selected to be replicated and added to the training dataset, in our work, we use synthetic oversampling algorithms, where new examples are generated based on a series of criteria and using the K nearest neighbors algorithm. We use the following algorithms:

- SMOTE: The Synthetic Minority Over-sampling Technique (SMOTE) algorithm generates new instances of the minority class by selecting instances of it, then find K nearest neighbors to that instance and synthetically generates new instances. These generated instances are located along the line segment between the real selected instance, and its K nearest neighbor [6].
- Geometric-SMOTE: The geometric-SMOTE algorithm generates new instances of the minority class based on its K nearest neighbors the same way as SMOTE algorithm, but instead of using a line segment, it uses a geometric region, which is usually a hyper-spheroid [9].
- ADASYN: The key idea behind this algorithm is to use a density distribution as a criterion to automatically decide the number of synthetic samples to generate for each minority sample, adaptively changing the weights of different minority samples to compensate for skewed distributions [15].

We use the Python implementation of these algorithms available in the imbalanced-learn library [20].

4.3 Classification

We built two different neural network architectures; the first architecture has an embedding layer to get a matrix representation of each text. We initialize the weights in this layer using the word2vec algorithm trained on the Spanish Billion Word Corpus [5]. This algorithm gets word vector representations on a continuous vector space whose objective is to preserve the semantic and syntactic similarity between the words [12]. After the embedding layer, we use a Long-Short Term Memory (LSTM) [26] to obtain a vector representation from the entire text. Finally, the output of this layer is connected to a dense layer to perform the text classification. This architecture uses a

Fig. 1 Block diagram for
the proposed recurrent neural
network architecture
(LSTM)

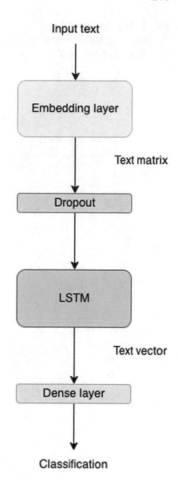

Dropout layer where a fraction of the input data is randomly assigned to zero in each
training update phase to avoid overfitting. The block diagram for this architecture is
shown in Fig. 1.

The second architecture is the same as the first one except for the Long-Short
Term Memory layer; instead of this recurrent layer, we use a convolutional net-
work followed by a pooling layer to get a vector representation for the entire text.
Figure 2 shows the block diagram for this architecture.

We compute the weight of each class in order to use a weighted cross-entropy as
a loss function for both architectures. We use 10 epochs and batch size of 32 and
K-fold cross-validation with $K = 5$ for all the experiments. For both cases, we used
the Keras library [7] to implement the network.

Fig. 2 Block diagram for
the proposed convolutional
neural network architecture

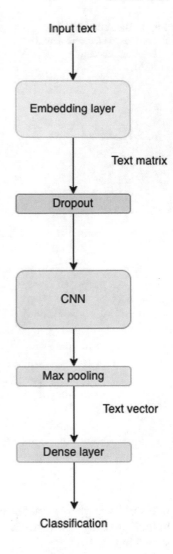

4.4 Comparison Experiments

In order to compare the performance of the proposed architectures, we carried out
additional experiments in which the traditional machine learning algorithms were
used: Support Vector Machines, Naive-Bayes, Logistic Regression, and Random
forest. We use two different inputs to those algorithms. First, we obtain a text matrix
using the pre-trained word2vec model trained on the Spanish Billion Word Corpus
[5], then we compute an average vector of each word on a text to obtain a vector
representation for the whole text that we use as inputs to the listed algorithms above.

We use the second input of a text vector representation obtained using a bag of words model.

Similarly, the Facebook library FastText [18] was used, which offers a tool that uses the ideas of *Mikolov et al.* [23] of efficient learning of word representations to train a linear classifier using a range constraint and a fast loss approximation. We also use the framework developed by *Socher et al.* [27] which is based on recursive auto-encoders for sentence-level prediction.

5 Results

Table 2 shows the results obtained with the deep learning architectures described in Sect. 4.3. We can observe that the convolutional neural network architecture outperforms the recursive neural network architecture both when oversampling techniques are used and when the original dataset (with no oversampling techniques applied) is used.

In the case of the recurrent network, it obtains the best performance when no oversampling algorithms are used with 0.59 for Accuracy and 0.46, 0.54, and 0.48 for precision, recall, and F1, respectively. The best-achieved result using an oversampling technique is with the SMOTE algorithm. However, there is about a 10% difference in terms of performance metrics with respect to results using the original dataset.

For the convolutional network, using the Geometric-SMOTE algorithm obtains the best performance with 0.64 of accuracy, 0.54 of precision, 0.56 of recall, and 0.54 of F1, which represent an improvement of 1% in Accuracy, recall, and F1 when no oversampling techniques are used.

Table 3 shows the results of the experiments using the classical machine learning algorithms with the average of the word vectors using the word2vec model as inputs

Table 2 Results with proposed architectures

LSTM	Accuracy	Precision	Recall	F1
Original dataset	0.59	0.46	0.54	0.48
SMOTE	0.47	0.37	0.47	0.38
Geometric-SMOTE	0.42	0.33	0.41	0.32
ADASYN	0.39	0.34	0.40	0.33
CNN	Accuracy	Precision	Recall	F1
Original dataset	0.63	0.54	0.55	0.53
SMOTE	0.61	0.50	0.56	0.51
Geometric-SMOTE	**0.64**	**0.54**	**0.56**	**0.54**
ADASYN	0.62	0.52	0.55	0.52

Table 3 Results of comparison experiments using average word2vec vectors as input

SVM	Accuracy	Precision	Recall	F1
Original dataset	0.65	0.61	0.45	0.49
SMOTE	0.64	0.59	0.47	0.49
Geometric-SMOTE	0.64	0.58	0.47	0.48
ADASYN	0.64	0.59	0.47	0.49
Naive-Bayes	Accuracy	Precision	Recall	F1
Original dataset	0.35	0.35	0.29	0.22
SMOTE	0.35	0.35	0.29	0.22
Geometric-SMOTE	0.35	0.35	0.29	0.22
ADASYN	0.35	0.35	0.29	0.22
Logistic Regression	Accuracy	Precision	Recall	F1
Original dataset	0.62	0.56	0.37	0.40
SMOTE	0.61	0.52	0.39	0.40
Geometric-SMOTE	0.60	0.53	0.38	0.39
ADASYN	0.61	0.54	0.39	0.40
Random forest	Accuracy	Precision	Recall	F1
Original dataset	0.59	0.67	0.34	0.40
SMOTE	0.60	0.68	0.35	0.41
Geometric-SMOTE	0.59	0.68	0.36	0.39
ADASYN	0.69	0.68	0.34	0.40
FastText	Accuracy	Precision	Recall	F1
–	0.60	0.36	0.27	0.31
RecursiveNN (Socher et al)	Accuracy	Precision	Recall	F1
–	0.12	0.09	0.11	0.10

to the classifiers as well as the results using FastText and RecursiveNN [27]. Support Vector machines outperform all other algorithms in any experimental setup and also outperform the recurrent network in any case. The Naive-Bayes classifier obtains the worst performance of all the classifiers. In addition, the oversampling techniques do not represent an improvement for this classifier since the performance remains constant.

Finally, Table 4 shows the results of the experiments with the classical machine learning algorithms using bag-of-words vectors as input. The best performing algorithm is Logistic Regression, followed by Support Vectors Machine, Random Forest, and Naive-Bayes. In this case, the oversampling techniques do not represent an

Table 4 Results of comparison experiments using bag-of-words vectors as input

SVM	Accuracy	Precision	Recall	F1
Original dataset	0.63	0.53	0.50	0.51
SMOTE	0.63	0.52	0.50	0.51
Geometric-SMOTE	0.63	0.53	0.51	0.51
ADASYN	0.63	0.52	0.50	0.51
Naive-Bayes	Accuracy	Precision	Recall	F1
Original dataset	0.58	0.60	0.28	0.32
SMOTE	0.58	0.60	0.29	0.33
Geometric-SMOTE	0.58	0.60	0.29	0.32
ADASYN	0.58	0.60	0.29	0.33
Logistic Regression	Accuracy	Precision	Recall	F1
Original dataset	**0.66**	**0.59**	**0.52**	**0.55**
SMOTE	0.66	0.58	0.52	0.54
Geometric-SMOTE	0.66	0.59	0.52	0.54
ADASYN	0.66	0.57	0.52	0.54
Random forest	Accuracy	Precision	Recall	F1
Original dataset	0.66	0.67	0.44	0.50
SMOTE	0.62	0.66	0.39	0.45
Geometric-SMOTE	0.66	0.69	0.44	0.51
ADASYN	0.63	0.66	0.39	0.45

improvement in the classifier's performance since the best result is obtained using the original dataset. Furthermore, this is the best result of all experimental setups; it outperforms the convolutional neural network when Geometric-SMOTE is used as an oversampling technique.

6 Conclusions

In this article, we present an automatic classifier of job offers using two neural network architectures to improve the operation of the DiCoDe system in Mexico City. For this classification task, we face the class imbalance problem in the field of multiclass classification. To tackle this problem, 3 synthetic oversampling algorithms were used, SMOTE, Geometric-SMOTE, and ADASYN. The results of the proposed architectures were compared with the classic machine learning algorithms Support

Vector Machine, Naive-Bayes, Logistic Regression, and Random forest using average word2vec vectors and bag-of-words vectors as input as well as the FastText library and a framework based on recursive auto-encoders.

The convolutional neural network architecture outperforms the recurrent neural network in all experimental setups. It also outperforms the FastText library, the recursive auto-encoder-based framework, and all machine learning algorithms when average word2vec vectors are used as inputs. The best result for this architecture is achieved when the Geometric-SMOTE algorithm is used as an oversampling method.

Nevertheless, our proposed convolutional neural network is outperformed by a logistic regression classifier when no oversampling techniques were used by 0.02 in Accuracy, 0.04 in precision, and 0.01 in F1.

Given the characteristics of our neural network architecture, future steps examine the use of a hybrid architecture that includes recurrent and convolutional neural networks through a [13] attention mechanism. On the other hand, in relation to oversampling methods, the effects of using methods based on Generative Adversarial Networks can be studied. These networks seek to approximate the real distribution of the data of the minority classes instead of using local information with the k-nearest neighbor algorithm [10].

Acknowledgements The work was done with partial support from the Mexican Government through the grants A1-S-47854 and CB A1-S-27780 of the CONACYT-Mexico, grants of PAPIIT-UNAM projects TA400121 and TA101722, and grants 20211784, 20211884, and 20211178 of the Secretaría de Investigación y Posgrado of the Instituto Politécnico Nacional, Mexico. The authors also thank the CONACYT for the computing resources brought to them through the Plataforma de Aprendizaje Profundo para Tecnologías del Lenguaje of the Laboratorio de Supercómputo of the INAOE, Mexico. The authors also acknowledge the support of the DiCoDe project.

References

1. Aditsania, A., Adiwijaya, Saonard, A.L.: Handling imbalanced data in churn prediction using adasyn and backpropagation algorithm. In: 2017 3rd International Conference on Science in Information Technology (ICSITech), pp. 533–536 (2017)
2. Ameer, I., Siddiqui, M.H.F., Sidorov, G., Gelbukh, A.: Cic at semeval-2019 task 5: simple yet very efficient approach to hate speech detection, aggressive behavior detection, and target classification in twitter. In: Proceedings of the 13th International Workshop on Semantic Evaluation, pp. 382–386 (2019)
3. Ameer, I., Sidorov, G., Gómez-Adorno, H., Nawab, R.M.A.: Multi-label emotion classification on code-mixed text: Data and methods. IEEE Access **10**, 8779–8789 (2022). https://doi.org/10.1109/ACCESS.2022.3143819
4. Barandela, R., Valdovinos, R.M., Sánchez, J.S., Ferri, F.J.: The imbalanced training sample problem: under or over sampling? In: Fred, A., Caelli, T.M., Duin, R.P.W., Campilho, A.C., de Ridder, D. (eds.) Structural, Syntactic, and Statistical Pattern Recognition, pp. 806–814. Springer, Berlin (2004)
5. Cardellino, C.: Spanish Billion Words Corpus and Embeddings (2019). https://crscardellino.github.io/SBWCE/

6. Chawla, N.V., Bowyer, K.W., Hall, L.O., Kegelmeyer, W.P.: Smote: synthetic minority over-sampling technique. J. Artif. Intell. Res. **16**, 321–357 (2002)
7. Chollet, F., et al. Keras (2015). https://github.com/fchollet/keras
8. Cording, P.H.: Algorithms for web scraping (2011). http://www.imm.dtu.dk/English.aspx, supervised by Associate Professors Inge Li Gørtz, ilg@imm.dtu.dk, and Philip Bille, DTU Informatics
9. Douzas, G., Bacao, F.: Geometric smote: effective oversampling for imbalanced learning through a geometric extension of smote (2017). arXiv preprint arXiv:1709.07377
10. Douzas, G., Bação, F.: Effective data generation for imbalanced learning using conditional generative adversarial networks. Expert Syst. Appl. **91** (09 2017). https://doi.org/10.1016/j.eswa.2017.09.030
11. Douzas, G., Bação, F., Fonseca, J., Khudinyan, M.: Imbalanced learning in land cover classification: Improving minority classes prediction accuracy using the geometric smote algorithm. Remote Sens. **11**, 3040 (2019). https://doi.org/10.3390/rs11243040
12. Ghannay, S., Favre, B., Estève, Y., Camelin, N.: Word embedding evaluation and combination. In: Proceedings of the Tenth International Conference on Language Resources and Evaluation (LREC'16). pp. 300–305. European Language Resources Association (ELRA), Portorož, Slovenia (2016), https://www.aclweb.org/anthology/L16-1046
13. Guo, L., Zhang, D., Wang, L., Wang, H., Cui, B.: Cran: a hybrid cnn-rnn attention-based model for text classification. In: International Conference on Conceptual Modeling, pp. 571–585. Springer (2018)
14. Hart, O.D.: Optimal labour contracts under asymmetric information: an introduction. Rev. Econ. Stud. **50**(1), 3–35 (1983)
15. He, H., Garcia, E.A.: Learning from imbalanced data. IEEE Trans. Knowl. Data Eng. **21**(9), 1263–1284 (2009)
16. He, H., Bai, Y., Garcia, E.A., Li, S.: Adasyn: Adaptive synthetic sampling approach for imbalanced learning. In: 2008 IEEE International Joint Conference on Neural Networks (IEEE World Congress on Computational Intelligence). IEEE (2008), pp. 1322–1328
17. Hossin, M., Sulaiman, M.N.: A review on evaluation metrics for data classification evaluations. Int. J. Data Mining Knowl. Manag. Process **5**, 01–11 (2015). https://doi.org/10.5121/ijdkp.2015.5201
18. Joulin, A., Grave, E., Bojanowski, P., Mikolov, T.: Bag of tricks for efficient text classification (2016). arXiv preprint arXiv:1607.01759
19. Lai, S., Xu, L., Liu, K., Zhao, J.: Recurrent convolutional neural networks for text classification. In: Proceedings of the Twenty-Ninth AAAI Conference on Artificial Intelligence, pp. 2267–2273. AAAI'15, AAAI Press (2015)
20. Lemaitre, G., Nogueira, F., Aridas, C.K.: Imbalanced-learn: A python toolbox to tackle the curse of imbalanced datasets in machine learning. CoRR (2016). http://arxiv.org/abs/1609.06570
21. Lemnaru, C., Potolea, R.: Imbalanced classification problems: systematic study, issues and best practices. In: International Conference on Enterprise Information Systems, pp. 35–50. Springer (2011)
22. Loper, E., Bird, S.: Nltk: The natural language toolkit. CoRR (2002). http://dblp.uni-trier.de/db/journals/corr/corr0205.html#cs-CL-0205028
23. Mikolov, T., Chen, K., Corrado, G., Dean, J.: Efficient estimation of word representations in vector space. In: Bengio, Y., LeCun, Y. (eds.) 1st International Conference on Learning Representations, ICLR 2013, Scottsdale, Arizona, USA, May 2–4, 2013, Workshop Track Proceedings (2013). http://arxiv.org/abs/1301.3781
24. Mohasseb, A., Bader-El-Den, M., Cocea, M., Liu, H.: Improving imbalanced question classification using structured smote based approach. In: 2018 International Conference on Machine Learning and Cybernetics (ICMLC), vol. 2, pp. 593–597. IEEE (2018)
25. More, A.: Survey of resampling techniques for improving classification performance in unbalanced datasets (2016)

26. Nowak, J., Taspinar, A., Scherer, R.: Lstm recurrent neural networks for short text and sentiment classification. In: International Conference on Artificial Intelligence and Soft Computing, pp. 553–562. Springer (2017)
27. Socher, R., Pennington, J., Huang, E.H., Ng, A.Y., Manning, C.D.: Semi-supervised recursive autoencoders for predicting sentiment distributions. In: Proceedings of the 2011 Conference on Empirical Methods in Natural Language Processing, pp. 151–161. Association for Computational Linguistics, Edinburgh (2011). https://www.aclweb.org/anthology/D11-1014

On Similarity Correlation of Probability Distributions

Maria Elena Ensastegui-Ortega, Ildar Batyrshin, and Alexander Gelbukh

Abstract In recent years there has been much attention to the analysis and comparison of different measures of similarity and dissimilarity of probability distributions. In this work, we consider such measures as functions defined on the set of probability distributions and satisfying given properties. Further, we consider a correlation function defined on the set of probability distributions and satisfying simple properties fulfilled for most of the known correlation and association coefficients considered in statistics. We discuss general methods of constructing correlation functions from similarity and dissimilarity functions. Finally, we introduced several new similarity, dissimilarity, and correlation functions on the set of probability distributions.

Keywords Similarity functions · Dissimilarity functions · Correlation functions

1 Introduction

Similarity measures and distances are widely used in information retrieval, data classification, machine learning, and decision making in ecology, computational linguistics, image and signal processing, financial data analysis, bioinformatics, and social sciences [1, 2]. The correlation and association coefficients are used in exploratory data analysis and data mining [2–5]. In recent years there has been much attention to the analysis of similarity and dissimilarity measures and operations over the set of probability distributions [6–9]. In this work, we consider the methods of construction of correlation functions from similarity and dissimilarity measures satisfying given properties [10–12]. The paper [6] has a dozen similarity and dissimilarity measures defined on the set of probability distributions. Our first goal was to determine which of these measures satisfies the required properties. We define correlation functions as symmetric and reflexive functions of two arguments taking values in the interval

M. E. Ensastegui-Ortega · I. Batyrshin (✉) · A. Gelbukh
Centro de Investigación en Computación, Instituto Politécnico Nacional, Av. Juan de Dios Bátiz S/N, Nueva Industrial Vallejo, CDMX, 07738 Ciudad de México, México

© The Author(s), under exclusive license to Springer Nature Switzerland AG 2023   249
Sh. N. Shahbazova et al. (eds.), *Recent Developments and the New Directions of Research, Foundations, and Applications*, Studies in Fuzziness and Soft Computing 422, https://doi.org/10.1007/978-3-031-20153-0_19

of real numbers $[-1, 1]$. These are the common properties fulfilled for correlation and association coefficients [3, 4]. We select from the survey [6] suitable similarity functions and, using these functions, construct correlation functions on the set of probability distributions.

The Sect. 2 considers the general properties of similarity, dissimilarity, and correlation functions and their relationships. The Sect. 3 explores measures of similarity like Cha [6]. It is shown that they comply with the necessary properties of similarity functions [10–12]. From them, new correlation functions are constructed. Section. 4 shows that some known similarity and dissimilarity measures are complementary and generate the same correlation functions. Section five contains conclusions.

2 Similarity, Dissimilarity, and Correlation Functions

Let Ω be the set of probability distributions $P = \{p_1, \ldots, p_n\}$ of the length n, satisfying the properties: $0 \le p_i \le 1$ and $\sum_{i=p}^{n} p_i = 1$. A function $S : \Omega \times \Omega \to [0, 1]$ is called a similarity function [12] on Ω if for all probability distributions P and Q in Ω the function S satisfies the properties of *symmetry*:

$$S(P, Q) = S(Q, P),$$

and *reflexivity*:

$$S(P, P) = 1.$$

Dually, a function $D : \Omega \times \Omega \to [0, 1]$ is called a *dissimilarity function* on Ω if for all probability distributions P and Q in Ω the function D satisfies *symmetry*:

$$D(P, Q) = D(Q, P)$$

and *irreflexivity*:

$$D(P, P) = 0.$$

Similarity and dissimilarity functions can be considered as *fuzzy relations* on Ω [12].

We will say that similarity function S, and dissimilarity function D are *dual or complementary* if for any two probability distributions functions P and Q in Ω the following holds:

$$S(P, Q) + D(P, Q) = 1.$$

We can rewrite this formula as follows:

$$S(P, Q) = 1 - D(P, Q), D(P, Q) = 1 - S(P, Q). \tag{1}$$

A function $A : \Omega \times \Omega \to [-1, 1]$ is called a *correlation (association) function* [10–12] on Ω if for all probability distributions P and Q in Ω, it satisfies the properties of *symmetry*:

$$A(P, Q) = A(Q, P),$$

reflexivity:

$$A(P, P) = 1,$$

and for some P and Q in Ω it is negative:

$$A(P, Q) < 0.$$

In [10, 11], such correlation functions are also called *weak correlation functions*.

Proposition 1 [11]. Similarity and dissimilarity functions S and D on Ω such that for some P and Q in Ω it is fulfilled: $S(P, Q) < D(P, Q)$, define a correlation function by:

$$A(P, Q) = S(P, Q) - D(P, Q). \tag{2}$$

If S and D are complementary and for some P and Q in Ω it is fulfilled: $S(P, Q) < 0.5$ or $D(P, Q) > 0.5$, then the function A will be a correlation function, and from Eqs. (1) and (2), we obtain:

$$A(P, Q) = 2S(P, Q) - 1, \tag{3}$$

$$A(P, Q) = 1 - 2D(P, Q). \tag{4}$$

The correlation function A related by (2)–(4) with complementary similarity S and dissimilarity D functions is called complementary to S and D and all three functions (S, D, A) together make up a *complementary (or correlation) triplet*. From (3), it is easy to see that a correlation function is nothing else but a rescaled similarity function. For this reason, such a correlation function is also referred to as a *similarity correlation function*.

It was shown [10, 11] that many classical correlation coefficients can be constructed from so-called *bipolar* [10, 11] similarity and dissimilarity functions using (3) or (4). Such correlation functions are called *invertible* o *strong correlation functions* [10, 11] because they satisfy an *inverse relationship* property $A(x, N(y)) = -A(x, y)$, for all elements x and y in the domain V of correlation triplet (S,D,A), where N is an involutive (reflection or negation) operation. For such correlation functions, we have $A(x, N(x)) = -1$, where $N(x)$ can be interpreted as an element "opposite" to x. In this paper, we do not consider invertible correlation functions.

3 Constructing Correlation Functions from Similarity Functions

Consider two probability distribution $P = \{p_1, \ldots, p_n\}$ and $Q = \{q_1, \ldots, q_n\}$ satisfying the properties: $0 \leq p_i \leq 1$, $\sum_{i=1}^{n} p_i = 1$, and $0 \leq q_i \leq 1$, $\sum_{i=1}^{n} q_i = 1$, respectively.

We will say that probability distributions P and Q are *orthogonal* if $p_i q_i = 0$ for all $i = 1, \ldots, n$. We will say that a similarity function S satisfies an *orthogonality property* if the following property is fulfilled on the set of probability distributions:

$$S(P, Q) = 0, \text{ if } P \text{ and } Q \text{ are orthogonal.} \tag{5}$$

From (1), (3) and (5) we obtain the following *orthogonality* property fulfilled for the dissimilarity function and correlation function complementary to similarity function satisfying orthogonality property:

$$D(P, Q) = 1, if \ P \text{ and } Q \text{ are orthogonal.}$$

$$A(P, Q) = -1, \text{ if } P \text{ and } Q \text{ are orthogonal.} \tag{6}$$

For simplicity of references, the functions satisfying orthogonality properties will also be called *orthogonal*.

Consider similarity measures of probability distributions described in [6]. If they satisfy the properties of similarity functions, we will construct their complementary correlation functions. We will suppose that the denominators in all formulas considered below are not equal to zero.

The *Bhattacharyya similarity coefficient* is defined as follows [6, 12]:

$$S(P, Q) = \sum_{i=1}^{n} \sqrt{p_i q_i}.$$

This function is symmetric. To show the reflexivity, consider the case when $P = Q$, then we have $p_i = q_i$ for all $i = 1, \ldots, n$, and:

$$S(P, P) = \sum_{i=1}^{n} \sqrt{p_i p_i} = \sum_{i=1}^{n} \sqrt{p_i^2} = \sum_{i=1}^{n} p_i = 1.$$

In [12], it was shown that the Bhattacharyya coefficient takes values in [0,1], and hence it is a similarity function. Applying (3) obtain correlation function:

$$A(P, Q) = 2 \sum_{i=1}^{n} \sqrt{p_i q_i} - 1.$$

Since the Bhattacharyya coefficient satisfies the orthogonality property, then the obtained correlation function also satisfies this property.

The *cosine similarity* of probability distributions P and Q is given by:

$$S_{cos}(P, Q) = \frac{\sum_{i=1}^{n} p_i q_i}{\sqrt{\sum_{i=1}^{n} p_i^2} \sqrt{\sum_{i=1}^{n} q_i^2}}$$

This function is a similarity function, and by (3), it defines a complementary correlation function:

$$A_{cos}(P, Q) = 2 \frac{\sum_{i=1}^{n} p_i q_i}{\sqrt{\sum_{i=1}^{n} p_i^2} \sqrt{\sum_{i=1}^{n} q_i^2}} - 1 = \frac{2 \sum_{i=1}^{n} p_i q_i - \sqrt{\sum_{i=1}^{n} p_i^2} \sqrt{\sum_{i=1}^{n} q_i^2}}{\sqrt{\sum_{i=1}^{n} p_i^2} \sqrt{\sum_{i=1}^{n} q_i^2}}$$

It is clear that both functions satisfy the orthogonality properties.

Consider *Czekanowski similarity measure* [6]:

$$S_{Cze}(P, Q) = \frac{2 \sum_{i=1}^{n} min(p_i, q_i)}{\sum_{i=1}^{n} (p_i + q_i)}.$$

The sum of probability values in denominator equals to 2, and we obtain:

$$S_{Cze}(P, Q) = \sum_{i=1}^{n} min(p_i, q_i).$$

It is clear that this function is a similarity function satisfying orthogonality property. It defines by (3) the following orthogonal complementary correlation function:

$$A_{Cze}(P, Q) = 2 \sum_{i=1}^{n} min(p_i, q_i) - 1.$$

The Ruzicka similarity measure [6] is given by:

$$S_{Ruz}(P, Q) = \frac{\sum_{i=1}^{n} min(p_i, q_i)}{\sum_{i=1}^{n} max(p_i, q_i)}.$$

It is easy to show that this function is an orthogonal similarity function. Using (3), we obtain an orthogonal complementary correlation function:

$$A_{Ruz}(P, Q) = 2\frac{\sum_{i=1}^{n} min(p_i, q_i)}{\sum_{i=1}^{n} max(p_i, q_i)} - 1.$$

The Jaccard similarity measure is defined in [6] as follows:

$$S_{Jac}(P, Q) = \frac{\sum_{i=1}^{n} p_i q_i}{\sum_{i=1}^{n} p_i^2 + \sum_{i=1}^{n} q_i^2 - \sum_{i=1}^{n} p_i q_i}.$$

It is clear that this function is symmetric and reflexive. Let us show that it takes values in [0,1]. The denominator is non-negative:

$$\sum_{i=1}^{n} p_i^2 + \sum_{i=1}^{n} q_i^2 - \sum_{i=1}^{n} p_i q_i \geq \sum_{i=1}^{n} p_i^2 + \sum_{i=1}^{n} q_i^2 - 2\sum_{i=1}^{n} p_i q_i$$

$$= \sum_{i=1}^{n} (p_i^2 + q_i^2 - 2p_i q_i) = \sum_{i=1}^{n} (p_i - q_i)^2 \geq 0.$$

Since the numerator is also non-negative, we obtain that $S_{Jac}(P, Q) \geq 0$. From the transformations above, we have:

$$\sum_{i=1}^{n} p_i^2 + \sum_{i=1}^{n} q_i^2 - 2\sum_{i=1}^{n} p_i q_i \geq 0,$$

hence:

$$\sum_{i=1}^{n} p_i^2 + \sum_{i=1}^{n} q_i^2 - \sum_{i=1}^{n} p_i q_i \geq \sum_{i=1}^{n} p_i q_i,$$

and $S_{Jac}(P, Q) \leq 1$, because the numerator of the similarity measure is less or equal to the denominator. Hence Jaccard similarity measure takes values in [0,1]. Applying (3), we obtain the correlation function complementary to the Jaccard similarity measure:

$$A_{Jac}(P, Q) = 2\frac{\sum_{i=1}^{n} p_i q_i}{\sum_{i=1}^{n} p_i^2 + \sum_{i=1}^{n} q_i^2 - \sum_{i=1}^{n} p_i q_i} - 1$$

$$= \frac{2\sum_{i=1}^{n} p_i q_i - \sum_{i=1}^{n} p_i^2 - \sum_{i=1}^{n} q_i^2 + \sum_{i=1}^{n} p_i q_i}{\sum_{i=1}^{n} p_i^2 + \sum_{i=1}^{n} q_i^2 - \sum_{i=1}^{n} p_i q_i}$$

$$= \frac{\sum_{i=1}^{n} p_i q_i - \sum_{i=1}^{n} (p_i - q_i)^2}{\sum_{i=1}^{n} p_i q_i + \sum_{i=1}^{n} (p_i - q_i)^2}.$$

It is easy to see that the Jaccard similarity measure and the complementary correlation function are orthogonal functions.

The following function is called the *Dice similarity measure* [6]:

$$S_{Dice}(P, Q) = \frac{2 \sum_{i=1}^{n} p_i q_i}{\sum_{i=1}^{n} p_i^2 + \sum_{i=1}^{n} q_i^2}$$

This function is symmetric and reflexive. As for the Jaccard similarity measure, we can show that it takes values in [0,1]. Using (3) obtain complementary correlation function:

$$A_{Dice}(P, Q) = 2 \frac{2 \sum_{i=1}^{n} p_i q_i}{\sum_{i=1}^{n} p_i^2 + \sum_{i=1}^{n} q_i^2} - 1 = \frac{4 \sum_{i=1}^{n} p_i q_i - \sum_{i=1}^{n} p_i^2 - \sum_{i=1}^{n} q_i^2}{\sum_{i=1}^{n} p_i^2 + \sum_{i=1}^{n} q_i^2}$$

$$= \frac{2 \sum_{i=1}^{n} p_i q_i - \sum_{i=1}^{n} (p_i - q_i)^2}{2 \sum_{i=1}^{n} p_i q_i + \sum_{i=1}^{n} (p_i - q_i)^2}.$$

The Dice similarity function and complementary correlation function satisfy the corresponding orthogonality properties.

4 Constructing Correlation Functions from Dissimilarity Functions

The construction of correlation functions from dissimilarity functions can be done similarly as in the previous section. Only we need to use an Eq. (4). One thing, looking for suitable dissimilarity functions we found that they are complementary to similarity functions considered before although they have another names. In such cases the complementary correlation function will be the same as we constructed from similarity functions. Consider an example.

Sorensen distance is formulated in [6] as follows:

$$D_{Sor}(P, Q) = \frac{\sum_{i=1}^{n} |p_i - q_i|}{\sum_{i=1}^{n} (p_i + q_i)}.$$

Let us show that it is dual (complementary) to Czekanowski similarity function considered in the previous section:

$$S_{Cze}(P, Q) = \frac{2 \sum_{i=1}^{n} min(p_i, q_i)}{\sum_{i=1}^{n} (p_i + q_i)},$$

that is, they are related by (1) as follows:

$$D_{Sor}(P, Q) = 1 - S_{Cze}(P, Q).$$

Consider:

$$1 - S_{Cze}(P, Q) = 1 - \frac{2 \sum_{i=1}^{n} min(p_i, q_i)}{\sum_{i=1}^{n} (p_i + q_i)} = \frac{\sum_{i=1}^{n} (p_i + q_i) - 2 \sum_{i=1}^{n} min(p_i, q_i)}{\sum_{i=1}^{n} (p_i + q_i)} =$$
$$= \frac{\sum_{i=1}^{n} (p_i + q_i - 2min(p_i, q_i))}{\sum_{i=1}^{n} (p_i + q_i)}.$$

Suppose that $p_i \geq q_i$ then

$$p_i + q_i - 2min(p_i, q_i) = p_i + q_i - 2q_i = p_i - q_i = |p_i - q_i|$$

Similarly, if $q_i \geq p_i$, obtain: $p_i + q_i - 2min(p_i, q_i) = |q_i - p_i| = |p_i - q_i|$. Finally, obtain:

$$1 - S_{Cze}(P, Q) = \frac{\sum_{i=1}^{n} (p_i + q_i - 2min(p_i, q_i))}{\sum_{i=1}^{n} (p_i + q_i)} = \frac{\sum_{i=1}^{n} |p_i - q_i|}{\sum_{i=1}^{n} (p_i + q_i)} = D_{Sor}(P, Q).$$

Since Czekanowski similarity function is complementary to Sorensen distance, both have the same complementary correlation function:

$$A_{Sor}(P, Q) = A_{Cze}(P, Q) = 2 \sum_{i=1}^{n} min(p_i, q_i) - 1.$$

The Czekanowski similarity function satisfies an orthogonality property and is complementary to Sorensen distance. Hence, the Sorensen distance is a dissimilarity function and it satisfies the orthogonality property.

Similarly, we can show that some dissimilarity functions considered in [6] are complementary to some of the similarity functions considered here and will generate correlation functions obtained above.

5 Conclusion

The paper introduces the correlation functions defined on the set of probability distributions. These functions are constructed using the known similarity measures of probability distributions. We selected similarity measures satisfying some simple properties defining similarity functions used for constructing correlation functions. All these functions can be used for the analysis of relationships between probability distributions and relative frequency distributions that often appear in data analytics.

Acknowledgements The work was done with partial support from the Mexican Government through grant A1-S-47854 of CONACYT, Mexico, and grants SIP 20220857, 20211874, 20220852, and 20220859 of the Instituto Politecnico Nacional, Mexico. The authors acknowledge the support of Microsoft through the Microsoft Latin America Ph.D. Award.

References

1. Clifford, H.T., Stephenson, W.: An introduction to numerical classification. Academic Press, New York (1975)
2. Tan, P.N., Kumar, V., Srivastava, J.: Selecting the right interestingness measure for association patterns. In: 8th ACM SIGKDD International conference on knowledge discovery and data mining, pp. 32–41 (2002)
3. Chen, P.Y., Popovich, P.M.: Correlation: Parametric and nonparametric measures. Sage, Thousand Oaks, CA (2002)
4. Gibbons, J.D., Chakraborti, S.: Nonparametric statistical inference, 4th edn. Dekker, New York (2003)
5. Mata, A.D.: Estadística aplicada a la administración y la economía. McGraw-Hill Interamericana, Mexico (2013)
6. Cha, S.H.: Comprehensive survey on distance/similarity measures between probability density functions. Int. J. Math. Model. Methods Appl. Sci. 1(4), 300–307 (2016)
7. Batyrshin, I.Z.: Contracting and involutive negations of probability distributions. Mathematics 9(19), 2389 (2021)
8. Batyrshin, I.Z., Kubysheva, N.I., Bayrasheva, V.R., Kosheleva, O., Kreinovich, V.: Negations of probability distributions: a survey. Computación y Sistemas 25(4), 775–781 (2021)
9. Koralov, L., Sinai, Y.: Theory of probability and random processes. Springer, Berlin (2007)
10. Batyrshin, I.Z.: Data science: Similarity, dissimilarity and correlation functions. In: Artificial intelligence, pp. 13–28. Springer, Cham (2019)
11. Batyrshin, I.Z.: Constructing correlation coefficients from similarity and dissimilarity functions. Acta Polytech. Hung. 16(10), 191–204 (2019)
12. Batyrshin, I.: Towards a general theory of similarity and association measures: similarity, dissimilarity, and correlation functions. J. Intell. Fuzzy Syst. 36(4), 2977–3004 (2019)

Fuzzy Model and Z-Information

Clustering of the State of Plant Species in the Urban Environment Under Z-Information

Olga Poleshchuk◉ and Evgeny Komarov

Abstract The paper developed a model for clustering the state of plant species in an urban environment using Z-information. The purpose of the paper is to analyze and identify groups of plant species that are most adapted to the harmful effects of large cities. This approach allows you to plan urban greening activities and reduce the costs associated with re-landscaping due to plant death. The state of plant species is assessed by experts with a certain degree of reliability, which must be taken into account in order to avoid the risk of expert errors. To formalize expert assessments of plants, the authors use Z-numbers, the components of which are the values of semantic spaces. The assessment of the state of each plant species is presented in the form of a set of Z-numbers. For a comparative analysis of the states of different plant species, indicators of difference and similarity are determined based on the distances between sets of Z-numbers. Pairwise similarity indicators are used to construct a fuzzy binary similarity relation on the set of states of plant species, which is transitively closed. The resulting fuzzy binary similarity relation is decomposed into equivalence relations and breaks down the states of plant species into clusters of similar ones with a certain level of reliability. A cluster analysis of the state of eight plant species growing in Moscow was carried out.

Keywords Expert information · Clustering · Plant species

1 Introduction

Urban plantings are part of urbanized areas that meet the needs of urban residents for environmentally sound improvement of the quality of the environment, health support and recreation.

Plant species adapt in different ways to the harmful effects of the climatic conditions of large cities. Based on this, they perform their functions in different ways for

O. Poleshchuk (✉) · E. Komarov
Bauman Moscow State Technical University, Moscow, Russia
e-mail: olga.m.pol@yandex.ru

© The Author(s), under exclusive license to Springer Nature Switzerland AG 2023   261
Sh. N. Shahbazova et al. (eds.), *Recent Developments and the New Directions of Research, Foundations, and Applications*, Studies in Fuzziness and Soft Computing 422,
https://doi.org/10.1007/978-3-031-20153-0_20

reduction of surface water runoff, purifying the air, dust-filtering capacity, carbon deposition, increasing the concentration of phytoncides in the air, protecting from heat, regulating air and soil moisture, creating recreation areas, etc. To assess the state of plants, not only physical instruments are used, but also experts are involved, whose assessments cannot be absolutely reliable. The possibility of taking into account the degree of reliability of expert assessments appeared in 2011 after the determination of the Z-number by Professor Zadeh [1]. To process information with Z-numbers (Z-information) over the past ten years, a mathematical apparatus has been developed that allows you to operate with Z-numbers, rank, compare and predict them. In papers [2–5], operations on Z-numbers are developed. Methods for ranking Z-numbers are developed in [6, 7].

In [4–7], the distances between Z-numbers are determined in different ways. In [4], the distance between Z-numbers is determined based on the aggregated Z-number segments. In [5], the authors defuzzify the second components of Z-numbers and use them to determine the weighted values of the first components, and then determine the distance, as for ordinary fuzzy numbers. In [6], to determine the distance between Z-numbers, both components of the Z-numbers and the first weighted components are used. In [7], the distance between Z-numbers is determined using the Jaccard similarity measure.

In [8–11] the developed methods of Z-information processing are used to support decision making. In [12–15], methods have been developed that allow one to formalize both components of Z-numbers and represent them as values of linguistic variables with the properties of completeness and orthogonality. Developed methods have been applied in [16] to determine the ratings of plant species under Z-information. Such an approach to finding rating points, taking into account the reliability of expert information, reduces the risk of errors when making decisions on the final results obtained.

For cluster analysis of fuzzy data, enough algorithms have been developed, among which the most famous and frequently used fuzzy clustering algorithm c-means [17]. In [18] this algorithm is combined with a genetic algorithm, in [19, 20] with an optimization algorithm for a swarm of particles [21]. In [22, 23], clustering is carried out based on fuzzy relations. In [24], a clustering procedure based on the transitive closure of the similarity relation is proposed. In [25], a Z-information clustering algorithm is described, which combines the c-means fuzzy clustering algorithm and the conversion of Z-numbers to fuzzy numbers. In [26], the foundations of the bimodal clustering problem are investigated, and an approach based on the connection between Type-II fuzzy sets and Z-numbers proved in [27] is proposed. In [28], an algorithm was developed for clustering Z-information using a transitive fuzzy binary relation. Z-information formalizes expert criteria of quality characteristics evaluations of objects.

The paper solves the problem of clustering the states of plant species growing in the climatic conditions of large cities, based on expert information that comes from experts with a certain level of reliability. The construction of a clustering model is carried out using theoretical developments [28]. The paper contains four sections. Section 2 includes the necessary basic concepts and definitions. Section 3 presents the model for clustering of plant species in urban environment under Z-information. Section 4 concludes this paper.

2   Basic Concepts and Definitions

$\{X, T(X), U, V, S\}$-is a linguistic variable with name X and names of values $T(X) = \{X_l, l = \overline{1, m}\}$ (term-set), which are determined by the rule V. The values of linguistic variable are fuzzy variables. A rule S defines for each fuzzy variable the corresponding fuzzy set of U[17].

A full orthogonal semantic space X is a linguistic variable with a fixed term-set and the following properties of continuous membership functions $\mu_l(x), l = \overline{1, m}$: $U_l = \{x \in U : \mu_l(x) = 1\} \forall l = \overline{1, m}$ is a point ore an interval; $\mu_l(x), l = \overline{1, m}$ does not increase and does not decrease correspondingly to the right and to the left of U_l; $\sum_{i=1}^{m} \mu_l(x) = 1, \forall x \in U$[14].

To formalize the values of a full orthogonal semantic space, a method based on statistical information was developed in [13, 15], which is based on the frequency approach and geometric probabilities. In the paper that method is used for formalization of expert assessments of the states of plant species. Trapezoidal fuzzy numbers that formalize the values of the full orthogonal semantic spaces are determined by four parameters, namely: the first two parameters are the abscissas of the left and right vertices of the upper base of the trapezoid, and the last two parameters are the lengths of the left and right wings of the trapezoid. Triangular fuzzy numbers are defined by three parameters, namely: the first parameter is the abscissa of the apex of the triangle, and the last two parameters are the lengths of the left and right wings of the triangle.

In [11], for a trapezoidal fuzzy number $\tilde{A} = (a_1, a_2, a_L, a_R)$ an aggregating indicator ϑ is determined:

$$\vartheta = \frac{a_1 + a_2}{2} + \frac{(a_R - a_L)}{12} \tag{1}$$

A Z–number is a pair of two fuzzy numbers $Z = \left(\tilde{A}, \tilde{R}\right)$, where $\tilde{A}$ is the value of a fuzzy variable, $\tilde{R}$ is a fuzzy value of the reliability of $\tilde{A}$ [1].

In [11], the distance between two Z–numbers $Z_1 = \left(\tilde{A}_1, \tilde{R}_1\right)$, $Z_2 = \left(\tilde{A}_2, \tilde{R}_2\right)$ with aggregating indicators $(a_1, r_1), (a_2, r_2)$ of fuzzy numbers $\tilde{A}_1, \tilde{R}_1$ and $\tilde{A}_2, \tilde{R}_2$ found by formula (1) is determined:

$$d(Z_1, Z_2) = \sqrt{\frac{1}{3}(a_1 - a_2)^2 + \frac{1}{3}(a_1 r_1 - a_2 r_2)^2 + \frac{1}{3}(r_1 - r_2)^2} \tag{2}$$

In the paper, formula (2) is used to construct a fuzzy binary similarity relation on the set of states of plant species.

3　Clustering of the State of Plant Species

To assess the state of plant species, experts use a scale A_l, $l = \overline{1,7}$: A_1–"old dead", A_2–"recently died", A_3–"drying up", A_4–"very weak", A_5–"mean weak", A_6–"moderately weak", A_7–"no signs of weakness" and to assess the reliability of information, experts use a scale R_k, $k = \overline{1,5}$:R_1–"Unlikely", R_2–"Not very likely", R_3–"Likely", R_4–"Very likely", R_5–"Extremely likely".

The scales A_l, $l = \overline{1,7}$, R_k, $k = \overline{1,5}$ are formalized using full orthogonal semantic spaces according to the method [13, 15]. The state of the i-th plant species is represented as a full orthogonal semantic space with fuzzy numbers $\tilde{A}_{il}$, $l = \overline{1,7}$, $i = \overline{1,n}$ and membership functions μ_{il}, $l = \overline{1,7}$, $i = \overline{1,n}$. A full orthogonal semantic space R with name "Reliability of information" is represented by fuzzy numbers $\tilde{R}_1 = (0, 0, 0.25)$, $\tilde{R}_2 = (0.25, 0.25, 0.25)$, $\tilde{R}_3 = (0.5, 0.25, 0.25)$, $\tilde{R}_4 = (0.75, 0.25, 0.25)$, $\tilde{R}_5 = (1, 0.25, 0)$, which correspond to the values R_1–"Unlikely", R_2–"Not very likely", R_3–"Likely", R_4–"Very likely", R_4–"Extremely likely".

Then the state of the i-th plant species taking into account the reliability of expert information is presented as a set of Z-numbers $Z_{il} = \left(\tilde{A}_{il}, \tilde{R}_{il} \right)$, $l = \overline{1,7}$, $i = \overline{1,n}$, where $\tilde{R}_{il} = (r_{il1}, r_{il2}, r_{il3})$, $l = \overline{1,7}$, $i = \overline{1,n}$ equals to one of fuzzy number $\tilde{R}_k$, $k = \overline{1,5}$ $r_{il1} = \max_m (r_{mil1})$, $_{il2} = \max_m (r_{mil2})$, $r_{il3} = \max_m (r_{mil3})$, if an expert assessed M plants of the i-th species by the level A_l, $l = \overline{1,7}$ with reliability $\tilde{R}_{iml} = (r_{iml1}, r_{iml2}, r_{iml3})$, $i = \overline{1,n}$, $m = \overline{1,M}$, $l = \overline{1,7}$ [15].

Using formula (2), we determine the distance between the states of the i-th and j-th plant species by the formula:

$$d_{ij} = \frac{1}{7} \sum_{l=1}^{7} \sqrt{\left[\frac{1}{3}(a_{ij} - a_{jl})^2 + \frac{1}{3}(a_{il}r_{il} - a_{jl}r_{jl})^2 + \frac{1}{3}(r_{il} - r_{jl})^2 \right]} \quad (3)$$

where $a_{il}, r_{il}, a_{jl}, r_{jl}$, $i = \overline{1,n}$, $j = \overline{1,n}$,$l = \overline{1,7}$ are aggregating indicators of fuzzy numbers $\tilde{A}_{il}, \tilde{R}_{il}, \tilde{A}_{jl}, \tilde{R}_{jl}$, $i = \overline{1,n}$, $j = \overline{1,n}$,$l = \overline{1,7}$, obtained by the formula (1).

Let $d_{max} = \max_{i,j} d_{ij}$, $i = \overline{1,n}$, $j = \overline{1,n}$. We define for states of plant species $Z_i = \left(\tilde{A}_{il}, \tilde{R}_{il} \right)$, $l = \overline{1,m}$, $i = \overline{1,k}$, a difference index δ_{ij}, $i = \overline{1,n}$, $j = \overline{1,n}$ and a similarity index ρ_{ij}, $i = \overline{1,n}$, $j = \overline{1,n}$:

$$\delta_{ij} = \frac{d_{ij}}{d_{max}}, i = \overline{1,n}, j = \overline{1,n} \quad (4)$$

$$\rho_{ij} = 1 - \delta_{ij}, i = \overline{1,n}, j = \overline{1,n}. \quad (5)$$

In [28] was proved that $\tilde{\Xi}$ with membership function $\mu_\Psi(Z_i, Z_j) = \rho_{ij}$ defines a fuzzy binary relation of similarity on the set $Z_i = \left(\left(\tilde{B}_{il}, \tilde{R}_{il} \right), l = \overline{1,m} \right)$, $i = \overline{1,n}$.

Using $\tilde{\Xi}$, its transitive closure was constructed and used for cluster analysis of $Z_i = \left(\left(\tilde{B}_{il}, \tilde{R}_{il} \right), l = \overline{1,m} \right), i = \overline{1,n}$.

The state of eight plant species (Norway maple, European white elm, Common lilac, Rough-bark poplar, Witch elm, Cottonwood, Large-leaved linden, European ash) growing in the Moscow was evaluated by experts. The results obtained for the relative number of plants $\left(\frac{k_i}{K}, i = \overline{1,7} \right)$, evaluated within each of the seven levels of the scale, as well as the reliability R of this information are entered in the Tables 1 and 2.

According to Tables 1 and 2, we construct full orthogonal semantic spaces with values $\tilde{A}_{il}, l = \overline{1,7}, i = \overline{1,8}$. For example, we have for the "Witch elm": $\tilde{A}_{51} = (0, 0, 0)$, $\tilde{A}_{52} = (0, 0.025, 0, 0.03)$, $\tilde{A}_{53} = (0.055, 0.03, 0.03)$, $\tilde{A}_{54} = (0.085, 0.42, 0.03, 0.3)$, $\tilde{A}_{55} = (0.72, 0.845, 0.3, 0.05)$, $\tilde{A}_{56} = (0.895, 0.05, 0.05)$, $\tilde{A}_{57} = (0.895, 0.05, 0.05)$.

Aggregating indicators for all fuzzy numbers using formula (1) and pairwise distances using formula (2) were calculated. For example, we have for "Witch elm" the obtained results presented in the Table 3.

Table 1 Frequency and reliability of plant species

№ Plant species	1. Norway maple	2. European white elm	3. Common lilac	4. Rough-bark poplar
$\frac{k_1}{K}, R$	0.06, R_5	0, R_5	0.01, R_5	0, R_5
$\frac{k_2}{K}, R$	0.05, R_3	0.08, R_3	0.02, R_2	0.16, R_3
$\frac{k_3}{K}, R$	0.03, R_4	0.05, R_4	0.06, R_5	0.04, R_5
$\frac{k_4}{K}, R$	0.09, R_4	0.19, R_4	0.23, R_3	0.3, R_3
$\frac{k_5}{K}, R$	0.46, R_5	0.36, R_5	0.41, R_4	0.4, R_4
$\frac{k_6}{K}, R$	0.24, R_5	0.27, R_4	0.26, R_3	0.07, R_4
$\frac{k_7}{K}, R$	0.07, R_5	0.05, R_5	0.01, R_4	0.03, R_4

Table 2 Frequency and reliability of plant species

No Plant species	5. Witch elm	6. Cottonwood	7. Large-leaved linden	8. European ash,
$\frac{k_1}{K}, R$	0, R_5	0,02, R_5	0, R_5	0, R_5
$\frac{k_2}{K}, R$	0.04, R_2	0, R_3	0.08, R_2	0.02, R_3
$\frac{k_3}{K}, R$	0.03, R_4	0.06, R_5	0.06, R_4	0.05, R_4
$\frac{k_4}{K}, R$	0.5, R_5	0.23, R_3	0.22, R_3	0.5, R_5
$\frac{k_5}{K}, R$	0.3, R_5	0.41, R_4	0.36, R_4	0.3, R_4
$\frac{k_6}{K}, R$	0.05, R_5	0.26, R_3	0.26, R_4	0.06, R_5
$\frac{k_7}{K}, R$	0.08, R_5	0.01, R_4	0.02, R_5	0.07, R_5

Table 3 Aggregating indicators for "Witch elm"

a_{5l}	r_{5l}	$a_{5l}r_{5l}$
0	0,979	0
0,015	0,250	0,004
0,055	0,750	0,041
0,275	0,979	0,269
0,799	0,979	0,782
0,895	0,979	0,876
0,930	0,979	0,910

Table 4 Clustering of the eight plant species

α-level	Plant species
0.512	{1, 2, 3, 4, 5, 6, 7, 8}
0.684	{1,4}, {2, 3, 5, 6, 7, 8}
0.813	{1}, {4}, {2, 8}, {3, 5, 6, 7}
0.844	{1}, {4}, {2}, {8}, {3, 5, 6, 7}
0.901	{1}, {4}, {2}, {8}, {3}, {5, 6, 7}
0.964	{1}, {4}, {2}, {8}, {3}, {6} {5, 7}
1	{1}, {4}, {2}, {8}, {3}, {6} {5}, {7}

The calculated pairwise distances made it possible to find the matrix of fuzzy binary relation of similarity and its transitive closure on the set of state of eight plant species, then cluster all plant states into groups of similar. The results obtained are presented in the Table 4.

From the results obtained, it can be concluded that all the plant species, except for plant species "Norway maple" and "Rough-bark poplar", have more similar states. This suggests that most of the plants of all the species belong to the same categories of state. It is assumed that the states of plant species "Norway maple" and "Rough-bark poplar" better than the rest of the species or worse. It is unambiguously determined from Table 1 that the state of these species cannot be the best, since the largest number of dead plants are in the plant species "Norway maple". One can make a preliminary conclusion that the state of these species is worse than the state of all other species. The states of species "European white elm" and "European ash" differ from species "Common lilac", "Rough-bark poplar", "Witch elm", "Cottonwood", "Large-leaved linden". It can be assumed that species "European white elm" and "European ash" have the best states of all the eight species. If we turn to Table 1 again, we can see that these species do not have old dead plants and most of the plants are rated by levels A_4–"very weak", A_5–"mean weak", A_6–"moderately weak", A_7–"no signs of weakness". The most similar are the states of plant species "Common lilac", "Witch elm", "Cottonwood", "Large-leaved linden" with an average state.

Table 5 Rating points of the eight plant species

Plant species	Rating points ρ_i
Norway maple	0,414
European white elm	0,608
Common lilac	0,517
Rough-bark poplar	0,428
Witch elm	0,524
Cottonwood	0,529
Large-leaved linden	0,538
European ash,	0,572

To confirm or refute the assumptions made, it was customary to calculate the ratings of plant species based on a comparative analysis with the ideal state of plants. We define the ideal state of a plant species as a set of seven Z-numbers $\left(\tilde{A}_l^{id}, \tilde{R}_l^{id}\right), l = \overline{1,7}$ where $\tilde{A}_1^{id} = \tilde{A}_2^{id} = \tilde{A}_3^{id} = \tilde{A}_4^{id} = \tilde{A}_5^{id} = \tilde{A}_6^{id} = (0,0,0), \tilde{A}_7^{id} = (0,1,0,0), \tilde{R}_l^{id} = \tilde{R}_5, l = \overline{1,7}$.

Rating points of the i-th plant species we determine by the formula:

$$\rho_i = 1 - \frac{1}{7}\sum_{i=1}^{7}\sqrt{\frac{1}{3}\left(a_{il} - a_l^{id}\right)^2 + \frac{1}{3}\left(a_{il}r_{il} - a_l^{id}r_l^{id}\right)^2 + \frac{1}{3}\left(r_{il} - r_l^{id}\right)^2} \quad (6)$$

where $a_{il}, r_{il}, i = \overline{1,8}, l = \overline{1,7}$ are aggregating indicators of fuzzy numbers $\tilde{A}_{il}, \tilde{R}_{il}, i = \overline{1,8}, l = \overline{1,7}$ and $a_l^{id}, r_l^{id}, l = \overline{1,7}$ are aggregating indicators of fuzzy numbers $\tilde{A}_l^{id}, l = \overline{1,7}$.

Using formula (6), we get the results listed in Table 5.

The results of Table 5 confirm the obtained results of the clustering of plant species. The best in the state are the species "European white elm" and "European ash", the worst in the state are the species "Norway maple" and "Rough-bark poplar". The most similar species are "Common lilac", "Witch elm", "Cottonwood", "Large-leaved linden". In paper [16], ratings of these plant species are also found. The definition of the distance between the states of plant species and the ideal state used here is different from the definition of distance in [16]. The distance used in [16] is less sensitive to the assessment of the reliability of information, since the reliability estimates are not used explicitly, but are used indirectly in the convolution with the first components of Z-numbers. However, the results obtained in the paper [16] and this paper are not contradictory. In both papers, the states of plant species "European white elm" and "European ash" are the best, but in ratings they change positions. The worst are the states of plant species "Norway maple" and "Rough-bark poplar", but in the ratings they also change positions. In [16], the plant species "Common lilac", "Witch elm", "Cottonwood", "Large-leaved linden" are the closest in terms of ratings. A similar result was obtained in this paper. The results of clustering the states of these plant species confirm the objectivity of the rating results, which means they allow including the best plant species in landscaping plans, thereby avoiding unjustified material costs, and receiving a full package of environmental services from these species for urban residents.

4 Conclusions

The purpose of this article was a cluster analysis of the state of plant species growing in megacities to identify the species most adapted to the harmful effects of the environment. It is advisable to include species with the best plant states in city plans for landscaping, reducing the costs associated with plant death and residents to receive from these species a full package of environmental services (softening the urban climate, creating recreation areas, reduction of surface water runoff, purifying the air, etc.).

The states of plant species are formalized using Z-numbers, the second component of which is the reliability of expert assessments, which reduces the risk of errors. Both components of the Z-numbers are the values of complete orthogonal semantic spaces. The state of each plant species is represented as a set of Z-numbers. For a comparative analysis of the states of different plant species, the distance is determined between them, in which the aggregating indicators of both components of the Z-numbers are considered, as well as their product. Based on pairwise distances between the states of plant species, indicators of difference and similarity of states are determined. Pairwise similarity indicators are used to construct a fuzzy binary similarity relation on the set of states of plant species, which is transitively closed. The resulting fuzzy binary similarity relation is decomposed into equivalence relations and breaks the states of plant species into clusters of similar ones with a certain level of reliability. The constructed model is applied for cluster analysis of the state of eight plant species growing in Moscow.

The results obtained in this case were confirmed using rating estimates, which were obtained based on a comparative analysis of the states of plant species with an ideal state. Thus, the model developed in the paper has shown its adequacy and effectiveness in a real practical task.

References

1. Zadeh, L.A.: A note on Z-numbers. Inform. Sci. **14**(181), 2923–2932 (2011). https://doi.org/10.1016/j.ins.2011.02.022
2. Aliev, R.A., Alizadeh, A.V., Huseynov, O.H.: The arithmetic of discrete Z-numbers. Inform. Sci. **1**(290), 134–155 (2015). https://doi.org/10.1016/j.ins.2014.08.024
3. Aliev, R.A., Huseynov, O.H., Zeinalova, L.M.: The arithmetic of continuous Z-numbers. Inform. Sci. **373**, 441–460 (2016). https://doi.org/10.1016/j.ins.2016.08.078

4. Poleshchuk, O.M.: Novel approach to multicriteria decision making under Z-information. In: Proceedings of the International Russian Automation Conference, (RusAutoCon-2019), 8867607 (2019). https://doi.org/10.1109/RUSAUTOCON.2019.8867607
5. Kang, B., Wei, D., Li, Y., Deng, Y.: A method of converting Z-number to classical fuzzy number. J. Inform. Comp. Sci. **9**(3), 703–709 (2012)
6. Wang, F., Mao, J.: Approach to multicriteria group decision making with Z-numbers based on topsis and power aggregation operators. Math. Probl. Eng. 1–18 (2019)
7. Aliyev, R.R., Talal Mraizid, D.A., Huseynov, O.H.: Expected utility based decision making under Z- information and its application. Comput. Intell. Neurosci. **3**, 364512 (2015). https://doi.org/10.1155/2015/364512
8. Aliev, R.A., Zeinalova, L.M.: Decision-making under Z-information. In: Human-centric Decision-Making Models for Social Sciences, pp. 233–252. (2013)
9. Kang, B., Wei, D., Li, Y., Deng, Y.: Decision making using Z-numbers under uncertain environment. J. Inform. Comp. Sci. **7**(8), 2807–2814 (2012)
10. Aliev, R.K., Huseynov, O.H., Aliyeva, K.R.: Aggregation of an expert group opinion under Z-information. In: Proceedings of the Eighth International Conference on Soft Computing, Computing with Words and Perceptions in System Analysis, Decision and Control, (ICSCW-2015), pp. 115–124. (2015)
11. Poleshchuk, O.M.: Object monitoring under Z-information based on rating point. Adv. Intell. Syst. Comp. **1197**, 1191–1198 (2021). https://doi.org/10.1007/978-3-030-51156-2_139
12. Zadeh, L.A.: The Concept of a linguistic variable and its application to approximate reasoning. Inform. Sci. **8**, 199–249 (1975). https://doi.org/10.1016/0020-0255(75)90036-5
13. Poleshchuk, O.M.: Creation of linguistic scales for expert evaluation of parameters of complex objects based on semantic scopes. In: Proceedings of the International Russian Automation Conference, (RusAutoCon–2018), 8501686. (2018). https://doi.org/10.1109/RUSAUTOCON.2018.8501686
14. Ryjov, A.P.: The concept of a full orthogonal semantic scope and the measuring of semantic uncertainty. In: Proceedings of the Fifth International Conference Information Processing and Management of Uncertainty in Knowledge-Based Systems, pp. 33–34. (1994).
15. Poleshchuk, O., Komarov, E.: The determination of rating points of objects with qualitative characteristics and their usage in decision making problems. Int. J. Comput. Math. Sci. **3**(7), 360–364 (2009)
16. Poleshchuk, O.M.: Monitoring stability of plant species to harmful urban environment under Z-information. Lect. Notes Netw. Syst. **308**, 863–870 (2022)
17. Bezdek, J.C.: Selected applications in classifier design. In: Pattern Recognition With Fuzzy Objective Function Algorithms, vol. 2, pp. 203 239
18. Bezdek, J.C., Hathaway, R.J.: Optimization of fuzzy clustering criteria using genetic algorithms. In: Proceedings of the IEEE Conf. on Evolutionary Computation, vol. 2, pp. 589–594. (1994)
19. Runkler, T.A., Katz, C.: Fuzzy clustering by particle swarm optimization. In: Proceedings of the IEEE Conf. on Fuzzy Systems, pp. 34–41. (2006)
20. Liu, H.C., Yih, J.M., Wu, D.B., Liu, S.W.: Fuzzy C-mean clustering algorithms based on Picard iteration and particle swarm optimization. In: Proceedings of the Int. Workshop on Geoscience and Remote Sensing (ETT and GRS-2008), vol. 2, pp. 75–84. (2008)
21. Kennedy, J., Eberhart, R.C.: Particle swarm optimization. In: Proceedings of the IEEE International Joint Conference on Neural Networks, vol. 4, pp. 1942–1948. (1995)
22. Ruspini, E.H.: A new approach to clustering. Inform. Control **15**, 22–32 (1969)
23. Ruspini, E.H.: Numerical methods for fuzzy clustering. Inform. Sci. **2**, 319–350 (1970)
24. Tamura, S., Higuchi, S., Tanaka, K.: Pattern classification based on fuzzy relations. IEEE Trans. Syst. Man Cybern. **1**, 61–66 (1971)
25. Jamal, M., Khalif, K., Mohamad, S.: The implementation of Z-numbers in fuzzy clustering algorithm for wellness of chronic kidney disease patients. J. Phys. Conf. Ser. **1366**, 012058 (2018)

26. Aliev, R.A., Pedrycz, W., Guirimov, B.G., Huseynov, O.H.: Clustering method for production of Z-numbers based if-then rules. Inform. Sci. **520**, 155–176 (2020)
27. Aliev, R., Guirimov, B.: Z-number clustering based on general Type-II fuzzy set. Adv. Intell. Syst. Comput. **896**, 270–278 (2018)
28. Poleshchuk, O.M.: Clustering Z-information based on semantic spaces. Lect. Notes Netw. Syst. **308**, 888–894 (2022)
29. Averkin, A.N., Batyrshin, I.Z., Blishun, A.F., Tarasov, V.B.: Fuzzy Sets in Models of Control and Artificial Intelligence. M: Nauka, Moscow (1986)

Multi-criteria Analysis of the Objects State Under Z-Information

Olga Poleshchuk

Abstract The paper developed a model for multi-criteria analysis of the objects state under Z-information. The scales used to evaluate the characteristics of objects are formalized using linguistic variables on the universal set [0,1]. Evaluations of objects are fuzzy numbers-values of linguistic variables, given with a certain level of reliability. The states of objects represented as a set of Z-numbers (the number of Z-numbers is equal to the number of evaluated characteristics) are analyzed using the ideal state of objects, which is also represented as a set of Z-numbers. For the components of Z-numbers that formalize the states of objects, aggregating indicators are defined, which are used to determine the distances between the real and ideal states of objects. When determining the distances, the indicators of both components are used, as well as their product. Based on these distances, the ratings of objects are determined for all the evaluated characteristics. The rating evaluation of objects is a special case of Z-number, the first component of which is an ordinary number, and the second component is a fuzzy evaluation of the reliability of this number. To analyze the states of objects in different periods of time, the first components of the rating Z-numbers are compared (larger number, higher rating), if they are equal, then their second components are compared (higher reliability, higher rating).

Keywords Z-number · Z-information · Multi-criteria analysis

1 Introduction

When assessing the states of objects, experts are often involved who, based on their experience and knowledge, evaluate the qualitative characteristics of objects or describe quantitative characteristics within their professional language.

O. Poleshchuk (✉)
Bauman Moscow State Technical University, Moscow, Russia
e-mail: Olga.m.pol@yandex.ru

© The Author(s), under exclusive license to Springer Nature Switzerland AG 2023
Sh. N. Shahbazova et al. (eds.), *Recent Developments and the New Directions of Research, Foundations, and Applications*, Studies in Fuzziness and Soft Computing 422,
https://doi.org/10.1007/978-3-031-20153-0_21

Even physical devices have an error in evaluations. Therefore, expert evaluations also have non-absolute reliability, which became possible to take into account after the determination of a Z-number by Professor Zadeh in 2011 [1].

The ability to take into account the reliability of expert information plays an essential role in decision-making problems, since it reduces the risk of error, especially in problems where expert information is the only information available.

In order to study the states of objects under multicriteria conditions, it was necessary to operate with Z-numbers, compare, rank and predict them, find the distances between them, and so on.

From the very beginning of the emergence of Z-numbers, it became clear that it is much more complicated to operate on them than just with fuzzy numbers. Naturally, this was due to the appearance of the second component, which is a fuzzy estimate of the reliability of the first component.

There are different approaches to operating on Z- numbers. In papers [2, 3] operations on Z-numbers are carried out on the basis of aggregating indicators of their components. In papers [4, 5] the second components of Z-numbers are considered to be elements of some probability distribution. How to rank Z- numbers is described in the papers [6, 7].

The authors of papers [2, 3, 6, 7] determine the distances between Z-numbers in different ways. In [2], the authors replace Z-numbers with ordinary fuzzy numbers, after which they determine the distances between these fuzzy numbers. Fuzzy numbers are obtained based on the first components of Z-numbers, which are multiplied by ordinary numbers, which are obtained after defuzzification of the second components. In [3], aggregating segments are defined for Z-numbers, which are used to determine the distance between Z-numbers. The authors of [6] use both components of the Z-numbers to determine the distance between the Z-numbers, as well as the first weighted components according to the principle described in [5]. In [7], the authors use the Jaccard similarity measure to determine the distance between Z-numbers.

To formalize expert information given with a certain level of reliability, methods have been developed in [8, 9] for representing it in the form of Z-numbers, the components of which are the values of linguistic variables [10] with special properties, which, as it turned out, make it possible to most accurately model the mental activity of experts in assessment procedures or descriptions of real processes and objects [11, 12].

In papers [13, 14], the authors developed methods for representing group expert information based on Z-numbers. Since in the papers the authors approached the representation of the second components of Z-numbers in different ways, the methods complement each other.

In papers [8, 9] methods were developed for determining the ratings of objects under Z-information, however, the ratings obtained in this case do not allow us to understand the degree of closeness of the object states to the ideal state, but nevertheless allow us to successfully rank objects according to several characteristics.

In paper [15], a particular case of the problem posed in this work is solved. In [15], a method was developed for determining the states of plant species in large cities based on a comparative analysis with their ideal state. The developed method cannot be applied in the general case, since there is a specific assessment of each individual plant of the corresponding species, followed by formalization of the state of this species.

Due to the lack of a general method for multi-criteria analysis of the objects state under Z-information based on an ideal state, the task is to develop such a method.

The paper contains four sections. Section 2 includes the necessary basic concepts and definitions. Section 3 presents the method for multi-criteria analysis of the objects state under Z-information. Section 4 concludes this paper.

2   Basic Concepts and Definitions

A linguistic variable X with name of values $T(X) = \{X_l, l = \overline{1, m}\}$ is called a five $\{X, T(X), U, V, S\}$. The rule V determines the names of the values, the rule S associates names $X_l, l = \overline{1, m}$ to the fuzzy sets of the universal set U [10].

The conducted studies have shown that linguistic variables with certain properties of the membership functions of their values make it possible to obtain models that most adequately reflect real processes and objects.

These properties consist in the fact that for each continuous membership function $\mu_l(x), l = \overline{1, m}$ there is a point or segment for which the value of the function is equal to one, that is: for $\forall \mu_l(x), l = \overline{1, m} \exists U_l = \{x \in U : \mu_l(x) = 1\}$. To the left of $U_l, l = \overline{1, m}$ the function $\mu_l, l = \overline{1, m}$ is increasing and to the right of $U_l, l = \overline{1, m}$ the function $\mu_l(x), l = \overline{1, m}$ is decreasing. The intersection points have only neighboring membership functions, the values of the functions at the intersection points are equal to 0.5. In addition, the sum of all membership functions at any point of the universal set U is equal to one.

Linguistic variables with such properties have been named full orthogonal semantic spaces [11].

Methods for constructing full orthogonal semantic spaces are developed and described in detail in the paper [16].

Trapezoidal fuzzy numbers that formalize the values of the full orthogonal semantic spaces are determined by four parameters, namely: the first two parameters are the abscissas of the left and right vertices of the upper base of the trapezoid, and the last two parameters are the lengths of the left and right wings of the trapezoid. Triangular fuzzy numbers are defined by three parameters, namely: the first parameter is the abscissa of the apex of the triangle, and the last two parameters are the lengths of the left and right wings of the triangle.

In this paper, we will use the method for the construction of full orthogonal semantic spaces based on a direct survey of an expert, which determines with absolute certainty (membership function value equal to unity) typical values $\left(\tau_l^1, \tau_l^2\right)$ for

each term $X_l, l =$ on the universal set $U = [0, 1]$. Then $\mu_1 = \left(0, \tau_1^2, 0, \frac{\tau_2^1 - \tau_1^2}{2}\right)$,
$\mu_l(x) = \left(\tau_l^1, \tau_l^2, \frac{\tau_l^1 - \tau_{l-1}^2}{2}, \frac{\tau_{l+1}^1 - \tau_l^2}{2}\right), l = \overline{2, m-1}, \mu_m(x) = \left(\tau_m^1, 1, \frac{\tau_m^1 - \tau_{m-1}^2}{2}, 0\right)$.

If an evaluating characteristic is quantitative, then the universal set U changes depending on the range of values of this characteristic. The approach to the construction of membership functions for terms of full orthogonal semantic space does not change. If the range of values of a quantitative characteristic is a limited set, then the universal set U can be transformed into a segment $[0, 1]$ by simple arithmetic operations.

In [12], for a fuzzy number $\widetilde{A} = (a_1, a_2, a_L, a_R)$ an aggregating segment $[\vartheta_1, \vartheta_2]$ is determined:

$$\vartheta_1 = a_1 - \frac{a_L}{6}, \vartheta_2 = a_2 + \frac{a_R}{6} \tag{1}$$

This segment is obtained by taking into account the α - cuts of all triangular numbers belonging to the fuzzy number $\widetilde{A} = (a_1, a_2, a_L, a_R)$. To obtain the ends of the segment, as integral indicators, the midpoints of α-cuts are taken with weight coefficients of 2α.

In [8], the midpoint of the segment $[\vartheta_1, \vartheta_2]$ is taken as an aggregating indicator ϑ of a fuzzy number $\widetilde{A} = (a_1, a_2, a_L, a_R)$:

$$\vartheta = \frac{a_1 + a_2}{2} + \frac{a_R - a_L}{12} \tag{2}$$

A Z- number is an ordered pair of fuzzy numbers $Z = \left(\widetilde{A}, \widetilde{R}\right)$, where $\widetilde{A}$ is the value of the fuzzy variable, and $\widetilde{R}$ is the fuzzy value of the reliability of the value $\widetilde{A}$ [1].

The paper will consider linguistic Z-numbers, where both components are the values of the corresponding full orthogonal semantic spaces. The first component is the value of the full orthogonal semantic space that formalizes the characteristic estimates, the second component is the value of the full orthogonal semantic space that formalizes the reliability estimates.

Consider two Z-numbers $Z_1 = \left(\widetilde{A}_1, \widetilde{R}_1\right), Z_2 = \left(\widetilde{A}_2, \widetilde{R}_2\right)$. Let us denote the aggregating indicators of fuzzy numbers $\widetilde{A}_1, \widetilde{R}_1, \widetilde{A}_2, \widetilde{R}_2$, respectively, as a_1, r_1, a_2, r_2.

According to [8], we define the distance between Z-numbers $Z_1 = \left(\widetilde{A}_1, \widetilde{R}_1\right), Z_2 = \left(\widetilde{A}_2, \widetilde{R}_2\right)$ as follows:

$$d(Z_1, Z_2) = \sqrt{\frac{1}{3}(a_1 - a_2)^2 + \frac{1}{3}(a_1 r_1 - a_2 r_2)^2 + \frac{1}{3}(r_1 - r_2)^2} \tag{3}$$

When determining the distance, the indicators of both components are taken into account, as well as their product with equal weighting coefficients.

3 Problem Formulation and Solution

Suppose that an expert evaluates n characteristics B_i, $i = \overline{1,n}$ of P objects using the scales with values B_{il}, $l = \overline{1,m_i}$ and gives their evaluations with certain levels of reliability. To evaluate these levels an expert uses the scale with values R_j, $j = \overline{1,k}$.

To formalize all the scales with the help of full orthogonal semantic spaces, method of [16] based on a direct survey of an expert have been used.

The scale for the characteristic B_i, $i = \overline{1,n}$ is represented as a full orthogonal semantic space with fuzzy values $\widetilde{B_{il}} = \left(b_{il}^1, b_{il}^2, b_{il}^L, b_{il}^R\right)$, $i = \overline{1,n}$, $l = \overline{1,m_i}$. The scale for the evaluating the reliability of expert information is represented as a full orthogonal semantic space with fuzzy values $\widetilde{R_j} = \left(r_j^1, r_j^2, r_j^L, r_j^R\right)$, $j = \overline{1,k}$.

The state of the p-th object in terms of the characteristic B_i, $i = \overline{1,n}$ is a Z-number $Z_{ip} = \left(\widetilde{B_{ip}}, \widetilde{R_{ip}}\right)$, $i = \overline{1,n}, p = \overline{1,P}$, where fuzzy number B_{ip}, $i = \overline{1,n}$, $p = \overline{1,P}$ equals to one of fuzzy number B_{il}, $i = \overline{1,n}$, $l = \overline{1,m_i}$ and fuzzy number $\widetilde{R_{ip}} = \left(r_{ip}^1, r_{ip}^2, r_{ip}^L, r_{ip}^R\right)$, $i = \overline{1,n}, p = \overline{1,P}$ equals to one of fuzzy number $\widetilde{R_j} = \left(r_j^1, r_j^2, r_j^L, r_j^R\right)$, $j = \overline{1,k}$.

The state of the p-th object in terms of all the characteristics B_i, $i = \overline{1,n}$ is a set of n Z-numbers $Z_{1p} = \left(\widetilde{B_{1p}}, \widetilde{R_{1p}}\right)$, ...$Z_{np} = \left(\widetilde{B_{np}}, \widetilde{R_{np}}\right)$, $p = \overline{1,P}$.

Let us define the ideal state of objects in terms of characteristics B_i, $i = \overline{1,n}$ as a set of n Z-numbers $\left(\widetilde{A_i^{id}}, \widetilde{R_i^{id}}\right)$, $i = \overline{1,n}$, where $\widetilde{A_i^{id}} = \widetilde{B_{im_i}} = \left(b_{im_i}^1, b_{im_i}^2, b_{im_i}^L, b_{im_i}^R\right)$, $\widetilde{R_i^{id}} = \widetilde{R_k}$, $i = \overline{1,n}$.

Determine the worst state of objects in terms of characteristics B_i, $i = \overline{1,n}$ as a set of as a set of n Z-numbers $\left(\widetilde{A_i^w}, \widetilde{R_i^w}\right)$, $i = \overline{1,n}$, where $\widetilde{A_i^w} = \widetilde{B_{i1}} = \left(b_{i1}^1, b_{i1}^2, b_{i1}^L, b_{i1}^R\right)$, $\widetilde{R_i^w} = \widetilde{R_1}$, $i = \overline{1,n}$.

We denote the aggregating indicators of $\widetilde{B_{ip}}, \widetilde{R_{ip}}$, $i = \overline{1,n}, p = \overline{1,P}$ as b_{ip}, r_{ip}, $i = \overline{1,n}$, $p = \overline{1,P}$ and aggregating indicators of $\widetilde{A_i^{id}}, \widetilde{A_i^w}, \widetilde{R_i^{id}}, \widetilde{R_i^w}$, $i = \overline{1,n}$ as $a_i^{id}, a_i^w, r_i^{id}, r_i^w$, $i = \overline{1,n}$.

Using the aggregating indicators, we determine the distance d_p, $p = \overline{1,P}$ between the state of the p-th object in terms of all the characteristics B_i, $i = \overline{1,n}$ and the ideal state of objects in terms of characteristics B_i, $i = \overline{1,n}$:

$$d_p = \frac{1}{n} \sum_{i=1}^{n} \sqrt{\left[\frac{1}{3}\left(b_{ip} - a_i^{id}\right)^2 + \frac{1}{3}\left(b_{ip} r_{ip} - a_i^{id} r_i^{id}\right)^2 + \frac{1}{3}\left(r_i^{id} - r_i^w\right)^2\right]} \quad (4)$$

Let

$$d = \frac{1}{n} \sum_{i=1}^{n} \sqrt{\left[\frac{1}{3}\left(a_i^{id} - a_i^w\right)^2 + \frac{1}{3}\left(a_i^{id} r_i^{id} - a_i^w r_i^w\right)^2 + \frac{1}{3}\left(r_i^{id} - r_i^w\right)^2\right]} \quad (5)$$

Using (2 and 3), determine ρ_p, $p = \overline{1, P}$:

$$\rho_p = 1 - \frac{d_p}{d}, p = \overline{1, P} \tag{6}$$

Rating point of the p-th object state is defined as a special case of a Z-number– $Z_p = \left(\rho_p, \widetilde{R}_p\right)$, $p = \overline{1, P}$, the first component of which is an ordinary number ρ_p, $p = \overline{1, P}$, and the second component is a fuzzy number $\widetilde{R}_p = \left(r_p^1, r_p^2, r_p^L, r_p^R\right)$, $p = \overline{1, P}$, which is obtained as follows:

$$r_p^1 = \max_i r_{ip}^1, r_p^2 = \max_i r_{ip}^2, r_p^L = \max_i r_{ip}^L, r_p^R = \max_i r_{ip}^R$$

Then the question arises of analyzing the states of objects in different periods of time. To do this, you need to rank the ratings represented by Z-numbers. When ranking, we will use the method developed in [2].

Let $Z_p^1 = \left(\rho_p^1, \widetilde{R}_p^1\right)$, $Z_p^2 = \left(\rho_p^2, \widetilde{R}_p^2\right)$, $p = \overline{1, P}$ be ratings of the p-th object for the periods 1 and 2, β_p^1, β_p^2, $p = \overline{1, P}$–aggregating indicator for the fuzzy numbers $\widetilde{R}_p^1, \widetilde{R}_p^2$, $p = \overline{1, P}$. If $\rho_p^2 > \rho_p^1$, then the state of the p-th object has improved, if $\rho_p^2 < \rho_p^1$, then the state of the p-th object has become worse. If $\rho_p^2 = \rho_p^1$, then consider the relationship between β_p^1, β_p^2. If $\beta_p^2 > \beta_p^1$, then the state of the p-th object has improved, if $\beta_p^2 < \beta_p^1$, then the state of the p-th object has become worse, if $\beta_p^2 = \beta_p^1$, then the state of the p-th object has not changed.

4 Conclusions

The paper has developed a method of multi-criteria analysis of the objects state under Z-information. The states of the objects are formalized as a set of Z-numbers, the number of elements of which is equal to the number of evaluated object characteristics.

The components of Z-numbers are the values of the corresponding full orthogonal semantic spaces (linguistic variables with properties of completeness and orthogonality). The first component is the value of the full orthogonal semantic space that formalizes the characteristic estimates, the second component is the value of the full orthogonal semantic space that formalizes the reliability estimates.

The paper defines the ideal state of objects in the form of a set of Z-numbers, the components of which are the largest values of full orthogonal semantic spaces that formalize the corresponding scales for the estimated characteristics and reliability of information.

The distance between the real and ideal states of the object is determined on the basis of the aggregating indicators of the components of the Z-numbers. Based on this distance, the rating point of the object state is determined.

Rating point of the object state is defined as a special case of a Z-number, the first component of which is an ordinary number, and the second component is a fuzzy number.

The analysis of the state of objects in different periods of time is carried out according to the first component of the rating Z-numbers according to the principle– the greater the value, the better the state. If the first components of Z- numbers are equal, then the aggregating indicators of the second components are considered according to the same principle as the first components.

The method of comparative analysis of the objects states developed in the paper under conditions of multicriteria based on the ideal state of objects significantly complements the existing methods of analysis and makes it possible to choose an alternative method for different problem formulations.

References

1. Zadeh, L.A.: A note on Z-numbers. Inform. Sci. **14**(181), 2923–2932. https://doi.org/10.1016/j.ins.2011.02.0222011
2. Kang, B., Wei, D., Li, Y., Deng, Y.: A method of converting Z-number to classical fuzzy number. J. Inform. Comp. Sci. **9**(3), 703–709 (2012)
3. Poleshchuk, O.M.: Novel approach to multicriteria decision making under Z-information. In: Proceedings of the International Russian Automation Conference, (RusAutoCon-2019), p 8867607. (2019). https://doi.org/10.1109/RUSAUTOCON.2019.8867607
4. Aliev, R.A., Huseynov, O.H., Zeinalova, L.M.: The arithmetic of continuous Z-numbers. Inform. Sci. **373**, 441–460 (2016). https://doi.org/10.1016/j.ins.2016.08.078
5. Aliev, R.A., Alizadeh, A.V., Huseynov, O.H.: The arithmetic of discrete Z-numbers. Inform. Sci. **1**(290), 134–155 (2015). https://doi.org/10.1016/j.ins.2014.08.024
6. Wang, F., Mao, J.: Approach to multicriteria group decision making with Z-numbers based on Topsis and Power Aggregation Operators. Math. Probl. Eng. 1–18 (2019).
7. Aliyev, R.R., Talal Mraizid, D.A., Huseynov, O.H.: Expected utility based decision making under Z- information and its application. Comput. Intell. Neurosci. **3**, 364512 (2015). https://doi.org/10.1155/2015/364512
8. Poleshchuk, O.M.: Object monitoring under Z-information based on rating point. Adv. Intell. Syst. Comp. **1197**, 1191–1198 (2021). https://doi.org/10.1007/978-3-030-51156-2_139
9. Poleshchuk, O.M.: Recognition of the state of objects under the initial Z-information. Lect. Notes Electr. Eng. **729**, 404–412 (2021). https://doi.org/10.1007/973-3-030-71119-1_40
10. Zadeh, L.A.: The Concept of a linguistic variable and its application to approximate reasoning. Inform. Sci. **8**, 199–249 (1975). https://doi.org/10.1016/0020-0255(75)90036-5
11. Ryjov, A.P.: The concept of a full orthogonal semantic scope and the measuring of semantic uncertainty. In: Proceedings of the Fifth International Conference Information Processing and Management of Uncertainty in Knowledge-Based Systems, pp 33–34. (1994)
12. Poleshchuk, O., Komarov, E.: The determination of rating points of objects with qualitative characteristics and their usage in decision making problems. Int. J. Comput. Math. Sci. **3**(7), 360–364 (2009).
13. Aliev, R.K., Huseynov, O.H., Aliyeva, K.R.: Aggregation of an expert group opinion under Z-information. In: Proceedings of the Eighth International Conference on Soft Computing, Computing with Words and Perceptions in System Analysis, Decision and Control, (ICSCW-2015), pp. 115–124 (2015).
14. Poleshchuk, O.M.: Expert group information formalization based on Z-numbers. J. Phys. Conf. Ser. **1703**, 012010 (2020). https://doi.org/10.1088/1742-6596/1703/1/012010

15. Poleshchuk, O.M.: Monitoring stability of plant species to harmful urban environment under Z-information. Lect. Notes Netw. Syst. **308**, 863–870 (2022)
16. Poleshchuk, O.M.: Creation of linguistic scales for expert evaluation of parameters of complex objects based on semantic scopes. In: Proceedings of the International Russian Automation Conference, (RusAutoCon–2018), 8501686. (2018). https://doi.org/10.1109/RUSAUTOCON. 2018.8501686

On One Boundary-value Problem in a Non-Classical Treatment for the Mangeron Equation

Shahnaz N. Shahbazova⊙ **and Ilgar G. Mamedov**

Abstract In this paper, on a rectangular domain, a boundary value problem is substantiated in a nonclassical interpretation for the generalized Mangeron equation with nonsmooth coefficients. Such a boundary value problem in a nonclassical interpretation has a number of advantages: in this formulation, no additional conditions of agreement are required; this problem can be regarded as a problem formulated by traces in the Sobolev space. It should be especially noted that the generalized Mangeron equation is one of the basic differential equations of mathematical biology. In addition, the equation under consideration is a generalization of many model equations of some processes (string vibration equation, telegraph equation, Laplace equation, heat conduction equation, Aller equation, Boussinesq-Love equation, Mangeron equation, etc).

Keywords Generalized Mangeron equation · Boundary value problem · Equation with nonsmooth coefficients

1 Introduction

One of the main model differential equations of mathematical biology with applicable values is the generalized Mangeron equation, and this equation has the following

Sh. N. Shahbazova (✉)
Institute of Control Systems, Ministry of Science and Education of the Republic of Azerbaijan, 68 Bakhtiyar Bahabzadeh Str., AZ1141 Baku, Azerbaijan
e-mail: shahbazova@cyber.az; shahnaz_shahbazova@unec.edu.az

Department of Digital Technologies and Applied Informatics, Azerbaijan State University of Economics (UNEC), Baku AZ1001, Azerbaijan

I. G. Mamedov
Ministry of Science and Education of the Republic of Azerbaijan, Institute of Control Systems of ANAS, Baku, Azerbaijan
e-mail: ilgar-mamedov-1971@mail.ru

© The Author(s), under exclusive license to Springer Nature Switzerland AG 2023   279
Sh. N. Shahbazova et al. (eds.), *Recent Developments and the New Directions of Research, Foundations, and Applications*, Studies in Fuzziness and Soft Computing 422,
https://doi.org/10.1007/978-3-031-20153-0_22

form:

$$\frac{\partial^4 u(x,y)}{\partial x^2 \partial y^2} + a_{2,1}(x,y)\frac{\partial^3 u(x,y)}{\partial x^2 \partial y} + a_{1,2}(x,y)\frac{\partial^3 u(x,y)}{\partial x \partial y^2} + a_{2,0}(x,y)\frac{\partial^2 u(x,y)}{\partial x^2} +$$

$$+a_{0,2}(x,y)\frac{\partial^2 u(x,y)}{\partial y^2} + a_{1,1}(x,y)\frac{\partial^2 u(x,y)}{\partial x \partial y} + a_{1,0}(x,y)\frac{\partial u(x,y)}{\partial x} +$$

$$+a_{0,1}(x,y)\frac{\partial u(x,y)}{\partial y} + a_{0,0}(x,y)u(x,y) = \Phi(x,y).$$

Now consider the following special equation, which is a special case of this generalized Mangeron equation:

$$\frac{\partial^4 u(x,y)}{\partial x^2 \partial y^2} - \frac{\partial^2 u(x,y)}{\partial x^2} + \frac{\partial^2 u(x,y)}{\partial y^2} = 0.$$

This equation is called the Boussinesq-Love equation. It is known from the mathematical literature that the Boussinesq-Love equation describes longitudinal waves on thin elastic rods, taking into account the effects of transverse inertia. It is known from mathematical physics that even if the direction of particle oscillations coincides with the direction of wave propagation, the waves remain waves on the surface. With the propagation of longitudinal waves in the medium, compression and tension deformations occur. Longitudinal waves propagate in all media due to the formation of elastic forces during compression and tension deformations both in solids and in liquids and gases. Theoretically and applied it is determined that sound is longitudinal, light is transverse, and earthquake is a seismic wave propagating in both transverse and longitudinal directions. Note that if the direction of particle oscillations is perpendicular to the direction of wave propagation, then such waves are transverse waves. Thus, since the Boussinesq-Love equation describes longitudinal waves, and longitudinal waves propagate in all media, this equation is one of the model differential equations of great practical importance. Now we write the Businesq-Love equation as follows:
$\frac{\partial^2}{\partial x^2}\left(\frac{\partial^2 u(x,y)}{\partial y^2} - u(x,y)\right) + \frac{\partial^2 u(x,y)}{\partial y^2} = 0$ and additionally assume that $\frac{\partial^2 u(x,y)}{\partial y^2} - u(x,y) = u(x,y)$ in other words, suppose that the condition $u(x,y) = \frac{1}{2}\frac{\partial^2 u(x,y)}{\partial y^2}$. Then, within the framework of this hypothesis, the Boussinesq-Love equation, which belongs to the class of hyperbolic equations, coincides with an elliptic equation, one of the basic model differential equations of mathematical physics: $\frac{\partial^2 u(x,y)}{\partial x^2} + \frac{\partial^2 u(x,y)}{\partial y^2} = 0$. As is known from the literature, this equation, called the Laplace equation in mathematical physics, describes the processes of stationary filtration of liquids. The Laplace operator $\frac{\partial^2 u(x,y)}{\partial x^2} + \frac{\partial^2 u(x,y)}{\partial y^2}$, which is self-adjoint in this equation, can be written as a composition of two differential operators as follows: $\frac{\partial^2 u(x,y)}{\partial x^2} + \frac{\partial^2 u(x,y)}{\partial y^2} = \left(\frac{\partial}{\partial x} - i\frac{\partial}{\partial y}\right)\left(\frac{\partial}{\partial x} + i\frac{\partial}{\partial y}\right)u(x,y)$, here $i-$ is an imaginary unit: $i^2 = -1$. What is related to the Laplace equation is the equation

$\left(\frac{\partial}{\partial x} + i\frac{\partial}{\partial y}\right)u(x, y) = 0$ is a differential equation of elliptic type of the first order and is called the Cauchy-Riemann equation. The first-order elliptic differential operator on the left-hand side of the Cauchy-Riemann equation can be written as a composition of two more differential operators:

$\left(\frac{\partial}{\partial x} + i\frac{\partial}{\partial y}\right)u(x, y) = \left(\frac{\partial^{\frac{1}{2}}}{\partial x^{\frac{1}{2}}} - i\sqrt{i}\frac{\partial^{\frac{1}{2}}}{\partial y^{\frac{1}{2}}}\right)\left(\frac{\partial^{\frac{1}{2}}}{\partial x^{\frac{1}{2}}} + i\sqrt{i}\frac{\partial^{\frac{1}{2}}}{\partial y^{\frac{1}{2}}}\right)u(x, y)$. As you can see, the equations of mathematical biology and mathematical physics have mathematical relations not only with each other, but also with differential equations of fractional order. Thus, in this case, the considered generalized Mangeron equation has not only theoretical, but also great practical application, and even in a particular case, describes real physical and biological processes.

2 Boundary-value Problem in a Classical Treatment

Consider the generalized Mangeron equation [5]:

$$
\begin{aligned}
(V_{2,2}u)(x) &\equiv \left(D_1^2 + a_{1,2}(x)D_1 + a_{0,2}(x)\right)D_2^2 u(x) + \\
&+ \left(a_{2,1}(x)D_1^2 + a_{1,1}(x)D_1 + a_{0,1}(x)\right)D_2 u(x) + \\
&+ \left(a_{2,0}(x)D_1^2 + a_{1,0}(x)D_1 + a_{0,0}(x)\right)u(x) = \\
&= \phi_{2,2}(x), \quad x = (x_1, x_2) \in G,
\end{aligned}
\tag{1}
$$

which is a generalization of the Boussinesq—Love equation from the theory of oscillations [1, formula (20)], which describes longitudinal waves in a thin elastic rod, taking into account the effects of transverse inertia. Here: $D_k^j = \partial^j/\partial x_k^j$, $(j = 0, 1, 2; k = 1, 2)$- generalized differentiation operator in the sense of S.L. Sobolev and D_k^0- operator of identity transformation. Equations of this type in [2, 3] are called pseudoparabolic. In addition, the considered equation is a generalization of many model equations of some processes (generalized equation of moisture transfer, telegraph equation, string vibration equation, heat conduction equation, Aller equation, etc.). The generalized Mangeron equation as the Aller equation is one of the basic differential equations of mathematical biology [5], p. 261]. In this case, an important fundamental point is that the equation under consideration has nonsmooth coefficients that satisfy only certain conditions such as p-integrability and boundedness, in other words, the coefficients belong to the Lebesgue class or are bounded to a particular argument, in other words, the considered pseudoparabolic differential operator has no adjoint operator in the sense of Lagrange:

$$(V_{2,2}^* \rho)(x) \equiv \frac{\partial^4 \rho(x)}{\partial x_1^2 \partial x_2^2} - \frac{\partial^3}{\partial x_1^2 \partial x_2}\left(a_{2,1}(x)\rho(x)\right) - \frac{\partial^3}{\partial x_1 \partial x_2^2}\left(a_{1,2}(x)\rho(x)\right)+$$

$$+\frac{\partial^2}{\partial x_1^2}\left(a_{2,0}(x)\rho(x)\right) + \frac{\partial^2}{\partial x_2^2}\left(a_{0,2}(x)\rho(x)\right) + \frac{\partial^2}{\partial x_1 \partial x_2}\left(a_{1,1}(x)\rho(x)\right)-$$

$$-\frac{\partial}{\partial x_1}\left(a_{1,0}(x)\rho(x)\right) - \frac{\partial}{\partial x_2}\left(a_{0,1}(x)\rho(x)\right) + a_{0,0}(x)\rho(x).$$

However, for this operator to make sense, the coefficients $a_{i_1,i_2}(x)$ must be sufficiently smooth. Therefore, the integral representation of the solution of Eq. (1) has so far been studied in the literature only under strict conditions on the coefficients. In other words, if the problem under consideration is related to an equation that has nonsmooth coefficients, then the question of constructing its formally conjugate differential problem in the sense of Lagrange is, generally speaking, an unsolvable problem.

Here $u = u(x)$ the required function defined on G; $a_{i_1,i_2} = a_{i_1,i_2}(x)$ given measurable functions on $G = G_1 \times G_2$, where $G_1 = (x_1^0, h_1)$, $G_2 = (x_2^0, h_2)$; $\phi_{2,2}(x)$ given measurable function on G.

It is known from the literature that if partial differential equations have real characteristics up to an order number, then these equations are called hyperbolic. If the characteristic is repeated, then such hyperbolic equations are called pseudoparabolic. Equation (1) is a hyperbolic equation that has two real characteristics $x_1 = const$, $x_2 = const$, the first and second of which are double. Therefore, Eq. (1) in a sense can be regarded as a pseudoparabolic equation. It should be especially noted that the Dirichlet and Neumann problem for the generalized Mangeron equation were investigated in [6–9] and others. In addition, in the literature so far, the Riemann function of Eq. (1) has been constructed only for the case when the functions are sufficiently smooth / those. when functions $a_{i_1,i_2} = a_{i_1,i_2}(x)$ continuous together with derivatives $D_1^{i_1} D_2^{i_2} a_{i_1,i_2}(x)$ in the area of $\overline{G}/$.

In this paper, Eq. (1) was first studied in the general case, when the coefficients $a_{i_1,i_2}(x)$ are nonsmooth functions satisfying only the following conditions:

$a_{i_1,i_2}(x) \in L_p(G), i_1 = 0, 1$, $i_2 = 0, 1$; $a_{i_1,2}(x) \in L_{p,\infty}^{x_1,x_2}(G)$, $a_{2,i_2}(x) \in L_{\infty,p}^{x_1,x_2}(G)$, $i_1 = 0, 1$, $i_2 = 0, 1$.

Under these conditions, we will seek a solution to Eq. (1) in the space of S.L. Sobolev.

$$W_p^{(2,2)}(G) = \left\{ u \in L_p(G) / D_1^{i_1} D_2^{i_2} u \in L_p(G); \ i_1 = 0, 1, 2; \ i_2 = 0, 1, 2 \right\}$$

where $1 \leq p \leq \infty$. The norm in space $W_p^{(2,2)}(G)$ will be defined by the equalit

$$\|u\|_{W_p^{(2,2)}(G)} = \sum_{i_1=0}^{2} \sum_{i_2=0}^{2} \left\| D_1^{i_1} D_2^{i_2} u \right\|_{L_p(G)}.$$

For Eq. (1), the conditions in the middle of a domain of the classical form can be specified in the form

$$
\begin{cases}
u/_{x_1 = \frac{x_1^0 + h_1}{2}} = \Phi_1(x_2), \\[2mm]
D_1 u/_{x_1 = \frac{x_1^0 + h_1}{2}} = \Phi_2(x_2), \\[2mm]
u/_{x_2 = \frac{x_2^0 + h_2}{2}} = \Psi_1(x_1), \\[2mm]
D_2 u/_{x_2 = \frac{x_2^0 + h_2}{2}} = \Psi_2(x_1)
\end{cases}
\tag{2}
$$

where $\Phi_1(x_2)$, $\Phi_2(x_2)$, $\Psi_1(x_1)$ и $\Psi_2(x_1)$ given measurable functions on G. Obviously, in the case of conditions (2), the functions $\Phi_1(x_2)$, $\Phi_2(x_2)$, $\Psi_1(x_1), \Psi_2(x_1)$ except conditions $\Phi_1(x_2) \in W_p^{(2)}(G_2)$, $\Phi_2(x_2) \in W_p^{(2)}(G_2)$, $\Psi_1(x_1) \in W_p^{(2)}(G_1)$, $\Psi_2(x_1) \in W_p^{(2)}(G_1)$ must also satisfy the following conditions:

$$
\begin{cases}
\Phi_1\left(\dfrac{x_2^0 + h_2}{2}\right) = \Psi_1\left(\dfrac{x_1^0 + h_1}{2}\right), \\[3mm]
\Phi_2\left(\dfrac{x_2^0 + h_2}{2}\right) = \Psi_1'\left(\dfrac{x_1^0 + h_1}{2}\right), \\[3mm]
\Psi_1\left(\dfrac{x_1^0 + h_1}{2}\right) = \Phi_1'\left(\dfrac{x_2^0 + h_2}{2}\right), \\[3mm]
\Phi_2'\left(\dfrac{x_2^0 + h_2}{2}\right) = \Psi_2'\left(\dfrac{x_1^0 + h_1}{2}\right),
\end{cases}
\tag{3}
$$

which are the terms of the agreement.

3 Boundary-Value Problem in a Non-Classical Treatment and its Justification

The presence of agreement conditions in the statement of problem (1), (2) means that conditions (2) also specify some redundant information about the solution of this problem. Therefore, the question arises of finding boundary conditions that do not contain unnecessary information about the solution and do not require the fulfillment of some additional conditions such as agreement. In this regard, consider the following boundary conditions

$$\begin{cases} V_{0,0}u \equiv u\left(\dfrac{x_1^0+h_1}{2},\dfrac{x_2^0+h_2}{2}\right)=\phi_{0,0}, \\[2mm] (V_{1,0}u)\equiv D_1 u\left(\dfrac{x_1^0+h_1}{2},\dfrac{x_2^0+h_2}{2}\right)=\phi_{1,0}, \\[2mm] (V_{0,1}u)\equiv D_2 u\left(\dfrac{x_1^0+h_1}{2},\dfrac{x_2^0+h_2}{2}\right)=\phi_{0,1}, \\[2mm] (V_{1,1}u)\equiv D_1 D_2 u\left(\dfrac{x_1^0+h_1}{2},\dfrac{x_2^0+h_2}{2}\right)=\phi_{1,1}, \\[2mm] (V_{2,0}u)(x_1)\equiv D_1^2 u\left(\dfrac{x_1^0+h_1}{2},\dfrac{x_2^0+h_2}{2}\right)=\phi_{2,0}(x_1), \\[2mm] (V_{2,1}u)(x_1)\equiv D_1^2 D_2 u\left(x_1,\dfrac{x_2^0+h_2}{2}\right)=\phi_{2,1}(x_1), \\[2mm] (V_{0,2}u)(x_2)\equiv D_2^2 u\left(\dfrac{x_1^0+h_1}{2},x_2\right)=\phi_{0,2}(x_2), \\[2mm] (V_{1,2}u)(x_2)\equiv D_1 D_2^2 u\left(\dfrac{x_1^0+h_1}{2},x_2\right)=\phi_{1,2}(x_2) \end{cases} \tag{4}$$

where $\phi_{0,0}\in R, \phi_{1,0}\in R, \phi_{0,1}\in R, \phi_{1,1}\in R$ is a given number, and the rest are given functions ϕ_{i_1,i_2} that satisfy the conditions: $\phi_{2,0}(x_1)\in L_p(G_1), \phi_{2,1}(x_1)\in L_p(G_1), \phi_{0,2}(x_2)\in L_p(G_2), \phi_{1,2}(x_2)\in L_p(G_2).$

If the function $u\in W_p^{(2,2)}(G)$ is a solution to the problem with conditions in the middle of the domain of the classical form (1), (2), then it is also a solution to problem (1), (4) for ϕ_{i_1,i_2} defined by the following equalities

$$\begin{cases} \phi_{0,0}=\Phi_1(\dfrac{x_2^0+h_2}{2})=\Psi_1(\dfrac{x_1^0+h_1}{2}), \\[2mm] \phi_{1,0}=\Phi_2(\dfrac{x_2^0+h_2}{2})=\Psi_1'(\dfrac{x_1^0+h_1}{2}), \\[2mm] \phi_{0,1}=\Psi_2(\dfrac{x_1^0+h_1}{2})=\Phi_1'(\dfrac{x_2^0+h_2}{2}), \\[2mm] \phi_{1,1}=\Phi_2'(\dfrac{x_2^0+h_2}{2})=\Psi_2'(\dfrac{x_1^0+h_1}{2}), \\[2mm] \phi_{2,0}(x_1)=\Psi_1''(x_1), \\[2mm] \phi_{2,1}(x_1)=\Psi_2''(x_1), \\[2mm] \phi_{0,2}(x_2)=\Phi_1''(x_2), \\[2mm] \phi_{1,2}(x_2)=\Phi_2''(x_2) \end{cases}$$

It is easy to prove that the converse is also true. In other words, if a function $u \in W_p^{(2,2)}(G)$ is a solution to problem (1), (4), then it is also a solution to problem (1), (2), for the following functions Φ_1, Φ_2, Ψ_1, Ψ_2:

$$
\left\{
\begin{aligned}
&\Phi_1(x_2) = \phi_{0,0} + (x_2 - \frac{x_2^0 + h_2}{2})\phi_{0,1} + \int_{\frac{x_2^0 + h_2}{2}}^{x_2} (x_2 - \tau)\phi_{0,2}(\tau)d\tau, \\[2mm]
&\Phi_2(x_2) = \phi_{1,0} + (x_2 - \frac{x_2^0 + h_2}{2})\phi_{1,1} + \int_{\frac{x_2^0 + h_2}{2}}^{x_2} (x_2 - \xi)\phi_{1,2}(\xi)d\xi, \\[2mm]
&\Psi_1(x_1) = \phi_{0,0} + (x_1 - \frac{x_1^0 + h_1}{2})\phi_{1,0} + \int_{\frac{x_1^0 + h_1}{2}}^{x_1} (x_1 - \eta)\phi_{2,0}(\eta)d\eta, \\[2mm]
&\Psi_2(x_1) = \phi_{0,1} + (x_1 - \frac{x_1^0 + h_1}{2})\phi_{1,1} + \int_{\frac{x_1^0 + h_1}{2}}^{x_1} (x_1 - \upsilon)\phi_{2,1}(\upsilon)d\upsilon
\end{aligned}
\right.
\tag{5}
$$

Note that, in this case, functions (5) have one important property related to the fact that for them the agreement conditions (3) are satisfied automatically for all ϕ_{i_1,i_2}, possessing the above properties. Therefore, equalities (5) can also be regarded as a general form of all functions

$$
\Phi_1(x_2) \in W_p^{(2)}(G_2), \quad \Phi_2(x_2) \in W_p^{(2)}(G_2), \quad \Psi_1(x_1) \in W_p^{(2)}(G_1), \Psi_2(x_1) \in W_p^{(2)}(G_1)
$$

satisfying the agreement conditions (3).

So, problems with conditions in the middle of the domain of the classical form (1), (2) and the form (1), (4) are generally equivalent. However, problem (1), (4) is more natural in its setting than problem (1), (2). This is due to the fact that in the formulation of problem (1), (4) on the right-hand sides of the boundary conditions, no additional conditions such as agreement are required. Therefore, problem (1), (4) can be regarded as a problem with conditions in the middle of a domain of a new type.

Note that integral representations of functions from Sobolev spaces with dominant mixed derivatives of general form were studied in the works of Amanov [10], Nikol'skii [11], Lizorkin and Nikol'skii [12], Besov et al. [13], Dzhabrailov [14], Akhiev [15], Nadzhafov [16] and others. To study the boundary value problem (1), (4), but among them we will use the fact that any function $u(x) \in W_p^{(2,2)}(G)$ is uniquely representable in the form [15]:

$$u(x) = u\left(\frac{x_1^0 + h_1}{2}, \frac{x_2^0 + h_2}{2}\right) +$$

$$\left(x_1 - \frac{x_1^0 + h_1}{2}\right) D_1 u\left(\frac{x_1^0 + h_1}{2}, \frac{x_2^0 + h_2}{2}\right) +$$

$$+\left(x_1 - \frac{x_1^0 + h_1}{2}\right)\left(x_2 - \frac{x_2^0 + h_2}{2}\right) D_2 u\left(\frac{x_1^0 + h_1}{2}, \frac{x_2^0 + h_2}{2}\right) +$$

$$+\left(x_1 - \frac{x_1^0 + h_1}{2}\right) D_1 D_2 u\left(\frac{x_1^0 + h_1}{2}, \frac{x_2^0 + h_2}{2}\right) +$$

$$+ \int_{\frac{x_1^0 + h_1}{2}}^{x_1} (x_1 - \alpha_1) D_1^2 u\left(\alpha_1, \frac{x_2^0 + h_2}{2}\right) d\alpha_1 +$$

$$+(x_2 - \frac{x_2^0 + h_2}{2}) \int_{\frac{x_1^0 + h_1}{2}}^{x_1} (x_1 - \alpha_1) D_1^2 D_2 u\left(\alpha_1, \frac{x_2^0 + h_2}{2}\right) d\alpha_1 +$$

$$+ \int_{\frac{x_2^0 + h_2}{2}}^{x_2} (x_2 - \alpha_2) D_2^2 u\left(\frac{x_1^0 + h_1}{2}, \alpha_2\right) d\alpha_2 +$$

$$+\left(x_1 - \frac{x_1^0 + h_1}{2}\right) \int_{\frac{x_2^0 + h_2}{2}}^{x_2} (x_2 - \alpha_2) D_1 D_2^2 u\left(\frac{x_2^0 + h_2}{2}, \alpha_2\right) d\alpha_2 +$$

$$+ \int_{\frac{x_1^0 + h_1}{2}}^{x_1} \int_{\frac{x_2^0 + h_2}{2}}^{x_2} (x_1 - \alpha_1)(x_2 - \alpha_2) D_1^2 D_2^2 u(\alpha_1, \alpha_2) d\alpha_1 d\alpha_2$$

through traces

$u(\frac{x_1^0 + h_1}{2}, \frac{x_2^0 + h_2}{2})$, $\qquad D_1 u\left(\frac{x_1^0 + h_1}{2}, \frac{x_2^0 + h_2}{2}\right)$, $\qquad D_2 u\left(\frac{x_1^0 + h_1}{2}, \frac{x_2^0 + h_2}{2}\right)$,

$D_1 D_2 u\left(\frac{x_1^0 + h_1}{2}, \frac{x_2^0 + h_2}{2}\right)$, $D_1^2 u\left(x_1, \frac{x_2^0 + h_2}{2}\right)$, $D_1^2 D_2 u\left(x_1, \frac{x_2^0 + h_2}{2}\right)$, $D_2^2 u\left(\frac{x_1^0 + h_1}{2}, x_2\right)$,

$D_1 D_2^2 u\left(\frac{x_1^0 + h_1}{2}, x_2\right)$ and the dominant mixed derivative $D_1^2 D_2^2 u(x_1, x_2)$.

4 Conclusion

This formulation of the problem has a number of advantages:

(1) in this formulation problem, no additional conditions such as the agreement condition are required;
(2) in this formulation, the considered equation is a generalization of many model equations of some processes(string vibration equation, telegraph equation, Laplace equation, heat conduction equation, Aller equation, Boussinesq-Love equation, Mangeron equation, etc.);
(3) it can be considered as a problem formulated by traces in the space of S.L. Sobolev.

Note. The article is dedicated to the 100th anniversary of the glorious son of the Azerbaijani people, world-famous Azerbaijani scientist Lotfi Zadeh. In the near future, it is planned to consider the fuzzy analogue of the generalized Mangeron equation.

References

1. Berezanskii, Yu.M.: On a Dirichlet type problem for the string vibration equation. Ukr. Mat. Zh. **12**(4), 363–372 (1960)
2. Colton, D.: Pseudoparabolic equations in one space variable. J. Differ. Equ **12**(3), 559–565 (1972)
3. Soldatov, A.P., Shkanukov ,M.Kh.: Boundary value problems with A.A.Samarsky general nonlocal condition for higher order pseudoparabolic equations. In: Doklady Akademii Nauk, vol.297, No.3, pp.547–552 (1987) (in Russian)
4. Nakhushev, A.M.: Equations of mathematical biology. M.: Visshaya Shkola, p. 301. (1995) (in Russian)
5. Mangeron, D.: New methods for determining solution of mathematical models governing polyvibrating phenomena. Bul. Inst. politehn. Jasi. Sectia **1**(14):1,2 433–436 (1968)
6. Utkina, E. A.: Dirichlet problem for a fourth-order equation. Differ. Equ. **47** (4), 599–603 (2011)
7. Mamedov, I. G.: On a nonclassical interpretation of the Dirichlet problem for a fourth-order pseudoparabolic equation. Differ. Equ. **50**(3), 415–418 (2014)
8. Mamedov, I.G.: On the well-posed solvability of the Dirichlet problem for a generalized Mangeron equation with nonsmooth coefficients. Differ. Equ. **51**(6), 745–754 (2015)
9. Mamedov, I. G., Mardanov, M. D., Melikov, T. K., Bandaliev, R. A.: Well-posed solvability of the Neumann problem for a generalized Mangeron equation with nonsmooth coefficients. Differ. Equ. **55**(10), 1362–1372 (2019)
10. Amanov, T. I.: Investigation of the properties of classes of functions with dominant mixed derivatives. Representation, embedding, continuation, and interpolation theorems, Extended Abstract of Doctoral (Phys–Math.) Dissertation, Novosibirsk (1967)
11. Nikol'skii, S. M.: Priblizhenie funktsii mnogikh peremennykh i teoremy vlozheniya (Approximation of functions of many variables and embedding theorems).Nauka, Moscow (1969)
12. Lizorkin, P. I, Nikol'skii, S. M.: Classification of differentiable functions based on spaces with a dominant derivative. Tr. Mat. Inst. im. V.A. Steklova Akad. Nauk SSSR. **77**, 143–167 (1965)

13. Besov, O. V., Il'in, V. P., Nikol'skii, S. M.: Integral'nye predstavleniya funktsii i teoremy vlozheniya (Integral representations of functions and embedding theorems). Nauka, Moscow (1975)
14. Dzhabrailov, A. D.: Investigation of differential–difference properties of a function defined on an n-dimensional domain, Extended Abstract of Doctoral (Phys–Math.) Dissertation, Moscow, (1971) (in Russian)
15. Akhiev, S. S.: On the general form of linear bounded functionals in a function space of the S.L. Sobolev type. Dokl. Akad. Nauk Azerbaijan SSR **35** (6), 3–7 (1979) (in Russian)
16. Nadzhafov, A. M., On integral representations of functions from spaces with a dominant mixed derivative. Vestn. Bakinskogo Gos. Univ. Ser. Fiz.-Mat. Nauk (3), 31–39 (2005) (in Russian)

Graphical Modeling of Information Coming from Numerous Sensors of Different Types

A. A. Akhundov and E. M. Akhundova

Abstract In this work, the results of graphical modeling of information coming from numerous of sensors of different types are presented in a form that is very convenient for observations. This seems to be important for the timely detection of possible deviations from the normal functioning of the observed objects.

1 Introduction

The basis for this work was our monograph [1, in electronic version 2], published in 2010. In [1], a new geometric version of the solution of the problem of simultaneous observation of the readings of a large quantity of sensors of different types is presented. Here, as we will see later, the expression "geometric version of the solution to a problem" in fact means a solution to the problem by means of graphical modeling. In [1], we paid a lot of attention to the rigorous mathematical substantiation of the legality of this version of the solution to the mentioned problem. This, on the one side, of course, proved the correctness of the proposed version of the solution to the problem. But, on the other side this made the material of the monograph seriously difficult to use in applied observation problems.

The purpose of the work is to present a version of the solution to the problem by means of graphical modeling in a form more accessible to applications. In particular, for this we abandoned the use of group time transformations used in [1]. Here we have developed a new approach, abandoning the use of somewhat more complicated coordinate systems from [1]. As it turned out, applying these changes greatly facilitates the observations process.

A. A. Akhundov (✉) · E. M. Akhundova
Institute of Control Systems, National Academy of Sciences of Azerbaijan, Baku, Azerbaijan
e-mail: akhundov_a@rambler.ru

© The Author(s), under exclusive license to Springer Nature Switzerland AG 2023
Sh. N. Shahbazova et al. (eds.), *Recent Developments and the New Directions of Research, Foundations, and Applications*, Studies in Fuzziness and Soft Computing 422, https://doi.org/10.1007/978-3-031-20153-0_23

2 Relevance of the Problem

Controlling the quality of functioning or deviations from a given trajectory of movement of various objects using the current readings of numerous sensors of different
types has always been an actual problem. In cases where the number of sensors is
large number and all sensors need to be monitored simultaneously, the problem is
usually solved by building a hierarchical monitoring system. Naturally, this increases
the number of people who are observing. Ultimately, this inevitably leads to complication, increase in cost and, which is extremely undesirable, loss of time during
observation.

In this work, we propose a version of the solution to the problem, in which in
some cases the need for a hierarchical observation system disappears. If the number
of sensors is not more than 120, then the observation can be carried out without much
difficulty and successfully by one observer.

3 Basic Ideas

We have two basic ideas that are used to solve the problem through graphical
modeling.

Idea 1. Since the readings of individual sensors, generally speaking, are not related
to each other, they can be considered independent variables. Therefore, in order to
be able to graphically depict n independent readings of sensors, in order to form a
geometric structure from them, for example, in the form of a graph, we certainly need
to have some kind of geometric place that has n independent measurements. So we
come to the need to consider Euclidean space R_n. But, the Euclidean space $R_n (n > 3)$
does not have geometric clarity and consequently is not suitable for conducting
observations. Therefore, *question* 1 naturally arises: Is it possible in a legal way to
map the coordinates of the sensor readings from the space $R_n (n > 3)$ to the plane
R_2. Moreover, to ensure a high quality of observation, such a mapping must be a
homeomorphism.

Note 1. Recall that a homeomorphism is a one-to-one and mutually continuous
mapping between two sets or spaces. Roughly speaking, the presence of homeomorphism would mean that to small changes in the coordinates of the sensor readings in
one space correspond to small changes in the displayed coordinates in another space.
This is very important for the high-quality implementation of observations. More
precisely, we note that two topological spaces in the presence of a homeomorphism
become the same in the topological sense.

Direct answer to question 1 is negative. This was proven back in 1913 year (L.E.J.
Brouwer [3]). For more details on this theorem, as well as on its further development,
see [4]. Namely, Brouwer proved that there is no homeomorphism between spaces
of different dimensions R_n and $R_m (n \neq m)$. By the way, apparently, it was after the

appearance of this theorem of Brouwer that the concept of homeomorphism began to be used as a mapping in all strict definitions of the important mathematical concept of a *manifold*. The dimensions of the spaces $R_n (n > 3)$ and R_2 are different; therefore, according to [3], it is impossible to have a homeomorphism between them.

The desire to solve the problem, as well as taking into account the fact that in such multidimensional spaces as $R_n (n > 3)$, almost all problems are solved in the "coordinate language" lead us to the following *question* 2: Can the coordinate axes themselves, defined in the space $R_n (n > 3)$, be mapped homeomorphically onto the space R_2. In this case, an important task for us is to position the coordinate axes of space $R_n (n > 3)$ on the R_2 plane in a form convenient for graphic modeling. The answer to this question was in the affirmative. This is the essence of idea 1 and this idea allows us to provide a homeorphic transition from space $R_n (n > 3)$ to space R_2 at the level of the coordinate axes.

In the next section, we'll go into more detail of Idea 1 and demonstrate its implementation.

For the convenience of presentation, before proceeding with the description of idea 2, we introduce some notation. Let there are N objects for observation, and all of them are numbered in arbitrary order with natural numbers: $1, 2, \ldots, N$. Suppose that information about the behavior of these objects comes from n sensors, $n \geq N$. According to the observation problem, one observer needs to simultaneously observe the behavior of all objects for some time: $T = [t_0, t_1]$, using the current readings of the sensors for this: $1_1(t), 1_2(t), \ldots, 1_{k_1}(t), 2_1(t), 2_2(t), \ldots, 2_{k_2}(t), N_1(t), N_2(t), \ldots, N_{k_N}(t)$, Here $i_j(t)$ is a scalar function equal at the moment of time $t \in T$ to the numerical value of the readings of the j sensor referred to the object i, $n = k_1 + k_2 + \ldots + k_N, i = \overline{1, N}, j = \overline{1, k_i}$. Thus, as can be seen from the notation, we assumed that the number of sensors involved for the object i is equal to $k_i, k_i \geq 1$.

We call *admissible* all those observed values $i_j(t)$ at which the object i functions normally (without emergency) at the moment of time $t \in T, i = \overline{1, N}, j = \overline{1, k_i}$.

Let us assume that all sensors participating in the observation process can take admissible values from some intervals. Let $I_{ij}(t) = \left(a_{ij}(t), b_{ij}(t)\right)$ be the minimum length interval containing at the moment of time $t \in T$ all admissible values of the function $i_j(t)$, so that the object $i, i = \overline{1, N}$, functions normally at the moment of time t, if $i_j(t) \in I_{ij}(t)$ for all values $j = \overline{1, k_i}$.

It can be seen from the notation that the value of the function $a_{ij}(t)$ is equal to the *lower bound* of the admissible values of the function $i_j(t)$. Accordingly, the value of the function $b_{ij}(t)$ determines the *upper bound* of the admissible values of the function $i_j(t), t \in T, i = \overline{1, N}, j = \overline{1, k_i}$.

Of course, cases are not excluded when the functions $a_{ij}(t)$ and $b_{ij}(t)$ can be constant. E.g., consider a coolant used in an internal combustion engine. Then the lower bound of admissible values will be equal to the freezing temperature of this liquid. In this case, the upper bound of admissible values is determined by the boiling temperature of this liquid. Both quantities are constant. In this example, one more quantity is important. We are talking about the optimal coolant temperature $c \in I_{ij}(t)$, at which the engine works in the best possible way. By observing the deviations of the values of the corresponding function $i_j(t)$ from the value of c, we can estimate the quality of the engine (this means the engine works with the least wear). Liquids are also used to create optimal temperatures in modern electric vehicle batteries.

Let's select in each interval $I_{ij}(t)$ one admissible point $c_{ij}(t) \in I_{ij}(t)$ and suppose that at each moment of time $t \in T$ the values $c_{ij}(t), j = \overline{1, k_i}$, correspond to the best (optimal, nominal, programmed) mode of functioning of the object $i, i = \overline{1, N}$. It's obvious that

$$a_{ij}(t) < c_{ij}(t) < b_{ij}(t), t \in T, i = \overline{1, N}, j = \overline{1, k_i} \tag{1}$$

For definiteness, let us agree to call the functions $c_{ij}(t)$ optimal functions $t \in T$, $i = \overline{1, N}, j = \overline{1, k_i}$.

In general, the dependence of these functions on time makes it possible to expand the varieties of the observed objects. In particular, various moving objects can be included in the observation process.

It is assumed that all three functions from inequalities (1) are given in advance, i.e. known before the start of the observation process. Thus, at each time moment $t \in T$, the readings of an arbitrary sensor j of object i can be estimated by four values: $a_{ij}(t)$, $c_{ij}(t)$, $b_{ij}(t)$ and $i_j(t)$, $i = \overline{1, N}, j = \overline{1, k_i}$. Indeed, the observation process for each separately taken observable function $i_j(t)$ is not complicated, namely it is required at each moment of time $t \in T$ to know two things:

(1) Is the inclusion fair $i_j(t) \in I_{ij}(t)$ (absence or presence of an emergency at the moment of time t);
(2) And if there is no emergency situation, then how close are the values $i_j(t)$ to the optimal values $c_{ij}(t)$ (for determine the quality of functioning of the corresponding node of the object i at time t).

This process turns into a very serious problem when the number n is large and it is required that one observer simultaneously observe for changes in the numerical values of all observed functions $i_j(t)$ over time $t \in T$.

For any fixed parameters i, j, t with the value of the function $a_{ij}(t)$, we one-to-one connect a point in some *adaptive coordinate system*, which we will construct in the next section. We will consider this point as the graph vertex. For each fixed time value, the values of the function $a_{ij}(t), i = \overline{1, N}, j = \overline{1, k_i}, n = k_1 + k_2 + \ldots + k_N$, can be used to determine n vertices in a one-to-one way. Connecting neighboring vertices with edges, we get a graph with n vertices in an adaptive coordinate system.

We call this graph the *left bound graph* and denote it by the symbol $GA(t), t \in T$. Obviously, in the general case, the vertices of the graph $GA(t)$ will be movable.

Proceeding in the same way with the functions $c_{ij}(t)$, $b_{ij}(t)$ and $i_j(t)$, we will construct three more graphs with n vertices in the adaptive coordinate system. They will be denoted, respectively, by the symbols $GC(t)$, $GB(t)$ and $GI(t), t \in T$.

The graph $GC(t)$ is called the *optimal graph*, which is determined in a one-to-one way by the values of the optimal functions $c_{ij}(t), t \in T$.

Similarly, the graph $GB(t)$ is called the *right bound graph*, which is determined in a one-to-one way by the values of the functions $b_{ij}(t), t \in T$.

And finally, we call the graph $GI(t)$ a *hal-graph* (or the *graph of current states*), which is determined in a one-to-one way by the current values of the functions $i_j(t)$, $t \in T$. Despite the fact that in the general case the vertices of the graphs $GA(t)$, $GC(t)$ and $GB(t)$ are movable, these three graphs can never intersect according to inequalities (1). Despite this, in the general case, the mobility of the vertices of all three graphs $GA(t)$, $GC(t)$ and $GB(t)$ create rather inconvenient conditions for making observations in an adaptive coordinate system.

We will prove that in the course of time, by applying some continuous group transformations such as translation group and scaling group, it is possible to greatly simplify the observation problem (sometimes, instead of the term "*translation group*", the term "*shift group*" is also used). Namely, with the help of these transformations, it is possible to ensure that the vertices of the three graphs are fixed and located in the course of time, respectively, on three static straight lines.

Consequently, as a result of the actions of these transformations, in the adaptive coordinate system, instead of three moving graphs $GA(t)$, $GC(t)$ and $GC(t)$, we get three fixed graphs in the process of observing. The vertices of these three graphs in the course of time will be motionless on three straight lines, respectively. We will make all three straight lines parallel to each other. One can hope that all this will create very convenient conditions for the implementation of the observation process.

The *question* arises 3, how to arrange these three fixed and parallel straight lines on the plane of the monitor screen?

Indeed, they can be arranged in different ways. We can have them positioned vertically or horizontally to the bottom of the monitor screen. What's better? We decided to position them vertically to the bottom of the monitor screen rectangle. It is curious that this consideration is quite convincingly consistent with the ancient and still very reliable experience of construction using a plumb line. Let us explain this.

Idea 2. The classic plumb by its design is a pendulum with an adjustable thread length. Such a pendulum at rest allows builders to reliably and very quickly determine even minor deviations from the vertical of the part of the wall being erected, which it touches. In our case, the wall corresponds to the current position of the hal-graph. And three fixed vertical lines symbolize three plumb lines.

The ease of detecting even the slightest changes in the positions of the hal-graph is as follows. Even minor changes in the position of the vertices of a hal-graph cause corresponding changes in the position of the edges of this graph. It is almost impossible not to notice such changes.

4 Adaptive Coordinate System

Let some n-dimensional Cartesian coordinate system defined in the space $R_n (n > 3)$. The axes of this coordinate system will be denoted by the symbols $y_1, y_2, \ldots, y_n$.

For convenience, the plane of the monitor screen will be called the *carrier plane*.

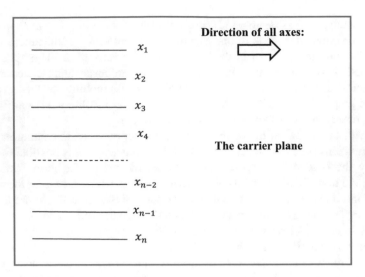

Fig. 1 Here, the rectangle represents the rectangle monitor screen. By depicting this rectangle, we wanted to show that all coordinate axes can be located in left half of the monitor screen

We place n one-dimensional coordinate axes $x_1, x_2, \ldots, x_n$ on the carrier plane based on the following 4 requirements (see Fig. 1):

1. Axes $x_1, x_2, \ldots, x_n$ must be parallel to the bottom of the rectangle of the monitor screen (the thought of using parallel coordinate lines is not new [4]).
2. All axes are oriented in the same direction from left to right.
3. For symmetry reasons, we assume that the distance between any two neighboring axes is the same.
4. The x_1 axis on the carrier plane is in the top position. The distance of each axis x_{i+1} from the bottom side of the rectangle of the monitor screen is less than the distance of the x_i axis from the bottom side of the same rectangle, $i = \overline{2, n-1}$.

Note 2. Of course, it is possible in this coordinate system to mapping any value of the quantity with two index $X_{ij}(t)$, $i = \overline{1, N}$, $j = \overline{1, k_i}$. If necessary, we can always transformation designations with two index to designations with one index. The transition formulas are as follows:

$$X_{ij} = \begin{cases} x_j, & i = 1, \; j = \overline{1, \, k_1}, \\ x_{k_1 + k_2 + \ldots + k_{i-1} + j}, & i = \overline{2.N}, \; j = \overline{1, k_i}. \end{cases} \tag{2}$$

For two index functions $a_{ij}(t)$ according to (2) we can pass to one index function: $a_{ij}(t) \rightarrow a_m(t), m = \overline{1, n}, i = \overline{1, N}, j = \overline{1, k_i}$.

Taking into account (2), in what follows we will use designations with one as well as two indices. In particular, we can equally well replace the designations of the single-indexed coordinate axes with their corresponding designations with a double index.

Note 3. In Fig. 1, we deliberately did not indicate the directions of the coordinate axes with arrows. Because it will be better this way, the arrows should not participate in the forthcoming "dance" of the coordinate axes. These movements in the general case will be realized over time under the influence of some continuous group transformations.

Note 4. By intuition, we can imagine that the coordinate axes $x_1, x_2, \ldots, x_n$ on the plane R_2 are "dismantled" coordinate axes $y_1, y_2, \ldots, y_n$ from the space $R_n(n > 3)$ [4]. From Sect. 2 we know that there is no homeomorphism between the spaces R_2 and $R_n(n > 3)$. However, in [1] we rigorously proved that a homeomorphism exists between the coordinate axes $x_1, x_2, \ldots, x_n$ on the plane R_2 and the coordinate axes $y_1, y_2, \ldots, y_n$ from the space $R_n(n > 3)$. This fact is useful for making observations.

The main difference of the coordinate system from Fig. 1 from many known coordinate systems will be that it does not have a common origin for all coordinate axes [4]. The absence of a common point for all axes creates opportunities for constructing an adaptive coordinate system in an ordinary two-dimensional Euclidean space as a kind of multidimensional coordinate system. The complete independence of the coordinate axes from each other allows us to simultaneously apply various group transformations to them.

Now we are ready to describe the transformations due to which our coordinate system is called adaptive.

Note 5. It is important to note that our further actions described in this section are more needed to substantiate the validity of the very idea of an adaptive coordinate system. In practice, we will not implement these actions, but will only use the final results of these actions.

On the carrier plane, we additionally introduce new coordinate axes $x_1', x_2', \ldots, x_n'$. Suppose that the new coordinate axes coincide with the old ones: $x_i' = x_i, i = \overline{1, n}$. We assume that before the start of the observation process, i.e. for $t < t_0$, each pair of coordinate axes x_i and $x_i', i = \overline{1, n}$, have common origins and units. In other words, before the start of the observation process, the new coordinate system $x_1', x_2', \ldots, x_n'$ is a complete copy of the coordinate system $x_1, x_2, \ldots, x_n$ (see Fig. 2). This identity stops only after the start of the observation process, namely at $t \geq t_0$. In the general case, after the start of observation, the new coordinate axes become movable $x_1' = x_1'(t), t \in T$, and participate in certain translation and scaling transformations groups. Consider now the optimal functions $c_{ij}(t)$ and based on (2) their one index equivalents: $c_{ij}(t) \to c_m(t), m = \overline{1, n}, i = \overline{1, N}, j = \overline{1, k_i}, t \in T$. In the general case, each function $c_m(t)$ in the course of time $t \in T$ will change its position on the coordinate axis x_m and $x_i', m = \overline{1, n}$.

Consider at time t each point having the coordinate $c_m(t)$ on the coordinate axis x_i' as the vertex of the graph, $m = \overline{1, n}, t \in T$. Connecting every two neighboring vertices with an edge, we get the already mentioned graph of optimal values $GC, t \in T$. In the general case, as already noted, due to the mobility of the vertices and the edges adjacent to them, the graph of optimal values can take on various configurations that are inconvenient for observation in the course of time. Starting to implement idea 2, we will greatly simplify the configuration of this graph.

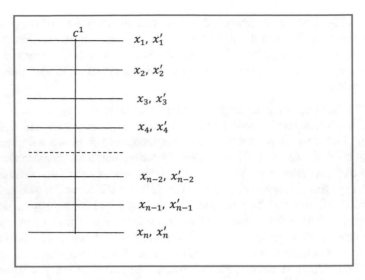

Fig. 2 The use of two notations for each coordinate axis means that the x_i and x_1' coordinate axes coincide before the start of the observation process. During the observation, all the x_i axes and the points c^i fixed on them remain motionless, $i = \overline{1, n}$. The transformations of the of the x_1' axes are defined relative to these fixed points

Note 6. Formally, from the this place we could go to the necessary transformations (11) and (12), using only some of the notation from the missing part of the text. We decided not to do this for the following reason. As we will see, transforms (11) and (12) are a composition of some translation and scaling transformations. These transformations each individually preserve the order relation between the positions of the points on any coordinate axis. In other words, if $a \mapsto a'$, $b \mapsto b'$ and $a < b$ then inequality $a' < b'$ is true, where a and b are the coordinates of any two points on any coordinate axis. Therefore, successive application of these transformations will not violate the order of the points. It is important. However, if we go straight to formulas (11) and (12), then we will have to give some reasoning showing that the order relation between points on the axes is an invariant of these transformations.

At each moment of time, we apply the following translation transformations on relation to the first coordinate axis:

$$x_1' \mapsto x_1' + c^1 - c_1(t), t \in T \tag{3}$$

where c^1 is some constant that we ourselves will choose. We choose the value of the constant c^1 on the fixed coordinate axis x_1 so that it is located closer to the center of the part of this axis drawn in Fig. 1. Let us draw a perpendicular from the location of this point to the bottom side of the rectangle of the monitor screen (see Fig. 2).

It can be seen from transformation (3) that the x_1' axis can move to the left or right, but so that the x_1' axis translations ensure a transformation $c_1(t) \to c^1$ at each moment in time Connecting every two neighboring vertices with an edge. Since a point with a coordinate c^1 is fixed on a non-moving x_1 axis, it will remain motionless in the course of time $t \in T$. But then the vertical from Fig. 2 will also be immobile in the course of time. The points of intersection of this vertical with the coordinate $x_2, x_3, \ldots, x_n$ axes will be denoted by symbols $c^2, c^3, \ldots, c^n$ respectively. If we assume that the coordinate origins of the axes $x_1, x_2, \ldots, x_n$ are on the same vertical, then the numerical values of the quantities $c^1, c^2, \ldots, c^n$ are the same. The superscript of the quantity is equal to the number of the axis to which this quantity belongs.

On each coordinate axis x_m', we implement the following translation transformations:

$$x_1' \mapsto x_m' + c^m - c_m(t), t \in T, m = \overline{2, n} \tag{4}$$

If now, in the interval of the text between formulas (3) and (4), we replace index 1 by index m, $m = \overline{2, n}$, then we will obtain a proof that the graph of optimal values $GC(t)$ with a movable configuration is transformed into a vertical from Fig. 2, not movable in the course of time $t \in T$. This graph, which is motionless in the course of time, will be called the *optimal graph* and denoted by the symbol GC.

Note 7. In [1], we got the same result in a much more complicated way.

Note 8. This is especially important. Here, in contrast to [1], we have abandoned the images of the coordinate axes of time in the adaptive coordinate system. By this, we have achieved a serious qualitative simplification. No confusion arises here, since we assume that it is imperative to use the following technique that exists in programming languages. Namely, the current data show on the coordinate system in the usual way, but we make the past data invisible.

Now we want to achieve the same result for the two graphs defined above GA (t) and GB (t), $t \in T$. In other words, we want to simplify the mutable configurations of these two graphs by applying certain group transformations.

Note 9. The use of double indices in designations is good because we can immediately determine the numbering of the sensor and the observed object using these indices. The advantage of single-index designations is that such indexes directly indicate the number of the coordinate axis to which they refer. For simplicity of presentation, according to (2), we will use the notation with one index. We did this above for the notation of optimal functions and left boundary functions. So, we will denote functions: $b_{ij}(t)$ and $i_j(t)$ as follows: $b_{ij}(t) \to b_m(t)$, $i_j(t) \to h_m(t)$, $t \in T$, $m = \overline{1, n}$, $i = \overline{1, N}$, $j = \overline{1, k_i}$.

Let us find out how the other three functions: $a_m(t), b_m(t)$ and $h_m(t), t \in T, m = \overline{1, n}$, change during the implementation of transformations (3) and (4). According to (3) and (4), we have:

$$\overset{\circ}{a}_m(t) = a_m(t) - c_m(t) + c^m, \tag{5}$$

$$b_m^{\circ}(t) = b_m(t) - c_m(t) + c^m,$$ (6)

$$h_m^{\circ}(t) = h_m(t) - c_m(t) + c^m$$ (7)

here $t \in T, m = \overline{1, n}$.

The inequalities are invariants of the translation transformations. This means that inequalities (1) remain valid after transformations (3)–(7), and we get:

$$a_m^{\circ}(t) < c^m < b_m^{\circ}(t), t \in T, m = \overline{1, n}.$$ (8)

We saw that the axes of the coordinate system $x_1', x_2', \ldots, x_n'$ are used to implement transformations (3) and (4), and the axes of the not mobile coordinate system $x_1, x_2, \ldots, x_n$ we used to fix the constants $c^1, c^2, \ldots, c^n$. When transforming the translation, the order of the points on the coordinate axes and the distance between the points do not change. These transformations allowed us to achieve an important result: $GC(t) \to GC \forall t \in T$. In order to achieve the same result for graphs $GA(t)$ and $GB(t), t \in T$, we will use scaling transformations. Scaling transformations do not change the order of the points on the coordinate axes, but the distances between the points change. We plan at each moment of time to achieve a simplification of the configuration of two graphs by means of a special change in the distance between the points. Clearly, we will have to apply two scaling transformations acting on the left and right sides of the optimal vertical, respectively. We will choose fixed points with coordinates $c^m, m = \overline{1, n}$, as the centers for some scaling transformations. This is possible because the scaling centers remained immobile during transformations (3) and (4). In addition, because the coordinates $c^1, c^2, \ldots, c^n$ in the general case are not equal to zero, in these transformations we will also have to carry out translation transformations by constant values.

When implementing translation transformations, we chose one point with a fixed coordinate c^1 on a non-moving coordinate axis x_1 as a landmark. Now, not forgetting about the optimal graph, as additional landmarks, we will choose two constant positive numbers: $\lambda > 0$ and $\mu > 0$. These two numbers determine the distances to the left and to the right of the optimal graph, respectively, of the future fixed two verticals. The case where $\lambda = \mu$ is certainly not excluded.

We will need the following transformations, which are implemented on the coordinate axes $x_1', x_2', \ldots, x_n'$:

$$x' \mapsto c^m - \left(\lambda\left(x' - c^m\right)/\left(a_m^{\circ}(t) - c^m\right)\right), \quad \text{if } x' \le c^m$$ (9)

$$x' \mapsto c^m + \left(\mu\left(x' - c^m\right)/\left(b_m^{\circ}(t) - c^m\right)\right), \quad \text{if } x' \le c^m, t \in T, m = \overline{1, n}.$$ (10)

Inequalities (8) exclude the possibility of division by zero here. Now we will create a composition of transformations (3), (4) and (9), (10). We have

$$x' \mapsto c^m - \left(\lambda\left(x' - c_m(t)\right)/(a_m(t) - c_m(t))\right) \text{ if } x' \le c_m(t) \tag{11}$$

$$x' \mapsto c^m + \left(\mu\left(x' - c_m(t)\right)/(b_m(t) - c_m(t))\right), \text{ if } x' \ge c_m(t), t \in T, m = \overline{1, n}. \tag{12}$$

Inequalities (1) exclude the possibility of division by zero here.

Note 10. It is obvious that even after obtaining (11) and (13) acting on the coordinate axes $x'_1, x'_2, \ldots, x'_n$ transformations, we cannot abandon the coordinate axes $x_1, x_2, \ldots, x_n$. From a geometric point of view, this non-movable coordinate system, or at least its first coordinate axis x_1, is necessary to fix the point with coordinate c^1 on the carrier plane.

From (11) it can be seen that for any value of time and for any value of the index m, point with coordinate $a_m(t)$ is mapped to point with coordinate $c^m - \lambda, t \in T$, $m = \overline{1, n}$. Since all the points with coordinates c^m, which are the vertices of the optimal graph, are on one fixed vertical, then the points with coordinates $c^m - \lambda$ will be on one fixed vertical, $m = \overline{1, n}$. Consider the points with coordinates $c^m - \lambda, m = \overline{1, n}$, as the vertices of the graph and denote it by the symbol **GA**. Thus, the composition of transformations (11) leads us to the desired result $GA(t) \mapsto GA \forall t \in T$. Now in the adaptive coordinate system, we obtained images of two parallel graphs **GA** and **GC** that do not move in the course of time. Acting by analogy, using transformations (12), we obtain a new graph that is fixed and parallel to graphs **GA** and **GC**. The vertices of the new graph, which we denote by the symbol **GB**, have the following coordinates: $c^m + \mu, m = \overline{1, n}$.

Note 10. We made sure that with the help of transformations (11) and (12) we can transform graphs **GC**(t), **GA**(t) and **GB**(t), which have a variable configuration, into graphs **GC**, **GA**, and **GB**, which have constant and simplest configurations (see Fig. 3).

Note 11. The configuration of graphs **GC**, **GA** and **GB** is so simple that we can draw them on an adaptive coordinate system without implementing transformations (11) and (12).

Therefore, we need transformations (11) and (12) only to determine the current positions of the hal-graph. To determine the current coordinates of the vertices of a hal-graph, based on (11) and (12), we obtain the following transformation formulas:

$$h'_m(t) = c^m - (\lambda(h_m(t) - c_m(t))/(a_m(t) - c_m(t))), \text{ if } h_m(t) \le c_m(t), \tag{13}$$

$$h'_m(t) = c^m + (\mu(h_m(t) - c_m(t))/(b_m(t) - c_m(t))), \text{ if } h_m(t) \ge c_m(t), t \in T, m = \overline{1, n}. \tag{14}$$

For us, these two formulas are *basic* when making observations. Functions: $a_m(t)$, $c_m(t)$ and $b_m(t)$, $t \in T$, $m = \overline{1, n}$, and constant parameters λ and μ are known to us even before the beginning of the observation process, $t < t_0$. This means that all expressions

$$\lambda/(a_m(t) - c_m(t)) < 0 \tag{15}$$

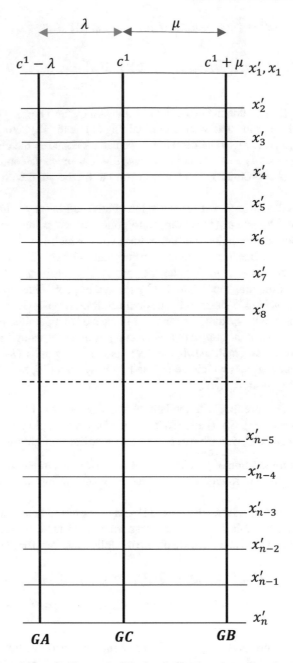

Fig. 3 According to Note 11, the graph *GC* of optimal values, as well as the graphs of left and right bound values *GA* and *GB*, respectively, can be drawn without carrying out any transformations before the start of the observation process. To do this, we need to select two positive numbers λ, μ and decide on the choice of the location of the point with coordinate c^1 All three graphs will remain motionless during the entire observation process

$$\mu/(b_m(t) - c_m(t)) > 0, t \in T, m = \overline{1, n} \tag{16}$$

included in (13) and (14) can be determined even before the beginning of the observation process. Performing preliminary calculations using expressions (15) and (16) have a special meaning. They will allow us to drastically reduce the amount of calculations by formulas (13) and (14) in a real observation process. Figures 4 and 5 show the possible, respectively, no emergency and emergency states of the hal-graph.

Note 12. By the way, if in the process of observation a drawn picture from Fig. 3 appears, then this will mean that the hal-graph coincides either with the graph of optimal values *GC* or with one of the bound verticals, *GA, GB*. The first case means that at the current moment all observed facilities are functioning flawlessly. The remaining two cases will mean that at the current time all observed objects are in emergency conditions. To avoid misunderstandings, we strongly recommend painting the image of the hal-graph in a different color.

Note 13. The readings of different sensors can have different physical dimensions. It is also possible that the readings of some sensors have no physical dimension. We can assign different physical dimensions to different coordinate axes. This is possible because the coordinate axes are independent and do not have intersection points. However, different physical dimensions cannot be attributed to the coordinate axes with the same indices x_i' and x_i, for some i. In addition, there is no need for unification of physical dimensions. E.g., the temperature can be measured in degrees Celsius on one coordinate axis at the same time, and in degrees Fahrenheit on the other axis, etc.

Note 14. Thus, as can be seen from Figs. 4 and 5 at each current moment of time, we receive information, which is not difficult for perception (in our opinion, of course), about the quality of functioning of the objects. Indeed, it is difficult to see the movements of the vertices of the hal-graph, but it is very difficult not to notice the movements of the edges of the hal-graph attached to these vertices. Figures 4 and 5 are a demonstration of the implementation of Idea 2 from Sect. 3. Graphs *GA*, *GC* and *GB* are motionless during the whole observation process. They are graph analogs of three fixed plumb lines. In this case, the hal-graph at each moment of time defines a certain configuration, which can be considered as an analogue of the wall mentioned in Idea 2.

Note 15. From Figs. 1, 2, 3, 4 and 5 we saw that the images these adaptive coordinate systems takes up little space on the carrier plane. In the free place of this plane, we can insert the new adaptive coordinate systems. They can be used either to implement observations of new objects or for observations corresponding to the past moments of time. The latter is important for monitoring the dynamics of the behavior of sensor readings. Also in the free space, we can place text boxes for entering information, as well as command buttons or various labels.

Note 16. If we abandon the need to substantiate the correctness of the above reasoning, then all of our material in the form of an instruction can be presented in two or three

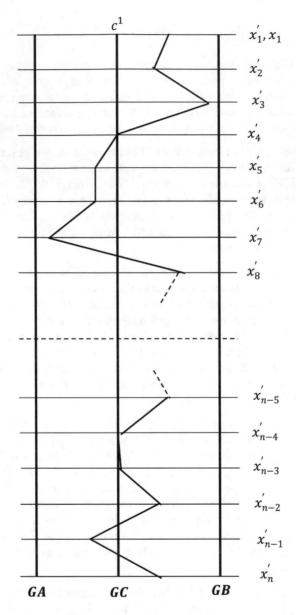

Fig. 4 The broken line shown here is a hal-graph. The positions of the vertices of the hal-graph are determined by the current values of the sensor readings and can change over time. The fact that these vertices are in the interval between the graphs **GA** and **GB** means that the received current values of the sensors are admissible, In other words, this means that in the observed objects at the current moment is no situation emergency. Also, we see that the vertices of the hal-graph on the coordinate axes x_3', x_7', x_8' at the current time moment are dangerously close to the bound graphs. At the same current moment, we see that the vertices of the hal-graph on the axes x_4', x_{n-4}', x_{n-3}' are on the desired optimal graph

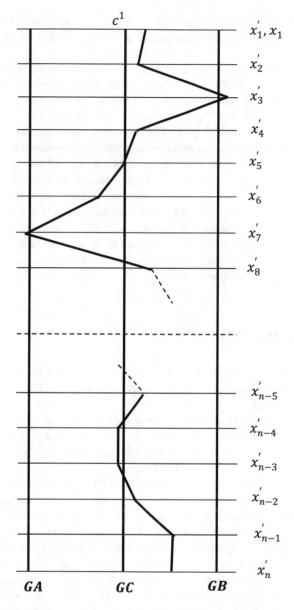

Fig. 5 Hal-graph signals the presence of an emergency at the current moment. The points on the coordinate axes x_3', x_7' have crossed or touched the "*forbidden*" boundary graphs at the current time

pages. Because after describing the adaptive coordinate system and the initial data, you can immediately go to the main formulas (13) and (14) in order to apply them in the program.

Example. If our system is used in the cockpits of modern airliners, then this can lead to an increase in the reliability of observation, since pilots will be able to easily perceive information about the dynamics of changes in the readings of all sensors from one place. In addition, this will lead to a drastic reduction in the number of instruments used in the cockpit (not as in Tesla cars, of course, but still close). An analysis of numerous aircraft accidents (demonstrated on the DISCOVERY channels) shows that some malfunctions occur almost immediately after an airliner takes off (far from all such accidents occur due to birds getting into the engine). In this case, the alarm is triggered when the aircraft is already in the air. We assure you that if the pilots had the opportunity to see the dynamics of dangerous changes in the readings of the sensors in a timely, they would refuse to take off. The same thing can happen in flight itself. Premature detection of danger, that is, before the alarm is triggered, can give additional time for an emergency landing. And the number of terrible aviation accidents would probably decrease.

5   System Advantages

1. Information about the current readings of all sensors is accumulated in one place (There is no need to try to keep numerous devices in sight. The devices themselves become redundant).
2. In the general case, the number of variable data is 4n, and of these, the observer must track only n variable data, that is, in the system, when observing, it becomes unnecessary to take into account changes 3n variable data (The result is achieved due to transformations (13) and (14).
3. The system allows you to see the dynamics of changes in the readings of all sensors (Additional seconds are won for timely intervention and the not very pleasant effect of an unexpected activation of the alarm is reduced).
4. The ease of observation follows from the extreme simplicity of using a plumb line when erecting vertical walls (The system uses three fixed plumb lines, that is, three lines perpendicular to parallel coordinate axes. The system adapts to incoming current data in such a way that at any time fixed vertical lines contain all information about invalid and optimal values. Only the current state graph (hal-graph) can moves in the system, the coordinates of the vertices of which are determined by the current values of the sensors. It is easy to observe the movement of these vertices and compare their position relative to three fixed verticals for two reasons:

   (a) Even the most insignificant movements of the vertices immediately cause movements of the edges of the hal-graph, which are hard not to notice.

(b) The "plumb line method" used in practice is so simple that it usually does not require of special efforts).

5. The reliability of observation is justified by Idea 2 about the use of a plumb line (Proved by time. Numerous buildings, both modern and old, with vertical walls prove this by their very existence. Apparently, such a result could not be obtained without the use of a plumb line (classical or using a laser beam)).
6. Some additional considerations that simplify the implementation of the system are given in the text and in some notes.

6 System Disadvantages

1. If we draw the location of units of measurement with small vertical lines (marks) on the coordinate axes, then, in the general case, these lines will be movable (In this case, the order of location of these marks does not change under transformations (13) and (14)).
2. We noted above that if the number of sensors is not more than 120, then the observation can be carried out without much difficulty and successfully by one observer. The system itself easily allows to increase the number of observed sensors up to 1000 and even higher. Unfortunately, we did not test for cases where the number of sensors is more than 120 (The number of parallel axes is limited by the height of the screen used).

References

1. Akhundov, A.A.: Geometrical solution of the problem of simultaneously tracking of the indications of the numerous sensors of different types, p. 256. Publishing house ELM Baku "ELM" (2010)
2. https://archive.org/search.php?query=Alakbar%20Akhundov
3. Brouwer, L.E.J.: Über den natürlichen dimensions-begriff. Journ. f. Math. **142**, 146–152 (1913)
4. https://en.wikipedia.org/wiki/Parallel_coordinates

Fuzzy Neural Networks

Advanced Trainings in the System of Teachers' Professional Development: The Empirical Study

T. Aliyeva and U. Rzayeva

Abstract This article presents a method for solving the algorithm of the model of recognition of visual and audio information of the environment by constructing an algorithmic model that simulates the human abilities of the motor (unconscious or non-intellectual) behavior. The article is devoted to the development of a machine intelligence system with a gradual increase in the complexity of the behavioral model of an artificial personality (IL) in order to experimentally study the issues of artificial intelligence.

Keywords Professional competence of teachers · University trainings · Analysis of impact of the teacher's training on their qualifications

1 Introduction

The current stage of development of Azerbaijani society is characterized by a rapid change in technologies, which determines the formation of a new education system, implying constant renewal. The success of the implementation of the concept of lifelong education adopted in the Republic of Azerbaijan depends on the ability of all subjects of the education system to maintain competitiveness, the most important conditions for which are such personality traits as activity, initiative, the ability to think creatively and find non-standard solutions.

Despite the fairly wide representation of the studied phenomenon of education in the scientific literature, there is still no unambiguity both in its rationalization and in determining its composition, and, consequently, in identifying the ways of its development. Thus, the relevance of the topic is due to the insufficient substantiation

T. Aliyeva · U. Rzayeva (✉)
Department of Digital Technologies and Applied Informatics, Azerbaijan State University of Economics (UNEC), Baku AZ1001, Azerbaijan
e-mail: ulviyya.rzayeva@unec.edu.az

T. Aliyeva
e-mail: tarana.aliyeva@unec.edu.az

© The Author(s), under exclusive license to Springer Nature Switzerland AG 2023
Sh. N. Shahbazova et al. (eds.), *Recent Developments and the New Directions of Research, Foundations, and Applications*, Studies in Fuzziness and Soft Computing 422,
https://doi.org/10.1007/978-3-031-20153-0_24

of the ways of developing the teachers' professional qualifications and the ever-increasing requirements of social practice in competence of workers.

This work is aimed at studying the influence of the means offered by the university management in order to increase the teacher's competence, and at identifying the features of these means.

Based on the foregoing, the purpose of the study was determined—to identify the nature of qualification changes after passing online training in specialties.

The object of the study was the opinions of the teachers of the Department of Digital Technologies and Applied Informatics of the Azerbaijan State Economic University (UNEC).

The subject of the research was the teacher's qualifications and their changes under the influence of online training.

2 Methodology

In the course of the research, the following methods were used: questioning, testing, including observation, ascertaining and shaping the experiment, and analysis of the results obtained using the SPSS software (the questionnaire is presented at the end of the paper). Components common to all teachers were used as an independent variable in the experiment, as a means of forming qualifications in professional activity (the presence and absence of an academic title, position, length of service, age, gender, etc.). The dependent variable is the teacher's competence, determined by the teacher's ability to carry out pedagogical activities, taking into account the specifics of the course.

3 Literature Review

The problem of the professional qualifications of a teacher has been studied by many philosophers, teachers, and psychologists.

The article [5] considers a number of factors that determine the degree of teacher motivation for effective innovative teaching at the university.

The article [3] analyzes ex post facto the results of a survey of 2,180 university professors from Andalusia (Spain) in order to study and compare the degree of digital competence of university teachers from different fields of knowledge and different age groups in accordance with the structure of DigCompEdu.

The article [1] describes a rather interesting approach to the preparation of students in pedagogical specialties. The study provides an analysis of postgraduate programs for teachers, in which video stories were assessed as an effective way to examine teaching competencies, their practical application, and their development over time.

Article [2] examines the competence of teachers in teaching a non-native language. The analysis of the opinions of three teachers in the context of teaching spoken English as a foreign language in Iran is carried out. By comparing teachers' views on how students should learn to speak and how teachers themselves learned to speak English, the study reveals the intertwined nature of the learning experience with teaching concepts.

The article [11] analyzes undergraduate students' discussions of the linguistic competence of their teachers after switching the classroom environment from German to English.

The aim of the article [8] is to analyze the impact of the professionalism of university lecturers on coordination processes in universities.

The article [9] discusses some of the classic indicators of academic professionalism, consisting of separate sub-indices of research and teaching activities.

In the current conditions, in order to be successful and in demand, a teacher must be ready for any changes, be able to quickly and effectively adapt to new conditions, show the desire to be a professional, constantly update their knowledge and skills, strive for self-development, show tolerance for uncertainty, be ready to risk, i.e., be professionally competent. However, as social practice shows, these characteristics are not formed in all teachers. On the contrary, a significant part of them experiences great difficulties in adapting to rapidly changing social, economic, and professional conditions, and then the lack of professional qualifications can cause serious social and psychological problems for the individual—from internal dissatisfaction to social confrontation and aggression.

4 Data and Analysis

This research is based on information about trainings initiated by the UNEC leadership with the involvement of trainers—teachers from leading universities in Turkey and Russia in order to improve the professional qualifications of UNEC teachers. Using the electronic form, questionnaires were sent to the teachers of the Department of Digital Technologies and Applied Informatics of the Azerbaijan State Economic University with a request to answer a number of questions on the assessment of online trainings conducted during 2020–2021. To fill out the questionnaire, it was enough to choose the answer option that most closely matches the respondent's opinion. The staff of the department was 45 people of teaching personnel. 30 completed questionnaires were received back, which was 66% of the total.

Teachers are encouraged to participate in the study, the purpose of which is to provide real help assess the impact of online training in order to improve the teachers' qualifications.

5 Experimental Results

For data analysis, we used the SPSS statistical data analysis package.

At the initial stage of data analysis, it was usually necessary to check whether they were subject to the normal distribution, as many statistical methods were based on the probability that the distribution of the studied data was normal. In the case under consideration, the analysis of the survey results was carried out using descriptive statistical methods.

The survey questions, which involved 30 people—26 women and 4 men—were classified into 3 sections.

For simplification of the analysis all the questionnaires were divided into 3 groups:

(1) General questions;
(2) Questions related to participation in trainings;
(3) Questions related to trainings.

5.1 Fitness Analysis

The article first analyzes the reliability of the second and third groups of questions (Fitness analysis).

(1) Analysis of the answers to 7 questions, one of which was multiple-choice in the second group of survey questions, showed that the average reliability value for this section was 0.620, and the data had a normal distribution. Based on this indicator, it can be concluded that teachers participated in the trainings, realizing the importance of trainings in increasing knowledge and experience (Table 1) (hereinafter we used the Cronbach alpha coefficient, which shows the internal consistency of characteristics describing one object, although it is not an indicator of the homogeneity of the object. The coefficient is often used in social sciences and psychology to design tests and to test their reliability [4]).

On the other hand, out of the 7 questions given, except for multiple-choice question options in the section under review, the question about the number of trainings had the highest average and the question about benefits from trainings had the lowest average.

The second group of questions showed that most teachers preferred to improve their skills and learn from the experience of trained teachers in answering a variety

Table 1 Fitness statistics for realiability of the second group questions

Fitness statistics		
Cronbach's alpha	Cronbach's alpha based on standardized items	Quantity of items
0.620	0.664	15

Source Authors' calculations

Table 2 Fitness statistics for realiability of the third group of questions

Fitness statistics		
Cronbach's alpha	Cronbach's alpha based on standardized items	Quantity of items
0.577	0.542	6

Source Authors' calculations

of questions related to the purpose of participation in the training, as well as to participate in more trainings and hoped that the training would be useful in future pedagogical activities.

(2) The analysis of the answers to 6 questions of the third group of questions showed that the average reliability score for this section was 0.577, i.e., more than 50% of the respondents had a positive attitude to the trainings at UNEC (Table 2).

On the other hand, out of 6 questions in the section under review, the question about the attitude to the exam had the highest average and the question about the trainings' languages had the lowest average.

This showed that the participants had a good attitude towards the training exams organized at UNEC and came to the conclusion that the language of training limited their opportunities.

5.2 Determining the Average for Participation in Trainings and Attitudes Towards Trainings

The auxiliary participation_score variable, which determines the average of "training participation" and "attitude to training", was used to analyze and compare the survey results.

5.3 Comparison of the Results of Some Survey Questions Related to the Third Section with the Participation_Score Variable

The relationship between the participation_score variable and the participants' responsibilities is analyzed based on the observation of the results of the answers to the question "Your position".

The calculations show that despite the high number of associate professors among the participants, the average of this number is lower than the average number of other teachers.

According to the Kolmogorov–Smirnov test, which is a nonparametric goodness-of-fit test designed to check simple hypotheses about the belonging of the analyzed sample to some known distribution law, since in our calculations $p > 0.05$, it meets the Kolmogorov–Smirnov criterion [6]. This can be seen again from the histograms obtained, based on all the probabilistic characteristics, it is judged that the data are distributed according to the normal law, and rule 3 σ is satisfied.

The use of Q-Q plots used to evaluate the conformity of a distribution of values to a standard normal (Gaussian) distribution [7], is especially productive in checking the nature of small data distributions. Such graphs reflect the quantiles of both the empirical (based on the analyzed data) distribution and the theoretically expected standard normal distribution. In the case of the normal distribution of the variable being tested, the points on the quantile graph are aligned in a straight line at a 45° angle.

Although the average of those with 20–25 years of experience is low, the number is high, with the vast majority of participants having more than 25 years of experience. This means that despite their high experience, teachers prefer lifelong learning and always focus on increasing their experience and professional development.

5.4 T-Test Independent of the Results of the Survey According to the Question "Your Basic Education" of the First Part of the Questions with the Variable Participation_Score

It is known that the t-test checks whether the average value in one group is different from the average in the other independent group. Let one of these groups be "Your Basic Education" and the other be the participation_score variable.

Hypotheses are accepted for testing the t-test:

H0: There is no difference between the average of basic education and the participation_score variable.

H1: There is a difference between the average of basic education and the participation_score variable.

H0 is rejected when $p < 0.05$, and H0 is accepted when $p > 0.05$.

As can be seen from the last table (Table 3), there is no difference between the averages of the two groups taken, since $0.430 > 0.05$ and $0.348 > 0.05$, the data in the first line are taken as the basis, i.e. H0 is assumed. Otherwise, the presence of the condition $0.318 > 0.05$ in the binary value column on the second line proves once again that the t-test is satisfied. Indeed, although 19 of the respondents had non-pedagogical basic education, their average participation rate was higher than the average participation rate of teachers with basic pedagogical education. We assessed these hypotheses using Levene's test, which is used to estimate the equality of variance for a variable calculated for two or more groups. Some general statistical procedures assume that the variances of populations from which different samples are taken to be equal. Levene's test evaluates this assumption [10].

Table 3 Hypothesis analysis

	Levene's test for equality of variances		t-test of equality of means				
	F	Sign	t	f.d	Sign. (2-tailed)	Av. dif.	Std. er.
Eq. var. are assumed	1.031	0.319	−1.220	28	0.233	−0.1285	0.11
Eq. var.is not assumed.			−1.324	26	0.197	−0.1285	0.08

Source Authors' calculations

In the same way, the average of the answers to the question "Gender" can be compared with the variable participation_score and the same considerations can be made.

5.5 Analysis of the Results of the Selection Questions Stating the Purpose of Participation in the Trainings

The question "The purpose of participation in trainings" was answered on a variety of criteria.

According to the frequency table of the variable participation general, which expresses the options, 31.1% of the participants accepted participation in trainings as a tool for professional development and 18.9% as a criterion for studying the experience of training teachers.

6 Discussion

During the period 2020–2021, UNEC implemented a significant variety of programs for retraining, preparation, and promotion of qualifications of scientific and pedagogical staff. The article analyzes the results of trainings conducted on the development of professional and pedagogical competence.

Teachers participated in the trainings, realizing the importance of trainings in increasing knowledge and experience. Most of the teachers preferred to participate in more trainings and hoped that the knowledge gained as a result of the training would be useful in future pedagogical activities. More than 50% of the respondents have a positive attitude to the trainings at UNEC. Participants had a good attitude towards the training exams organized at UNEC and concluded that the language of training limited their opportunities. Despite the high level of experience, teachers prefer lifelong learning and always focus on increasing their experience and professional development.

7   Conclusion

The article analyzes the results of the most common types of professional development of a teacher carried out—trainings, which are assessed as a type of training aimed at the formation and improvement of certain skills and their combinations. The purpose of the article is to demonstrate the advantages of using such an interactive form of teaching as pedagogical trainings in additional training of teachers. During training sessions, teachers are immersed in professional educational activities that contribute to the active development of their personal qualities and significantly increase their readiness for work.

This work may be of interest to the teaching staff of secondary and higher educational institutions, psychologists, teachers.

References

1. Admiraal, W., Berry, A.: Video narratives to assess student teachers' competence as new teachers. Teach. Teach. **22** (1), 21–34 (2015). https://doi.org/10.1080/13540602.2015.1023026
2. Baleghizadeh, S., Shahri Nasrollahi, M. N.: EFL teachers' conceptions of speaking competence in English. Teach. Teach. **20** (6), 738–754 (2014)
3. Cabero-Almenara, J., Guillén-Gámez, F. D., Ruiz-Palmero, J., et al.: Digital competence of higher education professor according to DigCompEdu. Statistical research methods with ANOVA between fields of knowledge in different age ranges. Educ Inf Technol. **26**, 4691–4708 (2021). https://doi.org/10.1007/s10639-021-10476-5
4. Cronbach, L.J.: Coefficient alpha and the internal structure of tests. Psychom. **16**(3), 297–334 (1951)
5. Fernández-Cruz, F.J., Rodríguez-Legendre, F.: The innovation competence profile of teachers in higher education institutions. Innov. Educ. Teach. Int. (2021). https://doi.org/10.1080/147 03297.2021.1905031
6. Hafner, R.: Kolmogorov-Smirnov-statistics under the alternative. Optim. **6**(5), 787–796 (1975)
7. Hinloopen, J., Wagenvoort, R.: Identifying all distinct sample P-P plots, with an application to the exact finite sample distribution of the L1-FCvM test statistic. SSRN Electron. J. (2010). https://doi.org/10.2139/ssrn.1669207
8. Macheridis, N., Paulsson, A.: Professionalism between profession and governance: how university teachers' professionalism shapes coordination. Stud. High. Educ. **44**(3), 470–485 (2019). https://doi.org/10.1080/03075079.2017.1378633
9. Nordkvelle, Y.T.: Professional development of higher education teachers, can distance education make a difference? Turk. Online J. Distance Educ.—TOJDE. **7**(1), (2006)
10. Pallmann, P., Hothorn, L.A., Djira, G.D.: A Levene-type test of homogeneity of variances against ordered alternatives. Comput. Statistics **29**(6), 1593–1608 (2014). https://doi.org/10. 1007/s00180-014-0508-z
11. Studer, P.: Coping with English: students perceptions of their teachers' linguistic competence in undergraduate science teaching. Int. J. Appl. Linguist. **25** (2), 183–201 (2014). 10.1111 / ijal.12062

Quality Estimation of Wine Data Using Improved Crow Search Algorithm Based Fuzzy Neural Networks Classifier

Zeinab N. Ali, I. N. Askerzade, and M. S. Guzel

Abstract This paper focuses on developing an efficient framework using Improved Principal Component Analysis (IPCA) and hybrid neural networks based machine learning techniques for estimating the quality of red wine and white wine datasets. IPCA is a dimensionality reduction technique based on cumulative sum improved PCA. The proposed machine learning classifier is introduced by integrating the Fuzzy Neural Networks (FNN) and Improved Crow Search Algorithm (ICSA). In this model, the parameters of the FNN are automatically tuned using the ICSA. The ICSA is developed by improving the dynamic awareness probability, local search and global search abilities of the standard crow search algorithm to overcome the slow convergence and local optimum problem. This hybrid classifier model of ICSA-FNN improves the classification accuracy for the wine data and provides highly accurate results within less computation time. Experiments are conducted using the Red Wine and White Wine datasets from UCI Machine Learning Repository. The results showed that the proposed quality estimation framework using IPCA and ICSA-FNN has higher performance than the existing models in terms of accuracy, precision and computation time.

Keywords Fuzzy neural networks · Crow search algorithm · Wine quality

1 Introduction

Wine is a globally used beverage since its values are considered vital in some societal customs [1]. The revenue associated with the wine market is greater and hence the quality of wine becomes of utmost importance for both the customers and the competing wine producers [2]. The wine quality has been traditionally determined only after the end of the production stage using suitable testing methods [3]. However, the wine quality cannot be determined using this approach each time as a large amount

Z. N. Ali (✉) · I. N. Askerzade · M. S. Guzel
Department of Computer Engineering, Ankara University, Ankara 06100, Turkey
e-mail: imasker@eng.kara.edu.tr

© The Author(s), under exclusive license to Springer Nature Switzerland AG 2023
Sh. N. Shahbazova et al. (eds.), *Recent Developments and the New Directions of Research, Foundations, and Applications*, Studies in Fuzziness and Soft Computing 422,
https://doi.org/10.1007/978-3-031-20153-0_25

of money and time has already been associated with the production process. When the produced wine quality is found to be poor or not sufficiently good, the producers will face greater financial troubles. The customers won't purchase the low quality or not-so-good category wine [4]. The high-quality wines fetch the highest profits for the producers in such cases. Hence the producers employ different strategies to estimate the quality of the wine at different stages of production based on the ingredients and the added chemical compositions to ensure high-quality wine production [5]. The wine manufacturers estimate the quality and quantity of the wine products at each stage of production and make sure the final wine is qualified to be among the best products. Since the taste opinions of every person differently, it cannot be suitable to select the quality of one person or a group of persons. Hence it is advisable to make sure there is a standard quality for the wine that is adequate for each wine consumer. In this manner, the wine producers can save more time and money while the consumers are also satisfied with the final product [6].

In recent decades, there have been multiple attempts to accumulate various data about estimating wine quality. Quantities as values of different chemicals, product ingredients, temperature, moisture, production time, etc. have all been considered as important elements in estimating the quality [7]. Various efforts have been made to analyse the available data from various databases regarding wine production [8]. Analysing and processing these data will help decide the quality of the wine and ensure the manufacturers focus on delivering quality wine to the customers [9]. Traditionally, these data were analysed using manual methods which are tedious and time-consuming. Also, the possibility of human errors is large. Hence the automated estimation strategies were employed using the statistical and data mining techniques [10]. However, the statistical and traditional data mining techniques are less effective in analysing the quality of the wine data. The efficiency of these methods was limited in terms of computational effectiveness and precision. Therefore, the need for advanced computation models has been increased in the past decade [11].

Machine learning (ML) algorithms have become the most sought algorithms for classification tasks like prediction, analysis and categorization tasks. ML methods are also commonly applied for estimation based studies [12]. ML algorithms can be used for wine data quality estimation by identifying the most important parameters that control the wine quality [13]. On studying various ML algorithms, it has been found that the shallow learning process of the ML algorithms limits their overall effectiveness. While deep learning (DL) algorithms solve this problem by learning the deep features of the dataset, they are not preferred for the wine datasets due to their moderate size [14]. Hence an ML algorithm that resembles the DL characteristics is required for this study. Additionally, it should also provide high prediction property with sparse samples. The standard neural networks (NN) are an ML algorithm resembling the DL characteristics. The NN are the base architectures for many DL algorithms like convolutional NN, recurrent NN and its variants. Fuzzy logic (FL) is a conventional method that has abilities to learn the deep features more efficient than the ML algorithms [15]. This research study focuses on integrating these two methods to form an effective ML algorithm as a solution to the wine quality estimation problem.

After thorough analysis, the NN and FL methods have been elected as the primary estimation methods. However, the analysis has also reflected some of the limitations of NN and FL. NN are dependent on the hardware characteristics and their ability to learn the problem and resolving the problem consumes more training time. FL methods are often inaccurate due to the fuzzy (vague) inputs and also they depend entirely on human knowledge. Additionally, the FL methods cannot recognize the automated learning approaches [16]. Yet, the advantages of these two methods over-power these limitations. Hence, NN and FL were combined to form the Fuzzy Neural Networks (FNN). Combining these two models also creates complexity in terms of training time and model architecture. To overcome this problem, the parameters of the FNN are tuned optimally using an Improved Crow Search Algorithm (ICSA). The proposed ICSA selects the best values of weights, width and center value of the network architecture in FNN that can improve the training speed of the classifier [17]. This tuning process also reduces the number of nodes in the layers of FNN, so that the architecture is simplified. The contributions of this paper are given as follows:

- Firstly, the Improved Crow Search Algorithm is developed by improving the dynamic awareness probability, local search and global search abilities of the standard crow search algorithm. This will increase the convergence rate and avoid the local optimum problem.
- Secondly, the parameters of the FNN classifier are tuned optimally using the proposed ICSA so that the model is simplified and training time is minimized.
- The evaluations of the proposed model are performed over the wine quality dataset available in UCI Machine Learning Repository.

This research article is arranged as: Related study in Sect. 1. The proposed ICSA-FNN model and its implementations are elucidated in Sect. 2. Performance outcomes and comparative investigations are demonstrated in Sect. 3. Conclusions and possible future directions of this research are specified in Sect. 4.

2 Wine Quality Estimation Methodology

The word "data" is plural, not singular. In American English, periods and commas are within quotation marks. The proposed wine quality estimation method includes the stages of pre-processing, feature extraction, feature dimensionality reduction and classification. The evaluation datasets have high-dimensional imbalance problems, including missing features, missing feature values and cumulative data which might be noisy, errors and outliers. These issues can be eliminated through the pre-processing and feature extraction tasks. The label transformation, redundancy removal and data normalization tasks are performed in the pre-processing stage to avoid noise, errors and outliers. In label transformation, the class label names are converted as numerical labels. Redundancy removal eliminates the duplicate samples or repeated samples and data normalization revises the sample values between certain

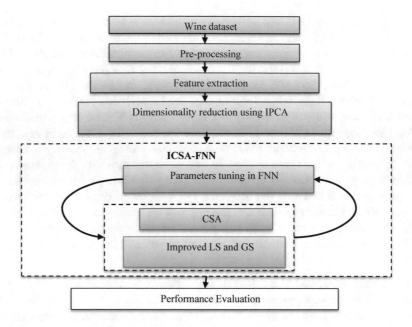

Fig. 1 Wine estimation methodology flow diagram

decimal values. Then the wine data features from the input dataset are extracted and the best features are selected using IPCA to feed the ICSA-FNN classifier to identify the good and bad samples [18]. Figure 1 illustrates the flow diagram of the proposed wine quality estimation framework.

2.1 Dataset Description

In presented study we use the publicly available wine quality dataset obtained from the UCL Machine Learning Repository [19]. There are two types of wine quality datasets is most popularly used for quality estimation. Both the red wine and white wine dataset contains 11 physicochemical parameters: fixed acidity (g[tartaric acid]/dm^3),volatile acidity (g[acetic acid]/dm^3), total sulphur dioxide (mg/dm^3), chlorides (g/dm^3), pH level, free sulphur dioxide (mg/dm^3), density (g/cm^3), residual sugar (g/dm^3), citric acid (g/dm^3), sulphates (g/dm^3), and alcohol (vol%). Additionally to these properties, a sensory score for quality was acquired from several different blind taste testers. Quality of wine graded with a score ranging from zero (poor) to 10 (excellent). The median was recorded and serves as the response variable. The red wine dataset contains the 1599 records and white wine contains 4898 random samples. A sample subset of the red wine dataset containing 20 samples taken from [18].

The analysis of red and white datasets provided few inferences:

- The sulphur dioxide (both free and total) and alcohol are the two important features among the 11 attributes.
- Alcohol is the most important ingredient. A higher concentration of alcohol leads to a better quality of wine and a lower density of the wine.
- Sulphates are used to prevent spoilage and have a positive correlation to wine quality.
- Volatile acidity adds to acidic tastes and has a negative correlation to wine quality.
- Citric acid is added to provide a freshness test and hence has a positive impact on wine quality.

2.2 Fuzzy Neural Networks

Fuzzy Neural Network (FNN) combines the knowledge expression ability of fuzzy logic and the self-learning ability of neural networks [20]. The practical design of FNN includes network structure and parameter identification. In traditional FNN, the Back Propagation algorithm was used to optimize the parameters in the parameter identification stage. But, the Back Propagation has the problem of providing the local optimum as the global solution and tends to limit the performance of the FNN. Hence, an advanced optimization algorithm has been sought which is fulfilled in this study using ICSA. Figure 2 shows the schematic structure of FNN.

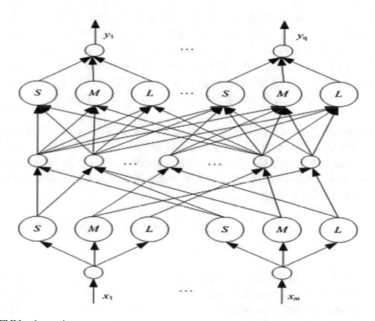

Fig.2 FNN-schematic structure

2.3 Improved Crow Search Algorithm

Crow Search Algorithm (CSA) is a new and advanced population-based meta-heuristic introduced by Askarzadeh [21]. It is based on the food search and storage behaviour of the crows. The pattern in which the crows hide their food away from the other crows is an optimization process. The position of the individual i-th crow at any generation in the search space is specified by vector and it denotes a possible solution to the optimization problem.

The FNN is designed as the multilayer FNN to accommodate the deep learning task. A total of five layers are used with each layer performing different functions.

$$x_i^{gen} = \left\{ x_{i,1}^{gen}, x_{i,2}^{gen}, \ldots, x_{i,n}^{gen} \right\}; i = 1, 2, \ldots, NP; gen = 1, 2, \ldots, \text{max\_gen} \quad (1)$$

Here, NP denotes the population size and max_gen denote the maximum number of generations.

Each crow has a memory m_i^{gen} that represents the position of its hiding food, and at the same time, it is considered as the best position of that crow. After each generation, crows move in the search space and try to steal the hidden food of others. At each generation gen, i-th crow moved to a new position,

$$x_i^{(gen+1)} = \begin{cases} x_i^{gen} + \alpha_i \times FL_i^{gen} \left(m_j^{gen} - x_i^{gen} \right) & if \ \alpha_j \geq AP_i^{gen} \\ random & otherwise \end{cases} \quad (2)$$

Here, α_i and α_j are the random numbers with uniform distribution in [0,1], AP_i^{gen} denotes the awareness probability of the j-th crow at generation gen and FL_i^{gen} denotes the flight length of the i-th crow at gen (also termed as fitness based on the problem instances) and m_j^{gen} denote the memory of i-th crow at gen. In this equation, the FL_i^{gen} has a significant impact on the search abilities such that when FL_i^{gen} values are low, local search is initiated and higher values initiate the global search.

At the initialization stage, the crows are randomly positioned at x_i^{gen} in the search space, which also represent the memories of the crows. After initialization, each crow is repositioned using Eq. (2). For each generation, the newly positioned crows are evaluated for fitness and the crows will update their memorized positions based on the best food hiding crows as

$$m_i^{gen+1} = \begin{cases} x_i^{gen+1} & if \ fit\left(x_i^{gen}\right) is \ bettter \ than \ fit\left(m_i^{gen}\right) \\ m_i^{gen} & otherwise \end{cases} \quad (3)$$

These three Eqs. (1), (2) and (3) represent the entire process of the standard CSA. Yet, the standard CSA has problems in maintaining the balance between exploitation and exploration, leading to slow convergence and local optimum problems [22]. On analysis, it has been found that the AP_i^{gen} parameter and the selection of crow for following by the other crows determine the convergence rate and search abilities. The main characteristic of the CSA is that it supports adaptability i.e. varying the parameters such as AP and FL. Therefore, an improved CSA (ICSA) is proposed using three modifications.

3 Results and Discussions

The performance of the proposed ICSA-FNN is evaluated over the two types of wine datasets using MATLAB R2016b under a controlled environment. Experiments are performed to compare the performance of the proposed model with the prominent existing methods. For comparison, the existing SVM, ANN, Naïve Bayes (NB), KNN, Decision tree (DT), Random Forests (RF), MLP, AB-SMOTE, AdaBoost-SVM and ARSKNN from the literature are also implemented to estimate the performance superiority. The evaluations are performed in three stages. Firstly, the IPCA is evaluated and compared with PCA. Secondly, the prominent existing models are compared with the proposed model using training/ testing set splitting. Finally, the K-fold cross-validation is applied to test the proposed model.

3.1 Evaluation Metrics

Accuracy, precision and time metrics are used in this study to evaluate the performance of the proposed ICSA-FNN model.

$$Accuracy = \frac{(TP + TN)}{(TP + TN + FP + FN)} \tag{4}$$

$$Precision = \frac{TP}{(TP + FP)} \tag{5}$$

TP-True Positive, TN-True Negative, FP-False Positive, FN-False Negative.

3.2 Evaluation of IPCA Model

To evaluate the proposed IPCA dimensionality reduction model, the model is implemented with the ICSA-FNN classifier and compared against the PCA based ICSA-FNN classifier and ICSA-FNN classifier without any feature selection. Table 1 shows the comparison results for the red wine dataset.

The results in Table 2 show that the proposed IPCA has outperformed the PCA for the red wine dataset. It should be noted that the accuracy of the ICSA-FNN

Table 1 Performance evaluation of IPCA for red wine

Metrics	ICSA-FNN	PCA with ICSA-FNN	IPCA with ICSA-FNN
Accuracy (%)	57.08	71.67	93.54
Precision (%)	60.72	73.12	90.99

Table 2 Performance evaluation of IPCA for white wine

Metrics	ICSA-FNN	PCA with ICSA-FNN	IPCA with ICSA-FNN
Accuracy (%)	82.37	84.68	97.14
Precision (%)	75.97	78.13	95.89

classifier is improved around 22% when the IPCA is applied compared with the traditional PCA and the classifier without the feature selection model. Table 2 shows the comparison results of IPCA for the white wine dataset.

The results in Table 2 show that the proposed IPCA has also outperformed the PCA for the white wine dataset. The accuracy of the ICSA-FNN classifier is improved around 13% when the IPCA is applied compared with the traditional PCA achieving 84.68% accuracy and the classifier without the feature selection model achieving 82.37%. Comparing the results of both the red wine and white wine results, it is clear that the white wine classification is performed better than the red wine. This is because of the quality of the data in both datasets.

3.3 Evaluation Based on Training/Testing Sets

The proposed ICSA-FNN classifier is also evaluated based on the training/testing set splitting technique. For this evaluation, the wine datasets (both white and red) are classified into training and testing sets in the ratio 70:30. Then the ICSA-FNN classifier is trained using the 70% dataset and the remaining data is used as testing data. Table 3 shows the evaluation results of the ICSA-FNN classifier for both the datasets using training/testing sets.

From Table 3, it can be seen that the training and testing results for both the red and white wine datasets are adequate. This indicates the effectiveness of the proposed model without over-fitting or under-fitting problems. The results of the

Table 3 Evaluation results of ICSA-FNN using training/testing sets

Metrics	Red wine		White wine	
	Training	Testing	Training	Testing
Accuracy (%)	96.99	93.67	99.72	97.75
Precision (%)	95.46	90.87	98.55	95.50

Table 4 Accuracy comparison based on training/testing sets

Methods	Red wine		White wine	
	Training	Testing	Training	Testing
SVM	88.76	69.88	92.56	88.50
ANN	91.89	71.22	91.25	83.44
NB	85.89	70.54	88.76	72.45
KNN	86.89	68.54	90.25	70.67
DT	90.87	73.23	90.10	75.32
RF	91.23	80.18	99.23	87.54
MLP	92.67	81.67	97.12	77.65
AB-SMOTE	95.84	83.65	97.67	83.56
AdaBoost-SVM	90.67	79.87	98.67	78.87
ARSKNN	91.44	76.65	90.12	85.61
FNN	90.33	88.45	97.66	90.34
ICSA-FNN	96.99	93.67	99.72	97.75

proposed model are also compared with the existing models from the literature in terms of training and testing accuracy. Table 4 shows the comparison of the proposed ICSA-FNN model with the existing methods.

From the results in Table 4, the proposed model has outperformed all the compared methods with higher training and testing accuracy. As some existing methods have over-fitting and under-fitting problems, their testing accuracy has been greatly reduced compared to the training accuracies. Yet, the proposed model achieved high accuracy of 93.67% for red wine and 97.75% accuracy for white wine.

3.4 K-fold Cross-Validation

To justify the efficiency of the proposed model achieved for the training/testing sets evaluation, the K-fold cross-validation is performed. K-fold cross-validation is conducted by splitting the wine dataset into K number of subsets with an equal number of instances. The set of instances in each subset is assigned randomly either from top–bottom or bottom-top splitting. In this study, fivefold validation is performed. Table 5 shows the fivefold validation results obtained for the proposed ICSA-FNN model over the red wine data.

Table 5 K-fold validation of red wine for ICSA-FNN

Metrics	Fold 1	Fold 2	Fold 3	Fold 4	Fold 5	Average
Accuracy (%)	93.88	94.01	92.06	94.25	93.5	93.54
Precision (%)	91.12	90.87	91.23	91.56	90.17	90.99

The results in Table 5 show that the fivefold validation of the proposed model over the red wine dataset also indicates the effectiveness of the proposed approach. It can be seen that the proposed model holds consistent performance values for all metrics. To attain the superiority of the proposed model over the existing models, the fivefold validation is compared with that of the existing methods. Table 6 shows the K-fold validation comparison of the proposed ICSA-FNN against the existing models over the red wine dataset.

From Table 6, it is evident that the accuracy of the proposed model for red wine data is significantly higher for all 5 folds of the cross-validation. These tables illustrates the performance of the proposed model compared against the existing methods in terms of accuracy, precision, recall, f-measure, chi-square and processing time. The proposed ICSA-FNN has achieved high values of accuracy, precision, recall, f-measure, and chi-square values and lesser processing time. This illustrates the effectiveness of the proposed approach for wine quality estimation.

Table 7 shows the fivefold validation results obtained for the proposed ICSA-FNN model over the white wine.

The results in Table 7 show that the fivefold validation of the proposed model over the white wine dataset also indicates the effectiveness of the proposed approach. It can be seen that the proposed model holds consistent performance values for all metrics. To attain the superiority of the proposed model over the existing models, the fivefold validation is compared with that of the existing methods. Table 8 shows the K-fold validation comparison of the proposed ICSA-FNN against the existing models over the white wine dataset.

From Table 8, it is evident that the accuracy of the proposed model for white wine data is also significantly higher for all 5 folds of the cross-validation. Compared with red wine, the results are higher for white wines.

Table 6 K-fold validation accuracy comparison (red wine)

Methods	Fold 1	Fold 2	Fold 3	Fold 4	Fold 5	Average
SVM	74.92	72.85	74.66	72.11	74.21	73.75
ANN	80.72	83.88	84.21	83.67	84.17	83.33
KNN	85.53	85.88	84.78	86.25	86.71	85.83
NB	75.18	76.15	77.6	75.12	74.05	75.62
RF	78.23	74.51	74.08	76.88	73.45	75.42
ICSA-FNN	93.88	94.01	92.06	94.25	93.5	93.54

Table 7 K-fold validation of white wine for ICSA-FNN

Metrics	Fold 1	Fold 2	Fold 3	Fold 4	Fold 5	Average
Accuracy (%)	96.35	96.91	97.12	97.33	97.99	97.14
Precision (%)	96.1	95.81	95.25	96.34	95.95	95.89

Table 8 K-fold validation accuracy comparison (white wine)

Methods	Fold 1	Fold 2	Fold 3	Fold 4	Fold 5	Average
SVM	91.21	91.24	92.56	92.8	90.99	91.76
ANN	93.10	93.56	94.67	94.87	93.5	93.94
KNN	93.20	94.88	95.32	94.94	95.11	94.69
NB	91.73	93.87	92.76	91.76	91.43	92.31
RF	88.44	91.5	91.82	92.59	92.10	91.29
ICSA-FNN	96.35	96.91	97.12	97.33	97.99	97.14

4 Conclusions

Wine quality estimation is performed using spectrometry and machine learning techniques. However, the ML algorithms have certain parameters that are not effectively tuned and results in attributes not adequately represented. This paper presented an efficient wine quality estimation model using the hybrid classifier of ICSA-FNN which is a combination of fuzzy logic, neural networks and an improved ICSA algorithm. This hybrid classifier effectively classifies the red and white wine types from the raw dataset available at the UCI repository. The evaluations are performed and the results have justified the theoretical effectiveness of the proposed model in classifying the good and bad wines accurately. In future, the possibility of minimizing the time complexity for large datasets will be investigated. Examinations will be conducted to study the impact of the proposed ICSA-FNN with the presence of high noise and model uncertainty. The possibility of applying the proposed model to other larger datasets with numerous attributes will also be explored in future researches.

Acknowledgements Authors thanks to Prof. F.V. Çelebi and Dr. G.E. Bostanci for usefull discussions.

References

1. Swami, S.B., Thakor, N.J., Divate, A.D.: Fruit wine production: a review. J. Food Res. Technol. **2**(3), 93–100 (2014)
2. Jones, G.V., White, M.A., Cooper, O.R., Storchmann, K.: Climate change and global wine quality. Clim. Change **73**(3), 319–343 (2005)
3. Charters, S., Pettigrew, S.: The dimensions of wine quality. Food Qual. Prefer. **18**(7), 997–1007 (2007)
4. Li, Y., Li, J., Jiang, Z.J.: Application of statistical analysis in the evaluation of grape wine quality. Liquor-Mak. Sci. Technol. **4**, 79–82 (2009)
5. Reynolds, A.G. (ed.): Managing Wine Quality: Viticulture and Wine Quality. Elsevier (2010)
6. Hosnedlova, B., Sochor, J., Baron, M., Bjørklund, G., Kizek, P.: Application of nanotechnology based-biosensors in analysis of wine compounds and control of wine quality and safety: a critical review. Crit. Rev. Food Sci. Nutr. **60**(19), 3271–3289 (2020)

7. Gambetta, J.M., Cozzolino, D., Bastian, S.E., Jeffery, D.W.: Towards the creation of a wine quality prediction index: Correlation of Chardonnay juice and wine compositions from different regions and quality levels. Food Anal. Methods **9**(10), 2842–2855 (2016)
8. Cortez, P., Cerdeira, A., Almeida, F., Matos, T., Reis, J.: Modeling wine preferences by data mining from physicochemical properties. Decis. Support Syst. **47**(4), 547–553 (2009)
9. Gawel, R., Godden, P.W.: Evaluation of the consistency of wine quality assessments from expert wine tasters. Aust. J. Grape Wine Res. **14**(1), 1–8 (2008)
10. Cortez, P., Teixeira, J., Cerdeira, A., Almeida, F., Matos, T., Reis, J.: Using data mining for wine quality assessment. In: International Conference on Discovery Science, pp. 66–79. Springer, Berlin, Heidelberg (2009)
11. Lingfeng, Z., Feng, F., Heng, H.: Wine quality identification based on data mining research. In: 2017 12th International Conference on Computer Science and Education (ICCSE), 358–361 (2017)
12. Balasubramanian, V., Ho, S.S., Vovk, V. (eds.): Conformal prediction for reliable machine learning: theory, adaptations and applications. Newnes (2014)
13. Trivedi, A., Sehrawat, R.: Wine quality detection through machine learning algorithms. In: 2018 International Conference on Recent Innovations in Electrical, Electronics & Communication Engineering (ICRIEECE), pp. 1756–1760 (2018)
14. Laughter, A., Omari, S.: A study of modeling techniques for prediction of wine quality. In: Science and Information Conference, pp. 373–399. Springer, Cham (2020)
15. Petropoulos, S., Karavas, C.S., Balafoutis, A.T., Paraskevopoulos, I., Kallithraka, S., Kotseridis, Y.: Fuzzy logic tool for wine quality classification. Comput. Electron. Agric. **142**, 552–562 (2017)
16. Adnan, M.M., Sarkheyli, A., Zain, A.M., Haron, H.: Fuzzy logic for modeling machining process: a review. Artif. Intell. Rev. **43**(3), 345–379 (2015)
17. da Costa, N.L., Valentin, L.A., Castro, I.A., Barbosa, R.M.: Predictive modeling for wine authenticity using a machine learning approach. Artif. Intell. Agric. (2021)
18. Bro, R., Smilde, A.K.: Principal component analysis. Anal. Methods **6**(9), 2812–2831 (2014)
19. UCI Repository: https://archive.ics.uci.edu/ml/datasets/wine+quality
20. Abbasbandy, S., Otadi, M.: Numerical solution of fuzzy polynomials by fuzzy neural network. Appl. Math. Comput. **181**(2), 1084–1089 (2006)
21. Askarzadeh, A.: A novel metaheuristic method for solving constrained engineering optimization problems: crow search algorithm. Comput. Struct. **169**, 1–12 (2016)
22. Hussien, A.G., Amin, M., Wang, M., Liang, G., Alsanad, A., Gumaei, A., Chen, H.: Crow search algorithm: theory, recent advances, and applications. IEEE Access **8**, 173548–173565 (2020)

Dow Jones Index Time Series Forecasting Using Feedforward Neural Network Model

Ramin Rzayev⊙ and Parvin Alizada⊙

Abstract A neural network-based method for modeling and forecasting the time series with regular periodic components is considered. As an example, the DJIA index time series has been selected that is distinguished by its volatility. The process of neural network modeling (restoring) of the DJIA index time series is carried out using the Neural Networks Toolbox within the MATLAB software package. The corresponding tools of this software package allowed users to focus on the application of neural network models in the process of empirical research and free themselves from the need, in fact, to design multilayer neural networks, and strain about their correctness, stability and performance under all sorts of conditions for forecasting time series. Therefore, in solving the problems of forecasting time series, the choice and justification of the declared type of neural network, adjustment (training) of its parameters and correct interpretation of the results have already become the main ones.

Keywords Dow Jones index · Time series · Neural network · Forecasting

1 Introduction

For the first time, the official DJIA (Dow Jones Industrial Average) publishing in the Wall Street Journal was 40.94. This value was calculated as the arithmetic average of the enterprise share value that make up the initial "basket" of the DJIA. In time, considering the rapid industrial growth of the US economy, forced rotations and changes in the composition of companies that make up the index basket, the values of the DJIA have changed cardinally. For example, the highest level of the indicator

R. Rzayev (✉)
The Ministry of Science and Education of the Republic of Azerbaijan, Institute of Control Systems, B. Vahabzadeh str. 9, AZ1141 Baku, Azerbaijan
e-mail: raminrza@yahoo.com

P. Alizada
Baku State University, Z. Xalilov str. 23, AZ1148 Baku, Azerbaijan

© The Author(s), under exclusive license to Springer Nature Switzerland AG 2023   329
Sh. N. Shahbazova et al. (eds.), *Recent Developments and the New Directions of Research, Foundations, and Applications*, Studies in Fuzziness and Soft Computing 422,
https://doi.org/10.1007/978-3-031-20153-0_26

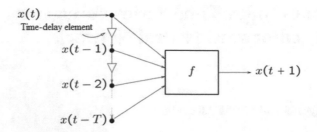

Fig. 1 TS forecasting based on approximation $f(\cdot)$ as a continuous function of $(T + 1)$ variables

in the amount of 1000 points was reached already in 1966, which did not change significantly over the next 15 years, i.e., during this period the DJIA dropped significantly (up to a third of the specified maximum) but managed to return to its primary state.

The DJIA is a multi-factor category. Therefore, its growth or fall is influenced by numerous factors, one of which is undoubtedly the global economic crisis. So, in the interval of the last two decades of the last century, there was a jump in the DJIA, which by January 2000 had grown to the level of 11,722.98. During this period, the index fell to 22%. However, this did not have a significant influence on the overall dynamics of the market value of the DJIA. The most significant default on the stock exchange was registered in 1987 on the so-called "Black Monday". At the end of 2017, the traded value of the DJIA was already within limits of 23,000 units (see Fig. 1). Thus, in the American and global economy, the DJIA is the main indicator that determines the strengthening or depreciation of the US dollar against other currencies. Therefore, the study of the DJIA time series (TS) is actual to the present.

Over the past decades, notable advances have been made in the field of modeling and forecasting TS by immersing historical data into the logical basis of neural networks. The existing standard algorithms have been studied in detail and tested in numerous periodic and fundamental sources, some of which are given in the references. The development of the neural network theory is organically and closely related to the names of Rosenblatt [1, 2], Rumelhart et al. [3], Kohonen [4], Hopfield [5] and many others. Among other scientists, it is necessary to note the works of Ivakhnenko [6] in the field of polynomial networks and the method of group consideration of arguments; T. Kussulya in the field of associative memory formalization based on a neural network; A.N. Gorban, who proposed new neural networks with complicated transmission functions; A.I. Galushkin in the fields of neuro-mathematics and applied aspects of neural networks.

2 Problem Definition

The general problem of forecasting TS is formulated as follows. Moving in the opposite direction in time t, we obtain TS in common form as: $\{x[t], x[t - 1], \ldots\}$.

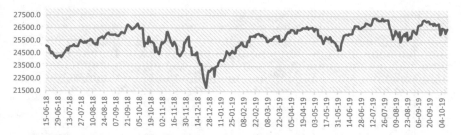

Fig. 2 DJIA TS

Then, based on this, it is necessary to estimate the future value of x on the base of the following equality $x[t + s] = f(x[t], x[t - 1], ...)$, where s is the forecast horizon. In our case, we will assume that $s = 1$. In the context of this reasoning, the process of finding predictions is identified with the problem of approximating the function $f(\cdot)$, the solution of which implies:

1. Assumption of the presence of a generative TS model.
2. Revealing cause-effect relations, reflecting the internal relationships of the k-th orders in the form: $(x[t_i - k], x[t_i - (k - 1)], ..., x[t_i]) \rightarrow x[t_i + 1]$, where $x[t_i + 1]$ is the desired model output. Based on such relations, $f(\cdot)$ is constructed as a function of k variables in a tabular form, and further it is approximated.
3. After approximating the function $f(\cdot)$, a generative model for forecasting the TS is launched, on the basis of which predicts $x[t + s]$ are established for the set of data $\{x[t], x[t - 1], ... \}$, that is, like this as shown in Fig. 1.

Neural networks are considered to predict the TS of the DJIA. The mathematical notation of neural networks, directly related to object-oriented programming, should be considered in a rather trivial form. It is necessary to predict the volatile TS of the DJIA, shown in Fig. 2 in graphical form and in Tables 1 and 2 in the form of historical data [7]. As the appropriate model, it is proposed the feedforward neural network (FNN) capable of reproducing the TS of the DJIA with a duration of 333 days, the historical data of which (see Table 2) were fixed by days on the stock exchange based on the results of trading closings.

3 Neural Network Forecasting of the Volatile DJIA TS

The Neural Networks Toolbox (NNT) of the MATLAB package is a kind of universal software shell, inside which there are all the procedures necessary for designing NNs of all known architectures and for analyzing and/or interpreting the results obtained. NNT is a powerful means for supporting mathematical modeling systems. Nevertheless, in some cases for the user it is not always an understandable tool for solving applied problems that cannot be solved by traditional methods and/or their solution is accompanied by significant technical difficulties. The explanation for this

Table 1 The set of training pairs for NN design

t	The NN model inputs						Desired output
	$x(t-10)$	$x(t-9)$	$x(t-8)$	...	$x(t-2)$	$x(t-1)$	$x(t)$
11	25090.5	24987.5	24700.2	...	24117.6	24216.1	24271.4
12	24987.5	24700.2	24657.8	...	24216.1	24271.4	24307.2
13	24700.2	24657.8	24461.7	...	24271.4	24307.2	24174.8
14	24657.8	24461.7	24580.9	...	24307.2	24174.8	24356.7
15	24461.7	24580.9	24252.8	...	24174.8	24356.7	24456.5
16	24580.9	24252.8	24283.1	...	24356.7	24456.5	24776.6
17	24252.8	24283.1	24117.6	...	24456.5	24776.6	24919.7
...							
328	27094.8	26935.1	26950.0	...	26573.0	26078.6	26201.0
329	26935.1	26950.0	26807.8	...	26078.6	26201.0	26573.7
330	26950.0	26807.8	26970.7	...	26201.0	26573.7	26478.0
331	26807.8	26970.7	26891.1	...	26573.7	26478.0	26164.0
332	26970.7	26891.1	26820.3	...	26478.0	26164.0	26346.0
333	26891.1	26820.3	26916.8	...	26164.0	26346.0	26496.7

is the complexity of the NN object created by NNT. Even under the condition that the multilevel NN model will work effectively, for example, by activating the *newelm* or *newff* functions, it is necessary to spend some time so that the user can fully master it: if necessary, edit his/her own objects and be able to adequately interpret the results obtained.

All commands given in the NNT documentation take on the nature of a network object by initiating the following function: $net = net(\cdot)$, which designs the NN without characteristic features in accordance with the specifics of the applied task. After that, the input, hidden and output layers of the NN, the connections between them are designed, the weights of the connections and thresholds of non-linear neurons from the hidden layer are initiated, and further down the list.

At the beginning, the trivial network consisting of one linear neuron is chosen as a neural network model (LNN) of the DJIA TS (see Fig. 3). The restoration of the DJIA TS with a duration of 333 days occurs by the following procedures:

```
>>time = 0:1:333;
```

```
>>T = [25090.5 24,987.5 24,700.2 ... 26,346 26,496.7];
```

```
>>Q = length(T);
```

For each day this neural network reduces the last 10 values of the DJIA, and its inputs P are determined by a delay from one day to ten days:

```
>>P = zeros(10, Q);
```

```
>>P(1,2:Q) = T(1,1:(Q-1));
```

Table 2 Results of forecasting the DJIA TS

Date	DJIA	Neural network		Date	DJIA	Neural network	
		LNN	FNN			LNN	FNN
15.06.2018	25090.5	17389	×	17.09.2019	27110.8	26143	27129
18.06.2018	24987.5	25226	×	18.09.2019	27147.1	26164	27146
19.06.2018	24700.2	24955	×	19.09.2019	27094.8	26183	27043
20.06.2018	24657.8	24910	×	20.09.2019	26935.1	26171	26962
21.06.2018	24461.7	24680	×	23.09.2019	26950.0	26120	26844
22.06.2018	24580.9	24784	×	24.09.2019	26807.8	26132	26915
25.06.2018	24252.8	24465	×	25.09.2019	26970.7	26093	26688
26.06.2018	24283.1	24554	×	26.09.2019	26891.1	26149	26819
27.06.2018	24117.6	24367	×	27.09.2019	26820.3	26122	26696
28.06.2018	24216.1	24512	×	30.09.2019	26916.8	26100	26693
29.06.2018	24271.4	25277	24153	01.10.2019	26573.0	26128	26770
02.07.2018	24307.2	25283	24222	02.10.2019	26078.6	26026	26485
03.07.2018	24174.8	25294	24313	03.10.2019	26201.0	25868	26042
05.07.2018	24356.7	25243	24109	04.10.2019	26573.7	25909	26238
06.07.2018	24456.5	25302	24352	07.10.2019	26478.0	26025	26508
09.07.2018	24776.6	25329	24414	08.10.2019	26164.0	25990	26410
10.07.2018	24919.7	25422	24865	09.10.2019	26346.0	25899	26250
11.07.2018	24700.5	25460	24987	10.10.2019	26496.7	25957	26365
12.07.2018	24924.9	25389	24754	MSE		616507	54170
				MAPE		2.0913	0.6608

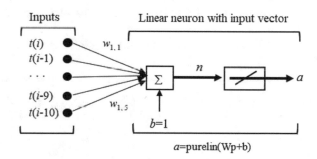

Fig. 3 The LNN consisting of one neuron

```
>>P(2,3:Q) = T(1,1:(Q-2));

>>P(3,4:Q) = T(1,1:(Q-3));

>>P(4,5:Q) = T(1,1:(Q-4));
```

\>\>P(5,6:Q) = T(1,1:(Q-5));

\>\>P(6,7:Q) = T(1,1:(Q-6));

\>\>P(7,8:Q) = T(1,1:(Q-7));

\>\>P(8,9:Q) = T(1,1:(Q-8));

\>\>P(9,10:Q) = T(1,1:(Q-9));

\>\>P(10,11:Q) = T(1,1:(Q-10));

By activating the following function in the workspace

\>\>plot(time, T);

the image of the volatile DJIA TS was obtained in the MATLAB notation (see Fig. 4).

Further, using the *newlind* function, corresponding NN is formed. It also has 10 inputs (10 delayed values of the DJIA) and one output as a subsequent value of the DJIA. Considering the minimization of the standard deviation, the *newlind* function adjusts the weights of the input and output synoptic connections, as well as the bias for the single neuron. As a result, the initiated NN reconstructs the TS as shown in Fig. 5, which also shows the dependence of the error on the number of iterations.

The restored DJIA TS is not satisfactory. It is necessary to apply the NN with a more complex topological design. As such, a three-layer FNN was chosen, which is shown in Fig. 6.

By activating the *newff* function, the FNN is formed, which has 10 inputs (10 delayed values of the DJIA) and one output as the subsequent DJIA value. Considering the minimization of the standard deviation, the *newff* function adjusts the weights of the input and output synoptic connections, as well as the thresholds of nonlinear neurons from the hidden layer. Based on the chosen concept of the FFN and the historical data of the DJIA TS, the training set consisting of pairs of the form:

$$f : \{[x(t-10), x(t-9), \ldots, x(t-1)] \rightarrow x(t)\}, t = 11 \div 333, \tag{1}$$

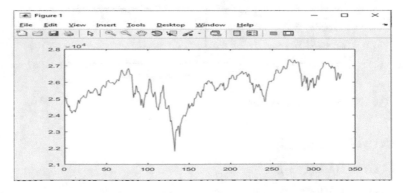

Fig. 4 Volatile DJIA TS in MATLAB notation

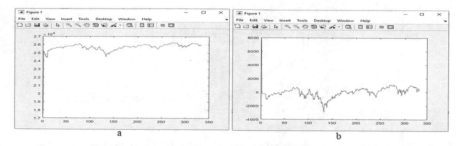

Fig. 5 *Newlind* function based volatile DJIA TS (**a**), and dependence of the error on the number of iterations (**b**)

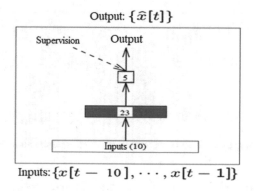

Fig. 6 FNN with one hidden layer

built in the form of Table 1.

The ten-factor function f is approximated by a three-layer FNN, which induces output signals of the following form

$$z_j = \sum_{k=1}^{r} c_k \varphi\big[\big(w_{ki} x_{ij}\big) - \theta_k\big], \tag{2}$$

where r is the number of nonlinear neurons in the hidden layer, selected by the user during the simulation; w_{ki} and c_k are the weights of the input and output synoptic connections, respectively; θ_i is a threshold (bias) of the k-th nonlinear neuron from the hidden layer; $\varphi(\cdot)$ is an activation function of the nonlinear neuron from a hidden layer, for example, sigmoidal type $\varphi(t) = 1/[1 + \exp(-(t - \theta))]$.

Figure 7 shows a scheme for training the NN (2) by identifying its parameters: weights of input and output synoptic connections, and thresholds of nonlinear neurons from the hidden layer. The NN receives 10 values from the segment [21500, 27500] as an input vector with 10 components. This is necessary to approximate the continuous function (1) presented in tabular form (see Table 1). The single output of the NN should represent the predict of the DJIA. To work correctly, the NN must respond, for

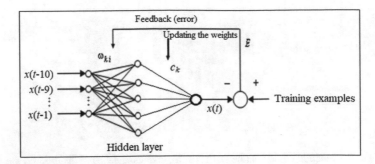

Fig. 7 Training the three-layer neural network

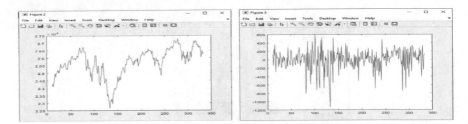

Fig. 8 DJIA TS restored by FNN and dependence of the error on the number of iterations

example, with a value of 25019.4 at the 20-th position of the input vector (24216.1, 24271.4, 24307.2, 24174.8, 24356.7, 24456.5, 24776.6, 24919.7, 24700.5, 24924.9).

The hidden layer of the NN includes 10 nonlinear neurons with sigmoid activation function $\varphi(\cdot)$, the range of which allows the output to be realized within the interval [21500, 27500]. After training and testing, products (results of pairs "input – output") are formed, which are summarized in Table 2. After training, the initiated NN restores the TS as shown in Fig. 8. The corresponding data of the DJIA TS restored by the FNN are summarized in Table 2.

4 Conclusion

The comparison results are shown in the following Table 2, and the geometric interpretation is shown in Fig. 9.

At the end of Table 2, the values of the statistical evaluation criteria MSE and MAPE are presented, which reflect the quality of the constructed predictive models, their adequacy and accuracy. Errors according to these criteria are calculated by the formulas:

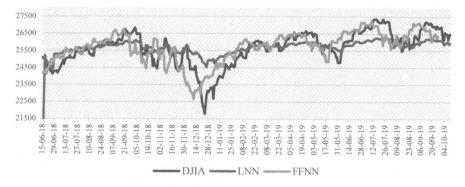

Fig. 9 Geometric interpretation of DJIA TS models

$$\text{MSE} = \sum_{t=1}^{m} (F_t - A_t)^2 / m, \quad \text{MAPE} = 100 \times \sum_{t=1}^{m} |F_t - A_t| / m A_t,$$

where m is the length of the TS; A_t is the DJIA at time t; F_t is the predict of A_t.

The results of restoring the volatile DJIA TS obtained using a three-layer FNN, comparing them with the results obtained using the LNN modeling showed that this method of TS modeling has a right to exist. In the article, we managed to consider only the ten-by-one training set, i.e., the set of the form $\{[x(t - 10), x(t - 9), ..., x(t - 1)] \rightarrow x(t)\}$, $t = 11 \div 333$. Nevertheless, in the search for a suitable sample of training pairs, there is also a considerable resource for building the most adequate NN model of volatile TS.

References

1. Rosenblatt, F.: The perceptron–a perceiving and recognizing automaton. Report 85–460–1, Cornell Aeronautical Laboratory, Buffalo, New York (1957)
2. Minsky, M., Papert, S.A.: Perceptrons: An Introduction to Computational Geometry. MIT Press, Cambridge, MA, USA (1969)
3. Rumelhart, D.E., Hinton, G.E., Williams, R.J.: Learning internal representations by error propagation: parallel distributed processing, vol. 1, pp. 318–362. MIT Press, Cambridge (1986)
4. Kohonen, T.: Self-organization and Associative Memory. Springer, Berlin (1984)
5. Hopfield, J.J.: Neurons with graded response have collective computational properties like those of two-state neurons. Proc. Natl. Acad. Sci. **81**(10), 3088–3092 (1984)
6. Ivakhnenko, A.G.: Polynomial theory of complex systems. IEEE Trans. Syst. Man Cybern. **4**, 364–378 (1971)
7. Dow Jones Industrial Index, https://ru.tradingview.com/symbols/DJ-DJI/

Artificial Intelligence and Expert Systems

Inverter Fault Diagnosis with AI at Edge

Aditya Anand, Priya G. Das, and K. Saneep

Abstract The paper presents the design, development, analysis of fault diagnosis system in inverter circuits using machine learning running in an edge computing device. A single phase inverter is used here. A supervised machine learning model is built using data from the circuit voltage under different conditions. The model is then optimized to run on a microcontroller board. The machine learning and optimizations are done on the Edge impulse platform. The training and testing data was generated via circuit simulations. The optimized machine learning model was run on a ARM based microcontroller to evaluate it's performance.

Keywords Machine learning · Edge · Inverter · Fault diagnosis · Artificial intelligence

1 Introduction

The paper aims to develop a simple, efficient and cost-effective method for diagnosing faults in an inverter circuit. Over the years the use of power electronics based inverters has become popular even in households. The compact, easy to design circuit and more accessible price for the components has helped as well. The use of these systems is

A. Anand (✉)
PG Scholar, Department of Electrical and Electronics Engineering, NSS College of Engineering, Kerala Palakkad, India
e-mail: adityaa4797@gmail.com

P. G. Das
Professor, Department of Electrical and Electronics Engineering, NSS College of Engineering, Kerala Palakkad, India
e-mail: priyadas@nssce.ac.in

K. Saneep
Asst Professor (Adhoc), Department of Electrical and Electronics Engineering, NSS College of Engineering, Kerala Palakkad, India
e-mail: saneepkarat@gmail.com

© The Author(s), under exclusive license to Springer Nature Switzerland AG 2023
Sh. N. Shahbazova et al. (eds.), *Recent Developments and the New Directions of Research, Foundations, and Applications*, Studies in Fuzziness and Soft Computing 422, https://doi.org/10.1007/978-3-031-20153-0_27

controlled via various industry and regulatory standards. This led to the development of robust control and fault protection schemes such as OCP, SCP, OPP etc. Still failing switches or gate drivers are the prime reasons for faults in the power electronic systems. With the increase in the use of PV systems for homes, offices and industries, the need for inverters has increased.

A large number of complex tasks such as fault diagnosis are now Artificial Intelligence assisted. Power electronics as a whole is undergoing a lot of changes due to AI. Hence, researchers are trying to build new and improved methods to detect, correct and at times prevent faults with the help of AI. A switch failure, gate driver failure or damage to the circuit board can cause an open circuit fault in the system. This issue becomes more prominent with an increasing number of switches. One of the challenges of using AI in any field was its relatively high computational requirement. But, this now changing with the projects such as tinyML, where the AI or ML libraries are shrunk down to run on low power microcontroller devices. Edge impulse is an online platform developed on this idea. The optimized model is used in the edge computing device, that is, in or near the control systems of the circuit.

2   Inverter and Its Faults

In the case of Inverters, multiple types of faults have different effects on their performance from open circuit faults to short circuit faults. A study of IGBT based inverter faults has been given in [2]. This article also provides several methods that can be used for fault diagnosis. One such method has been highlighted as a neural network-based, which is a part of AI.

Fault detection on multilevel inverters are discussed in [3, 4]. Most of these methods were developed on complex computer systems and tools, which reduces their deployability or cost-effectiveness with respect to a single-board microcontroller. Specialised FPGAs help in reducing the size, but they are not easy to reuse compared to a microcontroller board. Effects of such faults in real-world operation of PV systems is highlighted in [6, 7].

2.1   Single Phase Inverter

Here a single phase Voltage Source Inverter is selected as the power electronic system of choice. They can be in either half bridge or full bridge configuration. The full bridge configuration is widely used and hence selected. They have two more switches than half bridge configuration increasing the point of failures. The full bridge inverter is also known as the H bridge inverter due to it's switch layout. They are still easier to build and analyse with respect to multilevel inverters. A block diagram of the circuit built using switches and diodes are shown in Fig. 1.

Fig. 1 Single phase inverter
block diagram

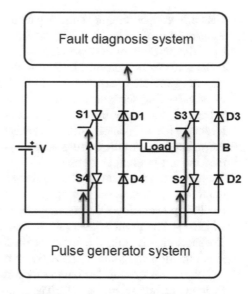

An array of two switches is termed as an "inverter leg". In an VSI, there are few rules to be followed to ensure it's safe operation [8]. These are

- Switches of the same leg cannot be turned on at the same time to avoid short circuit across the dc-link voltage source.
- Anti-parallel diodes should be connected to each switch, in order to provide a current path for inductive loads. If the commercial switch includes this diode (anti-parallel body diodes in MOSFETs, IGBTs), then the circuit is already complete.
- In practical implementation, a dead time, also known as blanking time, must be provided in the control signals of the leg switches, to avoid failing first rule.

If any of these rules are broken during design, the circuit will develop a fault which could eventually damage itself and/or connected devices as well.

The switches are driven via a pulse generator system. Usually the pulses are generated by low power devices such as flip flops or microcontrollers. These signals are used to drive the gate circuits (e.g., optocouplers) which in turn triggers the switches. Sine Pulse Width Modulation technique is used here, where a sine wave of desired frequency (e.g., power frequency of 50 Hz) is compared with a high frequency carrier signal such as triangular wave to form the required PWM signal.

2.2 Common Faults

These are the common fault types in inverters [2]:

- Gate driven open circuit fault: A gate-drive open-circuit fault may happen due to detaching of bonding wires caused by thermal cycling. It may be caused by a driver fault or a short-circuit-fault-induced IGBT failure. Open circuit faults leads to dc current offset in both the faulty and healthy phases. The dc currents generate unequal stress in the upper and lower switches. These may cause cascading secondary faults in the inverter, motor, or load. The current and voltage signals carry the fault signatures which can be analyzed to detect and locate the fault. Open-circuit faults generally do not cause complete system shutdown, but degrades the performance. Therefore, these diagnostic methods can be used in fault-tolerant systems.
- Transistor short circuit fault: A short-circuit fault may happen due to any of the following reasons. A wrong gate voltage, due to driver circuit malfunction, auxiliary power supply failure, or voltage change disturbance; an intrinsic failure, which may be caused by overvoltage/avalanche stress or a temperature overshoot. The short-circuit faults are difficult to deal with due to the short time between the fault initiation and failure. Therefore, most of the existing short-circuit detection and protection systems are based on hardware circuit.
- Intermittent gate-misfiring fault: The gate-misfiring faults can cause severe breakdown of the device, if the faults remain unchecked. It may be caused by a driver side open circuit, control circuit element degradation, deteriorated electromagnetic compatibility, etc. Inverters can work for a good period of time even with sustained gate misfiring. But, this is accompanied by reduced output voltage and higher stress on other switching devices and dc bus capacitors. Gate misfiring may also lead to a device short circuit fault. This may happen when the device fails to turnoff causing shoot-through. In most cases, gate misfiring is an intermittent issue.

3 AI Based Fault Diagnosis

Artificial Intelligence based fault diagnosis is proven to be a good solution [1–4]. But with any AI based technique proper data is required to form the model fit for the purpose. Good amount of dataset is required to properly design the fault diagnosis system. The required data is generated via MATLAB Simulink simulation. The simulation diagram of the circuit designed is shown in Fig. 2.

The circuit uses SPWM technique for pulse generation. An LC filter is added to the output to reduce harmonic distortion. The output voltage is saved to a file during the simulation. Normal and open circuit fault conditions are simulated. The gate triggers for one of the switches or its leg is removed to replicate the open circuit fault. The output voltage of each of these conditions are shown in Fig. 3. We can see that while in normal condition the output is sinusoidal, failure of any one switch can lead to loss of one half cycle of the output voltage, which will reduce the overall power being transferred.

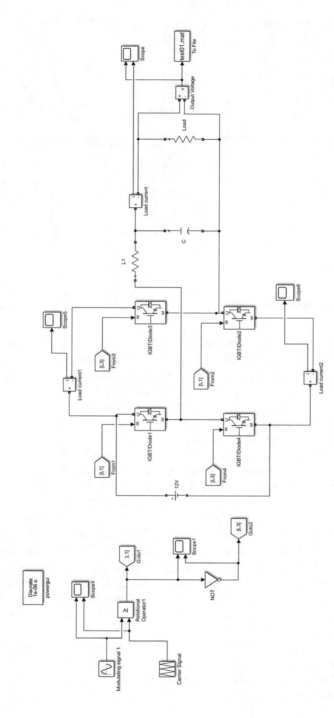

Fig. 2 Simulation diagram

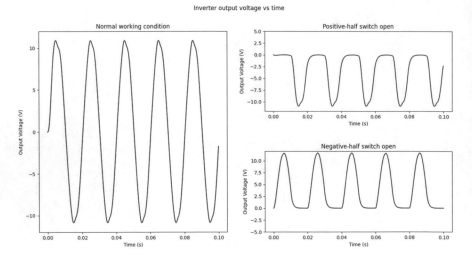

Fig. 3 Simulation output

3.1 Fault Classification Model

All the data collected is then sampled at a frequency of 1kHz with each of the data sets being appropriately labelled. The platform used for the machine learning, Edgeimpulse.com, requires the data to be in a specific format for processing. The output from the simulation is prepared with the sampling frequency, including the simulation time, as per the guidelines set by the platform in [5]. The data is uploaded to the cloud servers in JSON format with the help of a Python script through their ingestion services. After the upload, the data is then made part of the project created for this purpose. The Machine Learning model is then created inside the platform via its web GUI. Here we are using neural network based supervised ML classification model. The main objective of the model is to learn from the data available so that it can classify whether the output is in the normal or in the fault class.

Initially the voltage data is taken in as time series data since its working with AC voltage, where voltage variations are time dependent. This is then passed through the Spectral analysis block where the features are extracted. Various DSP functions are present inside this block, that will analyse the given data and generate the features. In our case the analysis gave frequency, peak height and other power spectrum features. After selecting the appropriate features the Neural Network classifier is trained. The classifier is a multi-layer sequential network with Adam optimisation. The basic network structure is shown in Fig. 4.

An anomaly detection model is also trained in this step. This is done to avoid one of the problems with NN classifiers. They will try and classify any given input to known classes (Normal and Fail) whether its actually true or not. Anomaly detection solves this issue by attaching an anomaly score to the output. If the given input deviates beyond a set threshold, it will flag the output as not trustworthy i.e., an anomaly. The

Input layer (5 features)

Dense layer (25 neurons)

Dense layer (12 neurons)

Add an extra layer

Output layer (2 classes)

Fig. 4 Neural network's basic structure

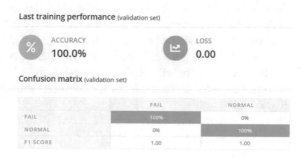

Last training performance (validation set)

%	ACCURACY		LOSS
	100.0%		**0.00**

Confusion matrix (validation set)

	FAIL	NORMAL
FAIL	100%	0%
NORMAL	0%	100%
F1 SCORE	1.00	1.00

Fig. 5 Validation accuracy

classifier model is generated using Keras, Tensorflow libraries in Python. The data taken is split in training and testing in 80:20 ratio and then fed to the model. The model is designed with two hidden layers apart from the input and output layers. The average validation and testing accuracy lies in between 98 and 99%. Validation results for a training session is shown in Fig. 5.

The model is now ready for live classification, where live data or data sets are given to the model for inference, so that we can verify if the model is correct. The model generated is in 32-bit floating point system, but such calculations are too expensive or not feasible in low power microcontrollers. Hence it is quantized to 8-bit integer, which is widely supported by most microcontrollers. The platform also optimizes the ML model according to the target device selected in the project window. This improves the performance, while reducing power consumption. The platform estimate for Arduino Nano 33 BLE sense was 4 ms of DSP, 1ms of ML inferencing with a peak RAM usage less than 2 kB and ROM usage less than 15 kB. These estimates are very good especially since running the non-optimised model with similar timings will require many times more computational power.

Fig. 6 Arduino performance
results

```
Edge Impulse standalone inferencing (Arduino)
run_classifier returned: 0
Predictions (DSP: 7 ms., Classification: 1 ms., Anomaly: 0 ms.):
[0.00391, 0.99609, 0.000]
    Fail: 0.00391
    Normal: 0.99609
    anomaly score: 0.000
```

3.2 Performance Evaluation

All the performance metrics obtained so far are simple estimates and they may or may not be true. In order test the same, the model was deployed into an Arduino Nano 33 BLE sense, an ARM Cortex M series low power microcontroller from Arduino. A static window of the data was loaded as buffer into the board during run time, emulating a sensor data load. The performance results were close to the estimates. The DSP runtime was around 6–7 ms while inference was steady at 1ms, through out different runs. The result captured at the serial communication terminal for one such run is shown in Fig. 6.

4 Advantages and Disadvantages

There are multiple advantages in using this method of fault diagnosis. They are:

– As the system is based on AI methods, its far more flexible to design, implement and maintain compared to a strict mathematical model or hardware based solutions.
– A very few additional components are necessary to use this system, in this case an output voltage sensor, which is already present many of the inverters.
– Being deployable at edge, improves the usability of the model without requiring any kind of network connectivity or other requirements in most cases.
– Since they can be deployed into low cost, low power microcontrollers, this method is better suited for cost sensitive applications as well as applications that require high scalability like a solar farm.
– All the data collection, inferencing, etc., happens on board, its more secure and private compared to network connected systems.
– The same system can be set to collect further data during run time to improve the model as it goes.
– As the model is more intuitive, it is easier for a different person or team to understand and improve upon it in the future.

There are few disadvantages to this system. They are:

- Even though support for microcontroller has improved, it is still limited to a few models from certain manufacturers, reducing the choices.
- Any AI method is only as good as the data used, so obtaining proper data or selecting proper features can quickly become tedious, especially in converters with large number of switches.

5   Conclusion

Implementation of AI methods in power electronics is expected to grow in the coming years, with more and more areas being identified over time. They will help power electronic systems to be incorporated into smart systems like smart grids etc. Here we explored one such possibility via Fault diagnosis in inverter. With recent advancements in ML like as tinyML and platforms such as Edge impulse has makes it easier than ever before to utilize these methods. The open circuit fault analysis in single phase inverter using AI at edge can be considered as a stepping stone to more complex and diverse systems in the future.

References

1. Zhao, S., Blaabjerg, F., Wang, H.: An overview of artificial intelligence applications for power electronics. IEEE Trans. Power Electr. **36**, 4633–4658 (2021). https://doi.org/10.1109/tpel.2020. 3024914
2. Lu, B., Sharma, S.K.: A literature review of IGBT fault diagnostic and protection methods for power inverters. IEEE Trans. Ind. Appl. **45**, 1770–1777 (2009). https://doi.org/10.1109/tia.2009. 2027535
3. Achintya, P., Kumar Sahu, L.: Open circuit switch fault detection in multilevel inverter topology using machine learning techniques. In: 2020 IEEE 9th Power India International Conference (PIICON). IEEE (2020)
4. Shahbazi, M., Zolghadri, M., Khodabandeh, M., Ouni, S.: Fast detection of open switch fault in cascaded H-bridge multilevel converter. Sci. Iran. **25**(3), 1561-1570 (2017). https://doi.org/10. 24200/sci.2017.4371
5. Edge Impulse Documentation. In: Edge Impulse. https://docs.edgeimpulse.com. Accessed 20 May 2021
6. Golnas, A.: PV system reliability: an operator's perspective. IEEE J. Photovolt. **3**, 416–421 (2013). https://doi.org/10.1109/jphotov.2012.2215015
7. Nagarajan, A., Thiagarajan, R., Repins, I.L., Hacke, P.L.: Photovoltaic Inverter Reliability Assessment. Office of Scientific and Technical Information (OSTI) (2019)
8. Vázquez, N., López, J.V.: Inverters. In: Power Electronics Handbook, pp. 289–338. Elsevier (2018)

Risk Management in Bank: Fuzzy Inference Based Process Model

Ramin Rzayev⑩ and Jamirza Aghajanov⑩

Abstract Any commercial bank (CB) is trying to determine the degree of acceptable risks of material and/or reputational losses, believing that its potential losses are inversely proportional to the volume of its capital. Therefore, the basis of successful banking is the presence of a Risk Management System (RMS) in the CB structure, the main task of which is to determine the best (or rational) strategy for concluding transactions that ensures maximum profit growth. Within the RMS the statistical analysis tools are used that allow to evaluate and compare the consequences and expediency of certain transactions by establishing a quantitative measure of banking risk from any transaction. Moreover, these tools make it possible to formalize various transactions and, thereby, ensure the accumulation of CB experience. To form heuristic knowledge from the history of the conclusion of various contracts, it is necessary to use and accumulate expert opinions, and, as a result, fuzzy methods of analysis and decision-making relative to banking transactions. There is a system of fuzzy models for assessing internal and external banking risks is considered. This approach involves the use a fuzzy cognitive model, based on which cause-effect relations for assessing banking risks at all levels of the hierarchy of their detailing are described using the Fuzzy Inference Systems. Qualitative evaluation criteria for the areas of localization of banking operations and external influences are used as inputs and outputs.

Keywords Risk management · Evaluation criteria · Fuzzy inference system

R. Rzayev (✉)
The Ministry of Science and Education of the Republic of Azerbaijan, Institute of Control Systems, B. Vahabzadeh str. 9, AZ1141 Baku, Azerbaijan
e-mail: raminrza@yahoo.com

J. Aghajanov
Baku State University, Z. Xalilov str. 23, AZ1148 Baku, Azerbaijan

© The Author(s), under exclusive license to Springer Nature Switzerland AG 2023   351
Sh. N. Shahbazova et al. (eds.), *Recent Developments and the New Directions of Research, Foundations, and Applications*, Studies in Fuzziness and Soft Computing 422,
https://doi.org/10.1007/978-3-031-20153-0_28

1 Introduction

Any banking activity is carried out under uncertainty, therefore, in the process of implementing its functions and providing financial and credit services, any commercial bank (CB) is inevitably forced to consider many various risks, both commercial and institutional. Since the publication of the book by J. Neumann and O. Morgenstern "Game Theory and Economic Behavior", the assessment of banking risks has become one of the main applied areas of modeling risk situations in economics and business. Each CB tries to determine the degree of acceptable risks of financial and/or reputational casualties. At the same time, the potential casualties of the CB are inversely proportional to the volume of its capital. Therefore, in the banking sector of the economy, special attention is paid to the study of the skills of risk management of possible casualties, which allows to assert that the basis of successful banking is the presence of risk management in the structure of the CB. The main task of risk management is to determine the best (or optimal) strategy for making deal, which ensures the maximum growth of the CB's profit due to the correct multi-criteria selection from all potential deals that are distinguished by high rates of profitability and reliability. In other words, risk management is designed to implement the optimal strategy for the allocation of free banking resources by determining the selected set of transactions, which can provide for CB to obtain the maximum average profit with the minimum risk.

To solve this problem relative to all banking operations, within the framework of risk management, the tools of the theory of statistical decisions are used, which, by establishing a quantitative measure of banking risk from a transaction, in each specific case allow to evaluate and compare the consequences and feasibility of certain transactions. Moreover, statistical analysis tools allow to formalize various transactions and, thereby, ensure the accumulation of CB experience. At the same time, for the formation of heuristic knowledge from the history of the conclusion of various kinds of transactions, it is necessary to use and accumulate expert opinions, and, therefore, fuzzy methods of analysis and decision-making relative to banking transactions.

The use of the apparatus of fuzzy logic in banking becomes possible since, along with quantitative measures of the reliability of banking transactions, it is increasingly necessary to apply their qualitative characteristics. As such an apparatus, it is proposed to use the fuzzy inference mechanism [1], which can combine and aggregate the statistical processing of the results of completed contracts with expert opinions relative to various conditions of their conclusion.

2 Problem Definition

The main functions of risk management are anticipation, prevention, localization, and elimination of banking solutions with high risk. At the same time, the defini-tion and assessment of banking risks are always relative, and the desire to assign them a numerical value is not always acceptable from the point of view of further interpretation of complex results. The acceptable level of risk that CB can consider acceptable for itself is a complex concept and cannot be considered as a simple set of its interrelated and/or interdependent components, since each of them is critical. Therefore, when assessing the aggregate banking risk, the numerical averaging of the results for all types of banking operations is not always acceptable. Distinctive features in the process of assessing the aggregate banking risk are: (1) incompleteness and uncertainty of the initial information on the composition and nature of factors affecting the magnitude of the risk; (2) the presence of multi-criteria problems of choosing alternatives associated with the need to consider many qualitative factors that determine the level of risk; (3) the impossibility of using classical optimization methods. Therefore, based on these considerations, it is necessary to develop an adequate model for the integral assessment of the aggregate banking risk.

3 Fuzzy Cognitive Model for Assessing Aggregate Banking Risk

According to [2], risk is the probability of unfavorable consequences or events, it is a situation that has the uncertainty of its outcome or the probability of a possible unwanted loss of something under an unfavorable concurrence of certain circum-stances. In the banking sector, risks, as the probability of manifestations of unde-sirable (sometimes dangerous) factors, manifest themselves not in isolation, but in aggregate. As a rule, one risk manifests itself in the composition of another or is its consequence or cause. Therefore, from the point of view of risk management strategy, the optimal hierarchy should demonstrate the relationships and interdependences between individual groups and types of banking risks.

In the banking sector of the economy, risk is considered as a category of entrepreneurial and institutional activity of CB. It reflects the hidden cause-effect relation between factors and outcomes. Therefore, the differentiation criteria and features of classification of banking risks that can be identified should be based on the reasons for their occurrence, in the entrepreneurial case, and, in the institutional case, on the differentiation of objects at risk, based on which the consequences of risk realization can be directly observed. Based on these considerations, a concept, and mechanisms for managing banking risks are being formed that meet the require-ments of risk management, which imply the need to consider the essential features of individual risks, which make it possible to form ways to influence them. Therefore, to streamline many banking risks, considering their interdependence by spheres of

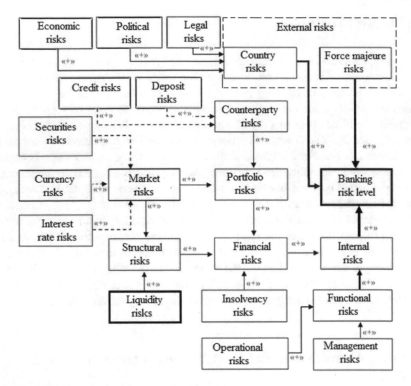

Fig. 1 FCM for assessing the aggregate banking risk

their occurrence and influence, by the nature of influence, by factors of influence and by areas of localization, a logical detailing of banking risks is proposed based on a fuzzy cognitive map (FCM) [3], shown in Fig. 1.

In practice, the "interaction" of even two elements of the FCM occurs according to more complex functional laws, which are very difficult to formalize in the traditional mathematical form. Therefore, it becomes necessary to apply the mechanism of fuzzy inference to describe the cause-effect relation between the terms of the aggregate banking risk, and to carry out the analysis based on the so-called fuzzy cognitive model (FCMd) [3]. In this case, the nodular factors (concepts) of the FCM are interpreted as fuzzy sets, and the cause-effect relations between them are established based on a bounded set of fuzzy linguistic rules, which are formed as follows:

$$\text{"If } x_{k1} \text{ is } A_{k1} \text{ and } x_{k2} \text{ is } A_{k2} \text{ and } \dots \text{ and } x_{kn} \text{ is } A_{kn}, \text{ then } y \text{ is } B_k \text{".} \qquad (1)$$

where x_{kj} $(j = 1 \div n; k = 1, 2, \dots)$ are input linguistic variables, characterizing the factors of influence; y is the output linguistic variable, which characterizes the level of consolidated risk; A_{kj} and B_k are terms (values) of the corresponding input and output linguistic variables, which can be described by appropriate fuzzy sets (FSs).

4 Description of Cause-Effect Relations by Fuzzy Inference

In [4], the so-called risk matrix is considered, which proposes to use as the information base for assessing and managing banking risks. The basis of this matrix is a scale of the probability of risk occurrence, which is characterized by the following terms: A–ALMOST CERTAINLY (risk situation expected under all circumstances); B–VERY PROBABLY (risk situation possible almost always); C–POSSIBLY (risk situation occurs from time to time); D–UNLIKELY (risk situation can sometimes happen); E–OCCASIONALLY (risk situation can occur under exceptional circumstances). The given terms are qualitative criteria for assessing risk situations, which can be described by appropriate fuzzy sets. For this purpose, choosing a discrete set $U = \{0, 0.25, 0.5, 0.75, 1\}$ as a universe, the corresponding FSs can be described as: $A=\{0/0, 0/0.25, 0/0.5, 0.5/0.75, 1/1\}$, $B=\{0/0, 0/0.25, 0.5/0.5, 1/0.75, 0.5/1\}$, $C=\{0/0, 0.5/0.25, 1/0.5, 0.5/0.75, 0/1\}$, $D=\{0.5/0, 1/0.25, 0.5.5/0.5, 0/0.75, 0/1\}$, $E=\{1/0, 0.5/0.25, 0/0.5, 0/0.75, 0/1\}$. To estimate the level of risk by areas of localization of banking operations, a scale of terms is used, which are described by fuzzy subsets of the universe $J=\{0, 0.1, 0.2, \ldots, 1\}$ with corresponding membership functions (MF) $\mu(j)$ ($j \in J$) in the following form: $TL = $ TOO LOW : $\mu_{TL}(j) = 1$, if $j < 1$ and $\mu_{TL}(j) = 0$, if $j = 1$; $VL = $ VERY LOW: $\mu_{VL}(j) = (1 - j)^2$; $\mu_{ML}(j) = (1 - j)^{(1/2)}$; $L = $ LOW : $\mu_L(j) = 1 - j$; $H = $ HIGH : $\mu_H(j) = j$; $MH = $ MORE THAN HIGH: $\mu_{MH}(j) = j^{(1/2)}$; $\cdot VH = $ VERY $\cdot$ HIGH : $\cdot\mu_{VH}(j) = j^2$, $TH = $ TOO HIGH: $\mu_{TH}(j) = 1$, if $j = 1$, if $j = 1$ and $\mu_{TH}(j) = 0$, if $j < 1$. So, shown in Fig. 1 the cause-effect relations can be described using a sufficient set of logically consistent rules in the form of (1). Thus, for each of the local concepts of FCM, the following typical fuzzy inference systems (FIS) can be chosen. All linguistic rules in the composition of FISs are quite trivial and can be easily realized in the notation of the MATLAB package. As a result, various banking risk situations can be simulated. For example, for localization "Market Risks" (see Fig. 1) we have:

e_{11}: "x_{11} is E and x_{12} is E and x_{13} is E, then y_1 is TL";

e_{12}: "x_{11} is D and x_{12} is C and x_{13} is E, then y_1 is VL";

e_{13}: "x_{11} is C and x_{12} is D and x_{13} is D, then y_1 is ML";

e_{14}: "x_{11} is D and x_{12} is C and x_{13} is C, then y_1 is L";

e_{15}: "x_{11} is C and x_{12} is C and x_{13} is B, then y_1 is H";

e_{16}: "x_{11} is B and x_{12} is B and x_{13} is C, then y_1 is MH";

e_{17}: "x_{11} is C and x_{12} is A and x_{13} is B, then y_1 is VH";

e_{18}: "x_{11} is A and x_{12} is A and x_{13} is A, then y_1 is TH",

where x_{11} is the securities risks, x_{12} is the currency risks (exchange operations), x_{13} is the interest rate risks; y_1 is the market risks level.

5 Classification of Banking Risks Using the FIS

Let's consider an approach to the classification of banking risks using the FIS on the example of assessing the levels of country risk (CR). Along with force majeure situations, country risks carry the dangers of political, legal, and socio-economic character. Therefore, to guarantee protection against such threats, it is necessary to consider the economic and political situation in the aggregate (especially in emerging markets), which predetermined the introduction of the concept of "Country Risk" (CR). CR is a multi-factor category characterized by a combined system of financial and economic, socio-political, and legal factors that distinguish the market of any country [5]. At present, many world rating agencies, and international institutions, such as Euromoney, Institutional Investor, Mood's Investor Service, the European Bank for Reconstruction and Development, the World Bank, etc., are currently ranking countries according to their CR-level. At the same time, existing approaches are conditioned by qualitative and/or quantitative, economic, combined, and structurally qualitative methods for estimating of CR. Well-known auditing firm Pricewaterhous Coopers uses a limited set of variables to formulate the ratings of the investment attractiveness of states. These variables are formulated and denoted in the following form: x_1–the presence of corruption; x_2–compliance of legislation; x_3–the level of economic growth; x_4–state policy on accounting and control; x_5–state regulation [5].

As a multi-criteria procedure the estimation of the CR-level implies the application of the compositional rule for aggregating the obtained results in each specific case. To assess the CR-level eight evaluative concepts (terms) are chosen: $u_1 =$ TOO LOW, $u_2 =$ VERY LOW, $u_3 =$ MORE THAN LOW, $u_4 =$ LOW, $u_5 =$ HIGH, $u_6 =$ MORE THAN HIGH, $u_7 =$ VERY HIGH, $u_8 =$ TOO HIGH. More simply, $C = \{u_1, u_2, ..., u_8\}$ is a set of criteria for the classification of CR-levels. Then, assuming the factors x_i ($i = 1 \div 5$) as linguistic variables (LV) that take their values in the form of denoted terms, the estimation of CR-levels can be carried out using a sufficient set of consistent implicative rules (1), based on which the appropriate scale of gradation of the final estimates of CR-levels can be established. In this case, basic judgments are formulated as follows:

d_1: "If there is no corruption in the country and economic development is observed, then the CR-level is acceptable",

d_2: "If, in addition to the above requirements, a state policy on accounting and control is carried out, then the CR-level is more than acceptable",

d_3: "If, in addition to the conditions specified in d_2, there is appropriate legislation and government regulation, then the CR-level is low",

d_4: "If there is no corruption, appropriate legislation and economic development are observed, a state policy on accounting and control is carried out, then the CR-level is very acceptable",

d_5: "If there is appropriate legislation, economic development is observed and a state policy on accounting and control is being implemented, but at the same time there is a manifestation of corruption, then the CR-level is still acceptable",

d_6: "If there is corruption in the country, economic development is not observed and government regulation is not carried out, then the CR-level is unacceptable".

In these judgments reflecting internal cause-effect relations, factors x_i $(i = 1 \div 5)$ are input LVs, and the output is the LV y, the terms of which reflect the CR-levels. Then, for specified terms of all LVs, the corresponding implicative rules are as follows:

d_1: "If x_1 is BE ABSENT and x_3 is OBSERVED, then y is ACCEPTABLE",

d_2: "If x_1 is BE ABSENT and x_3 is OBSERVED and x_4 is CARRIED OUT, then y is MORE THAN ACCEPTABLE",

d_3: "If x_1 is BE ABSENT and x_2 is EXIST and x_3 is OBSERVED and x_4 is CARRIED OUT and x_5 is IMPLEMENTED, then y is LOW",

d_4: "If x_1 is BE ABSENT and x_2 is EXIST and x_3 is OBSERVED and x_4 is CARRIED OUT, then y is VERY ACCEPTABLE",

d_5: "If x_1 is APPEARS and x_2 is EXIST and x_3 is OBSERVED and x_4 is CARRIED OUT, then y is ACCEPTABLE",

d_6: "If x_1 is APPEARS and x_3 is NOT VISIBLE and x_5 is NOT IMPLEMENTED, then y is UNACCEPTABLE".

Let the LV y is defined on the universe $J = \{0, 0.1, \ldots, 1\}$. Then $\forall j \in J$ its terms can be described by FSs using the corresponding MFs [5, 6]: $S =$ ACCEPTABLE, $\mu_S(j) = j$; $MS =$ MORE THAN ACCEPTABLE, $\mu_{MS}(j) = j^{(1/2)}$; $L =$ LOW, $\mu_L(j) = 1$ if $j = 1$ and $\mu_L(j) = 0$ if $j < 1$; $VS =$ VERY ACCEPTABLE, $\mu_{VS}(j) = j^2$; $US =$ UNACCEPTABLE, $\mu_{US}(j) = 1-j$. Fuzzification of terms in the left-hand sides of the rules is carried out using the Gaussian MF $\mu(u) = \exp[(u - u_k)^2/\sigma_i^2]$ $(i = 1 \div 5)$, which restore appropriate fuzzy subsets of the discrete universe C, where $u_k = (a_{k+1} + a_k)/2$, $a_k = 0.125k$ $(k = 1 \div 8)$; the density σ_i^2 for the i-th factor is selected individually under its criticality. Thus, the terms from the left-hand sides of the rules can be described as follows:

- BE ABSENT (the presence of corruption): $A = \{0.9070/u_1, 0.6766/u_2, 0.4152/u_3, 0.2096/u_4, 0.0870/u_5, 0.0297/u_6, 0.0084/u_7, 0.0019/u_8\}$;
- EXIST (compliance of legislation): $B = \{0.9070/u_1, 0.6766/u_2, 0.4152/u_3, 0.2096/u_4, 0.0870/u_5, 0.0297/u_6, 0.0084/u_7, 0.0019/u_8\}$;
- OBSERVED (economic development): $C = \{0.9394/u_1, 0.7788/u_2, 0.5698/u_3, 0.3679/u_4, 0.2096/u_5, 0.1054/u_6, 0.0468/u_7, 0.0183/u_8\}$;
- CARRIED OUT (state policy on accounting and control): $D = \{0.9497/u_1, 0.8133/u_2, 0.6282/u_3, 0.4376/u_4, 0.2749/u_5, 0.1557/u_6, 0.0796/u_7, 0.0367/u_8\}$;
- IMPLEMENTED (state regulation): $E = \{0.9575/u_1, 0.8406/u_2, 0.6766/u_3, 0.4994/u_4, 0.3379/u_5, 0.2096/u_6, 0.1192/u_7, 0.0622/u_8\}$.

Considering the introduced formalisms, the implicative rules in symbolic expression are described as follows:

d_1: $(x_1 = A)$ & $(x_3 = C)$ $\Rightarrow$ $(y = S)$,
d_2: $(x_1 = A)$ & $(x_3 = C)$ & $(x_4 = D)$ $\Rightarrow$ $(y = MS)$;
d_3: $(x_1 = A)$ & $(x_2 = B)$ & $(x_3 = C)$ & $(x_4 = D)$ & $(x_5 = E)$ $\Rightarrow$ $(y = L)$;

d_4: $(x_1 = A)$ & $(x_2 = B)$ & $(x_3 = C)$ & $(x_4 = D)$ $\Rightarrow$ $(y = VS)$;
d_5: $(x_1 = \neg A)$ & $(x_2 = B)$ & $(x_3 = C)$ & $(x_4 = D)$ $\Rightarrow$ $(y = S)$;
d_6: $(x_1 = \neg A)$ & $(x_3 = \neg C)$ & $(x_5 = \neg E)$ $\Rightarrow$ $(y = US)$.

Applying the rule of intersection of FSs [1] and Lukasiewicz's implication $\mu_{U \times J}(u, j) = \min\{1, 1 - \mu_U(u) + \mu_J(j)\}$, for each pair $(u, j) \in U \times J$ the corresponding fuzzy relations are obtained in the forms of corresponding matrixes, intersection of which generates the general functional solution in the form of following matrix R

$$
R = \begin{bmatrix}
 & 0 & 0.1 & 0.2 & 0.3 & 0.4 & 0.5 & 0.6 & 0.7 & 0.8 & 0.9 & 1 \\
\hline
u_1 & 0,0930 & 0,0930 & 0,0930 & 0,0930 & 0,0930 & 0,0930 & 0,0930 & 0,0930 & 0,0930 & 0,0930 & 0,9575 \\
u_2 & 0,3234 & 0,3234 & 0,3234 & 0,3234 & 0,3234 & 0,3234 & 0,3234 & 0,3234 & 0,3234 & 0,3234 & 0,8406 \\
u_3 & 0,5006 & 0,5848 & 0,5848 & 0,5848 & 0,5848 & 0,5848 & 0,5848 & 0,5848 & 0,5848 & 0,5848 & 0,6766 \\
u_4 & 0,7090 & 0,7904 & 0,7904 & 0,7904 & 0,7904 & 0,7904 & 0,7904 & 0,7904 & 0,6994 & 0,5994 & 0,4994 \\
u_5 & 0,8547 & 0,9130 & 0,9130 & 0,9130 & 0,9130 & 0,8379 & 0,7379 & 0,6379 & 0,5379 & 0,4379 & 0,3379 \\
u_6 & 0,9378 & 0,9703 & 0,9703 & 0,9096 & 0,8096 & 0,7096 & 0,6096 & 0,5096 & 0,4096 & 0,3096 & 0,2096 \\
u_7 & 0,9772 & 0,9916 & 0,9192 & 0,8192 & 0,7192 & 0,6192 & 0,5192 & 0,4192 & 0,3192 & 0,2192 & 0,1192 \\
u_8 & 0,9928 & 0,9622 & 0,8622 & 0,7622 & 0,6622 & 0,5622 & 0,4622 & 0,3622 & 0,2622 & 0,1622 & 0,0622 \\
\end{bmatrix}
$$

The matrix R reflects the cause-effect relations between the factors x_i ($i = 1 \div 5$) and the CR-levels. To determine the CR-level, the compositional rule $E_k = G_k \circ R$ is applied, where E_k is the degree of risk acceptability relative to the k-th CR-level ($k = 1 \div 8$); G_k is the mapping of the k-th CR-level as the fuzzy subset of the discrete universe J. Then, choosing the compositional rule as $\mu_{Ek}(j) = \max\{\min[\mu_{Gk}(j), \mu_R(j)]\}$ and, if in this case $\mu_{Gk}(j) = 0$, when $j \neq j_k$ and $\mu_{Gk}(j) = 1$, when $j = j_k$, then $\mu_{Ek}(j) = \mu_R(j_k, u)$, i.e., E_k is fuzzy subset of the universe J with values of corresponding MF from the k-th row of the matrix R.

To classify CR-levels by numerical criteria, the procedure of defuzzification of fuzzy outputs of the applied model is used. For example, for the evaluative concept of risk acceptability u_1, the fuzzy interpretation of the corresponding CR-level is the following fuzzy subset of the universe J: $E_1 = \{0.093/0, 0.093/0.1, 0.093/0.2, 0.093/0.3, 0.093/0.4, 0.093/0.5, 0.093/0.6, 0.093/0.7, 0.093/0.8, 0.093/0.9, 0.9575/1\}$. Establishing its level sets $E_{1\alpha}$ and calculating the corresponding cardinal number $M(E_{1\alpha}) = \sum_{r=1}^{m} x_r / m$ as:

- for $0 < \alpha < 0.093$: $\Delta\alpha = 0.0930$; $E_{1\alpha} = \{0, 0.1, 0.2, ..., 0.9, 1\}$. $M(E_{1\alpha}) = 0.50$;
- for $0.093 < \alpha < 0.9575$: $\Delta\alpha = 0.8645$; $E_{1\alpha} = \{1\}$, $M(E_{1\alpha}) = 1.0$,

to numerically estimate the fuzzy outputs E_k ($k = 1 \div 8$), the following formula is applied [6]: $F(E_k) = (1/\alpha_{max}) \int_0^{\alpha_{max}} M(E_{k\alpha}) d\alpha$. In particular, for E_1 we have:

$$
F(E_1) = (1/0.9575) \int_0^{0.9575} M(E_{1\alpha}) d\alpha = (0.5 \times 0.0930 + 1.0 \times 0.8645)/0.9575 = 0.9514.
$$

Similar actions are established point estimates for the others fuzzy outputs: for the evaluated concept of risk acceptability u_2–$F(E_2) = 0.8077$; u_3–$F(E_3) = 0.5741$;

Table 1 Gradation of CR-Levels using the FIS

Interval	CR-level	Interval	CR-level
(84.90, 100]	TOO LOW	(34.94, 41.66]	HIGH
(60.34, 84.90]	VERY LOW	(30.09, 34.94]	MORE THAN HIGH
(49.29, 60.34]	MORE THAN LOW	(27.11, 30.09]	VERY HIGH
(41.66, 49.29]	LOW	[0, 27.11]	TOO HIGH

u_4–$F(E_4) = 0.4689$; u_5–$F(E_5) = 0.3964$; u_6–$F(E_6) = 0.3324$; u_7–$F(E_7) = 0.2863$; u_8–$F(E_8) = 0.2579$. In this case, $F(E_8) = 0.2579$ is the smallest defuzzified output of the applied model of the multicriteria assessment of the CR-level. As the upper bound this value corresponds to the consolidated estimate of the CR-level "TOO HIGH OR UNACCEPTABLE". Further, from the point of view of the influence of CR-factors, the defuzzified output:

- 0.2863 is the upper bound of the qualitive estimate "VERY HIGH OR SIGNIFICANT";
- 0.3324 is the upper bound of the qualitive estimate "MORE THAN HIGH";
- 0.3964 is the upper bound of the qualitive estimate "HIGH";
- 0.4689 is the upper bound of the qualitive estimate "LOW";
- 0.5741 is the upper bound of the qualitive estimate "MORE THAN LOW";
- 0.8077 is the upper bound of the estimate "VERY LOW OR UNSIGNIFICANT";
- 0.9514 is the upper bound of the qualitive estimate "TOO LOW OR NONE.

Within the framework of the accepted assumptions, based on the criterion $E = 100 \times F(E_k)/F_{max}$, where $F(E_k)$ is the estimate of the k-th CR-level, $F_{max} = F(E_1) = 0.9514$, in the measure of the [0, 100] it is possible to construct a reasonable scale for assessing the CR-levels. The resulting scale is presented in the form of Table 1.

6 Conclusion

On the example of country risk, one can notice, that the FIS based classification of the final estimates of banking risks more confidence, since in this case the cause-effect relations between the influence factors and the risk levels are traced, even though these relations are formulated on the base of trivial, but consistent and sufficiently valid implicative rules. Using the scale obtained, within the proposed FCM, it is possible to assess risks for all localization of banking operations using various methods of multi-criteria assessment of alternatives, including scoring analysis methods, expert systems, as well as various fuzzy decision-making methods under uncertainty.

References

1. Zadeh, L.A.: Outline of a new approach to the analysis of complex systems and decision processes. IEEE Trans. Syst. Man Cybern. **3**(1), 28–44 (1973)
2. How to determine the acceptable amount of risk. http://fxgear.ru/kak-opredelit-dopustimyj-raz mer-riska/. (in Russian)
3. Kosko, B.: Fuzzy cognitive maps. Int. J. Man Mach. Stud. **1**, 65–75 (1986)
4. Prigodich, I.A.: Bank risk: essence, estimation, procedure, management. Econ. Manag. **4**, 85–90 (2012). (in Russian)
5. PwC Global Risk podcast series. https://www.pwc.com/gx/en/services/advisory/consult-ing/risk/cosoerm-framework/podcasts.html
6. Rzayev, R.R.: Analytical Support for Decision-Making in Organizational Systems. Palmerium Academic Publishing, Saarbruchen (2016). (in Russian)

Analysis of the Methods for Constructing Membership Functions Using Expert Data

K. R. Aida-Zade©, **P. S. Guliyeva, and R. E. Ismibayli**

Abstract The paper analyzes methods for constructing membership functions of fuzzy sets using expert information. Methods of approximation and interpolation are used. An analysis of the influence of noises in expert data on the accuracy of the obtained membership functions was carried out.

Keywords Expert data · Membership function · Fuzzy sets · Interpolation · Approximation

1 Introduction

The theory of fuzzy sets is widely used in mathematical modeling of complex techno-logical processes and technical objects functioning. The main characteristic of fuzzy sets is their membership functions. In practice, such functions as triangular, trape-zoidal, Gaussian, bell-shaped and some other types of functions are widely used as membership functions of fuzzy one-dimensional numerical sets. Each of these classes of functions is characterized by a number of parameters that can be used to construct the membership function (MF) of a particular fuzzy set. The number of parameters of the known classes listed above is in many cases few to construct a sufficient adequate MF for specifically given fuzzy sets [1].

In this paper, an analysis of methods for constructing an MF from various classes of known functions is carried out. In this paper, we suggest to carry out the analysis of functions from the class of piecewise continuous functions. To construct functions we use available expert information about the membership degree of some given elements from a universal set to a fuzzy set.

K. R. Aida-Zade (✉) · P. S. Guliyeva
The Ministry of Science and Education of the Republic of Azerbaijan, Institute of Control Systems, Baku, Azerbaijan
e-mail: kamil_aydazade@rambler.ru

R. E. Ismibayli
University of Architecture and Construction, Baku, Azerbaijan

© The Author(s), under exclusive license to Springer Nature Switzerland AG 2023
Sh. N. Shahbazova et al. (eds.), *Recent Developments and the New Directions of Research, Foundations, and Applications*, Studies in Fuzziness and Soft Computing 422, https://doi.org/10.1007/978-3-031-20153-0_29

To construct membership functions, namely, to determine the values of their parameters, mathematical methods such as various interpolation methods, approximation methods based on conditional and unconditional optimization methods are used [2–5].

The paper presents the results of numerical experiments using various approaches to the construction of the MF, the analysis and comparison of the results are carried out.

2 Statement of the Problem

Let there are expert information about the membership degree of the elements $\overline{x}_i \epsilon X, i = 1, \ldots, n$ to the fuzzy set A, here X is a universal set. Thus, $\overline{x}_i$ and $\overline{\mu}_i$ are given and:

$$\overline{\mu}_i = \mu_A(\overline{x}_i; P), i = 1 \ldots, n. \tag{1}$$

Here the $\mu_A(\overline{x}_i; P)$—is a membership function of the fuzzy set A; P—is a vector of parameters of this membership function.

It is required, firstly, to determine the type of MF $\mu_A(x, p)$, i.e. to determine what class it is from (we call this problem the problem of structural identification). Secondly, it is required to determine the values of the r-dimensional vector of parameters $P \in R^r$, participating as coefficients in the function $\mu_A(x; P)$, r— is the dimension of the vector of parameters (we call this problem the problem of parametric identification).

There are no mathematical methods to solve the problem of structural identification, i.e. to choose the form of the function $\mu_A(x_i, p)$ when choosing a class of functions, it is necessary to consider:

(1) the specifics of the fuzzy set itself; (2) number r—the dimension of the vector of function parameters, participating in the membership functions of a given class; (3) n—the number of elements, which have expert information about the membership degree to the fuzzy set [6–9].

To carry out parametric identification, mathematical methods such as the Lagrange interpolation method can be used, when the condition $r = n$ is satisfied. In the case, if $n < r$, then the interpolation problem is not correct, since the condition of uniqueness of the constructed membership function is not satisfied. The use of interpolation methods is also incorrect for $n > r$, due to the impossibility of constructing such a membership function [10, 11].

The application of approximation methods using methods, such as minimizing the standard deviation leads the original problem to the following:

$$S(P) = \frac{1}{n} \sum_{i=1}^{n} \left[\mu_A(\overline{x}_i; P) - \overline{\mu}_i \right]^2 \to \min_{P \in R^r}. \tag{2}$$

Here it is assumed that condition $r < n$ is satisfied. This is necessary for the uniqueness of the solution to the problem (2).

The problem (2) belongs to finite-dimensional optimization problems. To solve it, you can use well-known effective optimization methods such as conjugate gradient methods, variable metrics, and others. There are ready-made application packages to solve it [12].

As a rule, membership functions are required to fulfill the following conditions:

$$\mu_A(x; P) \le 1, x \in X, \tag{3}$$

$$\mu_A(x; P) \ge 0, x \in X. \tag{4}$$

Condition (3), is called the normalization condition and is not necessary in some problems. However, after solving the problem of parametric identification and determining the parameters P, the obtained function can be normalized by the formula:

$$\overline{\mu}_A(x) = \frac{\mu_A(x; P)}{\max\limits_{y \in X} \mu_A(y; P)}, x \in X. \tag{5}$$

It is easy to check that the function $\overline{\mu}_A(x)$ in (5) fulfills the condition (3). If the obtained function violates condition (4), then it can be transformed as follows:

$$\overline{\overline{\mu}}_A(x) = \frac{\mu_A(x; P) - \min\limits_{y \in X} \mu_A(y; P)}{\max\limits_{y \in X} \mu_A(y; P) - \min\limits_{y \in X} \mu_A(y; P)}, x \in X. \tag{6}$$

It is not so hard to check that the function $\overline{\overline{\mu}}_A(x)$ in (6) satisfies the conditions (3) and (4).

It can be done in another way. Namely, to solve the problem of minimizing function (2) under constraints (3) and (4). For this, for example, methods of penalty functions can be used. But in this case, constraints (3) and (4) on the parameters P must be discretized with respect to $x \in X$. Namely, to introduce, for example, a grid with a given step $h : x_j = x_0 + jh, j = 1, \ldots, M$, and to require the fulfillment of the conditions (3) and (4) at all points of the grid:

$$\mu_A(x_j; P) \le 1, j = 1, \ldots, M, \tag{7}$$

$$\mu_A(x_j; P) \ge 0, j = 1, \ldots, M. \tag{8}$$

To determine the vector of parameters P minimizing function (2), it is necessary to add $2M$ number of constraints (7) and (8).

3 Results of Numerical Experiments

Let us present some of the results of computer experiments carried out on examples of hypothetical expert data.

Tables 1 and 2 show exact expert data on the degree of membership of elements to some fuzzy sets. The fourth rows of these tables contain data on the degree of membership of the same elements with noises $\xi = 10\%$, i.e. $\overline{\mu}_i^{\xi} = (1 + \xi)\overline{\mu}_i$. Noises are random variables that have a uniform distribution on the interval $[-1; 1]$.

Table 1 and Fig. 1 show graphs of membership functions from a triangular class of functions, the parameters of which were obtained using the above methods.

Table 2 and Fig. 2 show similar results for the trapezoidal membership function. There are many other results obtained from computer experiments.

Table 1 The result of solving the problem for triangle type MF

№	1	2	3	4	5	6	7	8	9	10
$\overline{x}_i$	0.1	0.21	0.3	0.38	0.48	0.65	0.71	0.83	0.88	0.9
$\overline{\mu}_i$	0	0.2749	0.4999	0.7	0.95	0.625	0.475	0.175	0.05	0
$\overline{\mu}_i^{\xi}$	0	0.2954	0.5372	0.7522	1	0.6716	0.5104	0.188	0.0537	0

Table 2 The result of solving the problem for trapezoid type MF

№	1	2	3	4	5	6	7	8	9	10
$\overline{x}_i$	0.1	0.21	0.3	0.38	0.48	0.65	0.71	0.83	0.88	0.9
$\overline{\mu}_i$	0	0.36	0.66	0.93	1	1	0.95	0.35	0.1	0
$\overline{\mu}_i^{\xi}$	0	0.39	0.71	1	1	1	1	0.37	0.1	0

Fig. 1 The result of solving the problem after adding the error for triangle type MF

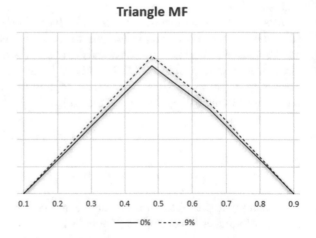

Fig. 2 The result of solving the problem after adding the error for trapezoid type MF

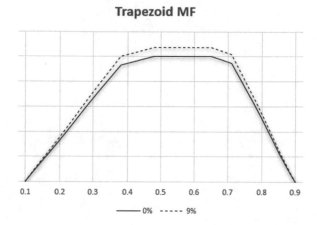

Trapezoid MF

As can be seen from the figures, the quality of the approximation of expert data significantly depends on the chosen class of the membership function. Noises in expert data have a comparatively small effect, since they are smoothed out in the process of approximation.

References

1. Aida-Zade K.R., Alieva N.T.: Study of one class of membership functions of fuzzy sets. Autom. Control Comput. Sci. **45**(3), 142–152 (2011)
2. Aida-Zade, K.R., Alieva N.T.: On class of smooth membership functions. J. Autom. Inf. Sci. **2**, 60–71324 (2012). (Beggel House, Inc., New York)
3. Guliyeva, P.S.: Analysis of the application of interpolation and approximation methods of constructing membership functions of fuzzy sets. In: News Bulletin Omsk Scientific and Educational Center of OmskTU and IM SB RAS in Mathematics and CS, pp. 22, 23. (in Russian)
4. Gottwald, S.: Universes of Fuzzy Sets and Axiomatizations of Fuzzy Set Theory. Part I: Model-Based and Axiomatic Approaches. Studia Logica, pp. 211–244 (2006)
5. Zadeh, L.A.: Fuzzy sets as a basis for a theory of possibility. Fuzzy Sets Syst. 9–34 (1999)
6. Zadeh, L.A.: Fuzzy Set. Inform. Cont. **8**, 338–353 (1965)
7. Liou, T.S., Wang, M.J.J.: Ranking fuzzy numbers with integral values. Fuzzy Sets Syst. **55**, 247–255 (1992)
8. Kaufmann, A., Gupta, M.M.: Introduction to Fuzzy Arithmetic Theory and Applications. Van Nostrand Reinhold, New York (1991)
9. Jang, S., Sun, C.-T., Mizutani, E.: Neuro-Fuzzy and Soft Computing. Prentice Hall, New York (2001)
10. Gill, P.E., Murray, W., Wright, M.H.: Practical Optimization. Academic Press, London (1981)
11. Baruah, H.K.: The Theory of fuzzy sets: beliefs and realities. Int. J. Energy Inf. Commun. **2**(2), 1–22 (2011)
12. Vasiliev, F.P.: Methods of optimization M, pp. 824. Factorial Press (2002). (in Russian)

Fuzzy Logic and Fuzzy Sets

Fuzzy Model of Defence of the Defensive Line by a Group of Dynamic Objects

Azad Bayramov, Adalat Pashayev, and Elkhan Sabziev

Abstract In the paper, the defence of a certain barrier in the plane by a group of guarded objects is considered. A model of the process of meeting a controlled attacking object by a group of controlled guard objects is offered. The mathematical model of the problem is based on the theory of differential games. As one of the issues of defence of the object, a strategy to neutralize the attacking object in the danger zone was developed and evaluated. The evaluation had been carried out by using Fuzzy theory.

Keywords Fuzzy model · Games theory · Attacking object · Defence object · Differential equation

1   Introduction

The problem of protecting defense lines from an attacking object is urgent. In the paper, this problem is considered on the basis of the theory of differential games described by ordinary differential equations, the main model for which is the process of pursuing one controlled object by another controlled object [1–3]. Mobile guard means are taken as an object.

Let in the process of movement the objects continuous observe each other and at each moment of time, with the help of control parameters, correct their movement depending on the information received about the enemy's behavior.

The following modified pursuit-escape problem is formulated: using information about the behavior of the attacking object, to choose self-control at each moment of time t in such a way that to prevent the attacking object from crossing the line of defense.

A. Bayramov (✉) · A. Pashayev · E. Sabziev
The Ministry of Science and Education of the Republic of Azerbaijan, Institute of Control Systems, 68 B. Bakhabzade str., AZ1141 Baku, Azerbaijan
e-mail: azad.bayramov@yahoo.com

2   The Methodological Essence of the Problem

Let us denote the number of controlled moving objects by k. The movement has carried out in $n = 2$ dimensional Euclidean space. We will mark these objects by numbers from 1 to k. The movement of i object is described by following equation:

$$\dot{x}_i(t) = u_i, \quad u_i \in P = \left\{ u : u \in \mathfrak{R}^n, \; \|u\| \le a \right\}$$

The movement of object is considered in duration $[0, \; T]$ interval. At the initial time, below condition is satisficed:

$$x_i(0) = x_i, \quad x_1, x_2, ..., x_n \in \mathfrak{R}^n$$

Moreover, there is attacking object and the movement of this object is expressed as below

$$\dot{y}(t) = v, \quad v \in Q = \left\{ v : v \in \mathfrak{R}^n, \; \|v\| \le b \right\}$$

$$y(0) = y_0, \quad y_0 \in \mathfrak{R}^n$$

The purpose of the guard group is to prevent the attacking object to cross the security barrier, and the aim of the attacking object is to cross the defensive barrier.

A point is an 0-dimensional object as a mathematical idealization. In real life, objects have a positive dimension. For example, the size of the attacking object can be $1 \times 2 \; m$ and the size of the guard object can be $1 \times 1.5 \; m$. Let give a Fuzzy definition of the concept that a defending object "collides" with an attacking object. For this purpose, let denote

$$\rho_i(t) = \|x_i(t) - y(t)\|$$

Then, the distance of the guard group from the attacking object is as below

$$S(t) = \min_{i=1, \, 2, \, ..., \, k} \{\rho_i(t)\}$$

Let for the given objects we have a parameter $m > 0$ which characterized a size of object.

Depending on the distance of the guard group to the attacking object, the function that determines its level of "damage" by a percentage can be given as the following "damage" $\{\mu, \; [0, \; \infty)\}$ Fuzzy set:

$$\mu(S) = \begin{cases} 100, & S \leq m, \\ 100 \cdot \left(1.2 - 0.2 \cdot \frac{S}{m}\right), & m \leq S \leq 2m, \\ 100 \cdot \left(1.6 - 0.4 \cdot \frac{S}{m}\right), & 2m \leq S \leq 4m, \\ 0, & 4m \leq S. \end{cases}$$

Definition For each $t_* \in [0, \ T]$ moment if $\mu(S(t_*)) > 80$ then attacking object collides with guard group; if $60 < \mu(S(t_*)) \leq 80$ then attacking object meets with guard group; if $40 < \mu(S(t_*)) \leq 60$ then attacking object approaches with guard group; if $\mu(S(t_*)) \leq 40$ then there isn't approaching.

On the other hand, it should be noted that the vulnerability varies depending on the strength of the defense system of the attacking object from external influences. The function that characterizes the weakness can be defined as the following "weakness" $\{v, \ [0, \ \infty)\}$ Fuzzy set:

$$v(S) = \begin{cases} 100, & S \leq 0.5\,m, \\ 100 \cdot \left(1.08 - 0.16 \cdot \frac{S}{m}\right), & 0.5\,m \leq S \leq 3\,m, \\ 100 \cdot \left(1.8 - 0.4 \cdot \frac{S}{m}\right), & 3\,m \leq S \leq 4.5\,m, \\ 0, & 4.5\,m \leq S. \end{cases}$$

It is assumed that the attacking object is completely neutralized by the guard group if the product of $\mu(S)v(S)$ is more than 50%, and partially neutralized if the product of $\mu(S)v(S)$ is more than 30%.

The purpose of the guard group is to prevent the attacking object from crossing the defence barrier so that it is $\mu(S)v(S) \geq 50$, and the aim of the attacking object is to cross the defence barrier so that it is $\mu(S)v(S) \leq 50$. The $\mu(S)v(S)$ product can be estimated depending on the strategy of the attacking object and the guard group.

Closed-loop control class [4, 5] ("позиционное управление"—positional control) will be considered as a controlling class, ie information about $x_i(t)$ and $y(t)$ can be used at any t time during the u_i and v generation of controlleds to determine the current status of the both attacking and guard dynamic group [5].

To study the task, let first look at the following ancillary issue of crossing the "barrier on the street".

Let D_0 and D_1 objects are placed face to face on some R street with $0 \leq y \leq R$ range (set). The question is D_0 object on the $y \in [0, \ R]$ strip condition, which on $\{x = -l\}$ line should be allocated and what should be the width of the street at least so that it can move and cross the $\{x = 0\}$ barrier from left to right?

3 Mathematical Assessment of the Problem of the "Street Barrier"

For the simplicity of mathematical formalization, we will consider the problem in the given $\{0 \leq x \leq l; \ 0 \leq y \leq R\}$ set with respect to the rectangular coordinate system (Fig. 1). The worst case of the scenario for the D_0 object at the origin to cross the $\{x = l\}$ barrier from left to right using its maximum speed advantage over D_1, is when they are in a straight line. Considering the above, suppose that at the initial $t = 0$ moment, in the $(0, \ 0)$ coordinate allocated D_0 object must move at least at the $l > 0$ distance from the D_1 object at the starting $(l, \ 0)$ point by moving to $\{x = l\}$ line at the beginning.

It is assumed that in absolute terms the maximum speeds of D_0 and D_1 objects are b and a, respectively. From the above, it is intuitively clear that each object must make maximum use of its capabilities, which means that the speed of movement of these objects must be maximum during the entire period of movement.

Depending on the adopted strategy, let us denote the movement trajectories of D_0 and D_1 objects as $(x_0(t), \ y_0(t))$ and $(x_1(t), \ y_1(t))$, respectively, $t \geq 0$. Then, the initial conditions can be written as follows:

$$\begin{cases} x_0(0) = 0, \\ y_0(0) = 0, \end{cases} \tag{1}$$

$$\begin{cases} x_1(0) = l, \\ y_1(0) = 0. \end{cases} \tag{2}$$

The conditions for the maximum velocity of objects to be equal to b and a, respectively, are written as follows:

Fig. 1 Rectangular coordinate system

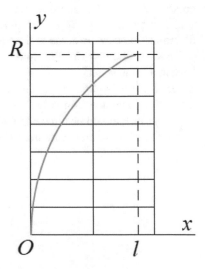

$$\left(\frac{dx_0}{dt}(t)\right)^2 + \left(\frac{dy_0}{dt}(t)\right)^2 = b^2, \tag{3}$$

$$\left(\frac{dx_1}{dt}(t)\right)^2 + \left(\frac{dy_1}{dt}(t)\right)^2 = a^2, \tag{4}$$

During movement the condition can be written as the distance between objects:

$$(x_1(t) - x_0(t))^2 + (y_1(t) - y_0(t))^2 = l^2, \tag{5}$$

Thus, a relationship was obtained to cross the barrier within the condition of maintaining the distance between the objects. To estimate the width of the lane ("street" width) through which the barrier can be crossed, it is necessary to find the point where it crosses the $\{x = l\}$ line of the trajectory determined by the conditions (1)–(5).

4   Solution of Passing the Barrier of the Street

According to the terms of the issue, the D_1 object ("guard") defends the $\{x = l\}$ barrier. In the worst case, it can be considered that the D_1 and D_0 objects are facing each other on the $\{y = 0\}$ line. From a strategic point of view, the D_0 object's motion in the upper or lower half-plane is the same. Without breaking the generality, it can be assumed that the D_0 object (attacking object) will try to get to the $\{x = l\}$ line by moving in the $\{y > 0\}$ semi-plane and overtaking it due to its high speed. Therefore, in order to get in front of the D_0 object, the D_1 object must move as fast as possible along the $\{x = l\}$ line. Based on the (2) and (4) conditions we can write that,

$$\begin{cases} x_1(t) = 1, \\ y_1(t) = a \cdot t. \end{cases}$$

Given the above, Eq. (5) can be write by the following form:

$$(l - x_0(t))^2 + \left(y_0(t) - a \cdot t\right)^2 = l^2, \tag{6}$$

From equality (6), it can be founded $x_0(t) = l - \sqrt{l^2 - \left(y_0(t) - a \cdot t\right)^2}$, and then it can be taking into account in (3). The $y_0(t)$ equation is obtained with respect to the function and its derivative:

$$\frac{\left(y_0(t) - a \cdot t\right)^2 \left(\frac{dy_0}{dt}(t) - a\right)^2}{l^2 - \left(y_0(t) - a \cdot t\right)^2} + \left(\frac{dy_0}{dt}(t)\right)^2 = b^2, \tag{7}$$

As a result of simplifications (7) can be written in the form of a nonlinear ordinary differential equation as follows:

$$l^2\left(\frac{dy_0}{dt}(t)\right)^2 - (y_0(t) - a \cdot t)^2\left(2a\frac{dy_0}{dt}(t) - a^2\right) - [l^2 - (y_0(t) - a \cdot t)^2]b^2 = 0$$

(8)

The (8) equation can be solved according to the $\frac{dy_0}{dt}(t)$ derivative:

$$\frac{dy_0}{dt}(t) = F(t, \ y_0(t)),$$

(9)

Here,

$$F(t, \ y_0(t)) \equiv \frac{1}{l^2}\left[a(y_0(t) - a \cdot t)^2\right] + \left\{(y_0(t) - a \cdot t)^4 a^2 - \right.$$
$$\left. - l^2\left[(y_0(t) - a \cdot t)^2(a^2 + b^2) - l^2 b^2\right]\right\}^{\frac{1}{2}}.$$

Thus, founding of the $y_0(t)$ function (9) leads to solution of (1) Cauchy task solving. The study of operations that express the $F(t, \ y_0(t))$ function shows that, (9), (1) have single solution and this solution can be calculated numerically, for example, by the Runge–Kutta method.

Thus, all functions included on the right side of the (7) formula are defined and the $y_0(t)$ trajectory. This means that in initially $(0, \ 0)$ located the width of the D_0 object is

$$R = y_0(l)$$

(10)

And when moving in the lane determined by the relationship (10) cannot move without "colliding" with the D_1 object.

Note, that attitude (9) is a strategy to overcome the barrier of the attacking object and can be considered as a management strategy in a closed-loop control class [4].

5 "Defence" Task

One of the issues of defence of the object is the strategy of neutralizing the threats to it in the "frontier" zone. The traditional task of defence consists in tracking (chasing) and destroying an impending danger with a higher speed. The solution of such problems has been the subject of research in game theory and has been commented on in the works of various authors.

Fig. 2 The behavior strategy of defenders: D_0 is attacking object; D_1, D_2, D_3, D_4, and D_5 are the guard group; DL is the defence line

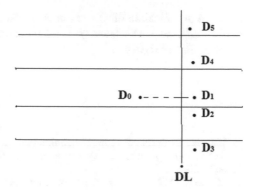

This study examines the issue of preventing a high-speed moving object from crossing a dangerous "boundary" with the use of relatively low-speed mobile security devices.

It is believed that both the object of danger and the objects of mobile security are controlled by regulating their speed. Mobile guard devices are required to be placed and operated in front of the dangerous "boundary" in such a way that the approaching source of danger must pass through the neutralization zone, which to some extent covers at least one of them, in order to achieve its goal.

Without breaking the generality, let us assume that the object of the attack is moving the $\{x = 0\}$ axis from the left hemisphere to the right one. To solve the problem, we divide the "barrier" into "streets". The width of each "street" should not be greater than the R determined by formula (10). In this case, let's define the behavior strategy of defenders as follows (see Fig. 2).

- If the attacking object is on the same street as the guarded object, the guard tries to stand in front of the object attacking the object. According to the solution of the auxiliary problem, in this case, the attacking object cannot cross the security barrier without "collision" $l = m$.
- If the attacking object is on another "street", the guard object is located on the border of the street in the direction of the attacking object. When the attacking object changes the street, the situation satisfies the condition of the auxiliary issue. In this case, the attacking object still can not cross the barrier without "collision".

6   Conclusion

So, in the paper, the defence of a certain barrier in the plane by a group of guarded objects is considered based on the differential equations. A model of the process of meeting a controlled object ("attacker") by a group of controlled guard objects is proposed. The model is based on the theory of differential games. As one of the issues of organization of protection of the facility, a strategy to neutralize the threats in the border zone was developed and evaluated. The evaluation had been carried out by using Fuzzy theory.

References

1. Vidal, R., Shakernia, O., Kim, H.J., Shim, D.H., Sastry, S.: Probabilistic pursuit-evasion games: theory, implementation, and experimental evaluation. IEEE Trans. Robot. Autom. **18**(5), 662–669 (2002)
2. Mamatov, M.Sh.: About the theory of differential game persecution in systems with distributed parameters. Autom. Comput. Technol. **1** 5–14 (2009)
3. Mamatov, M.Sh., Tuxtasinov, M.: About the task of persecution in distributed control systems. Cybern. Syst. Anal. **45**(2), 153–158 (2009)
4. Vlad, M., Mircea, Ş, Petru, D.: Krasovskii: passivity and μ-synthesis controller design for quasi-linear affine systems. Energies **14**, 5571 (2021)
5. Красовский, Н.Н., Субботин, А.И.: Позиционные дифференциальные игры. Moskva., «Nauka» (1974)

Comparison of Artificial and Convolutional Neural Networks in Recognition of Handwritten Letters of Azerbaijani Alphabet

Rustam Azimov🆔

Abstract Industrial and business life requires development of intelligent systems. Implementation of such systems and performing some automation tasks dictates building effective models for entering information. In the case the information is graphical, image recognition has a great potential to accelerate the increase of intelligence of machines. Known approaches for image recognition definitely include using Artificial Neural Networks with feature extraction and Convolutional Neural Networks which extracts a feature based on the training set. In the study these two models are compared on the problem of recognizing handwritten letters of Azerbaijani alphabet.

Keywords OCR · Handwritten alphabet recognition · Feature extraction

1 Introduction

Image recognition has a variety of applications including remote sensing, bar codes/QR codes reading, automated clinic diagnosis, fingerprint and face recognition, digital video processing, signature recognition, optical character recognition (OCR) [1, 2].

As a well-known direction of pattern recognition image recognition includes stages—preprocessing, feature extraction and classification [3]. Finding informative features for recognition is one of the key steps. In the study we experimentally compared three features. With two of them classification is achieved with artificial neural networks (ANN), the other one is founded during the learning process of a convolutional neural networks (CNN). With these networks and features recognition of handwritten letters of Azerbaijani alphabet was examined.

R. Azimov (✉)
Institute of Control Systems of the Azerbaijan NAS, Baku, Azerbaijan
e-mail: rustemazimov1999@gmail.com

© The Author(s), under exclusive license to Springer Nature Switzerland AG 2023
Sh. N. Shahbazova et al. (eds.), *Recent Developments and the New Directions of Research, Foundations, and Applications*, Studies in Fuzziness and Soft Computing 422, https://doi.org/10.1007/978-3-031-20153-0_31

2 Datasets

There are two kinds of graphics used worldwide: raster and vector graphics. In vector graphics an image is constituted by parametric equations of some primitives like lines, curves, circles. But in raster graphics an image is constructed with a color bitmap represented by a pixel array [4]. There are models to represent colors with a pixel array: RGB (Red, Green, Blue), HSV (hue, saturation, value/brightness), etc. Each pixels of an image in RGB format is expressed with three numbers. There are known ways to convert RGB format to grayscale which requires just a number to dye a pixel. Images in the dataset used in the study are in gray scale [5, 6].

Azerbaijani alphabet has 32 letters. In the study we used the dataset [7] with samples of 28 letters (see Fig. 1):

{ABCÇDEƏFGHIJKLMNOPQRSŞTUVXYZ}.

Remaining {ĞİOÜ} letters are not given in the dataset, because they can be recognized with much simpler approach: If a letter is a one of {GIOU}, then a "lightweight" algorithm can be used to identify class of that image—letter precisely by searching dots or strokes at upper part of the letter.

Training set has 500 samples for each letter, which means original training size is 14000 samples. Training size is enlarged to 28,000, 42,000, 70,000 samples with an augmentation technique. Augmentation includes operations rotation, shear, zoom. Test dataset has overall 2800 samples—100 samples for each letter. Images original training and test set have random, different sizes. They are all converted to a standard 20 × 20.

Fig. 1 Example of handwritten characters

3 Features

In the study three features are examined: Pixels of the image, Peripheral Directional Contributivity of black pixels, convolution kernels. First two are used with ANNs, the other with CNNs.

3.1 Pixels

Images in training/test sets are 20×20 sized arrays. Each array is flattened. As a result each image is characterized by a vector with 400 elements. These 400 numbers are given to an ANN as input.

3.2 Peripheral Directional Contributivity (PDC)

PDC reflects relative positioning of black pixels. This is achieved by looking for the first and second black pixels points in several directions from outside into inside from different segments. Then from each of these points connected pixels in multiple dimensions are counted [8, 9]. Total number of the feature vector can be calculated as the following:

- Dimension DC = 4,
- Directions quantity = 4 (two each in the horizontal and vertical directions),
- Depth quantity = 2,
- Number of segments = 8.

As a result, the dimension of the PDC feature will be 256.

Fig. 2 Extraction of PDC feature

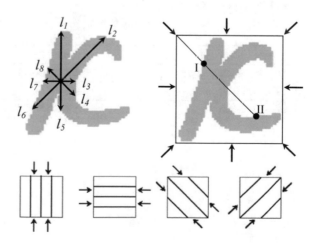

3.3 Convolutional Kernels Obtained from Training CNNs (Conv)

Images in training set are directly used as it is in the training process of a CNN. Elements of kernels that define convolution filters in the first and second convolutional layers are defined during the training. Input image has a 20 × 20 sized image of the letter in the center and total size of the image is 32 × 32.

4 Model Structures

4.1 Structures of the Models of Pixels and PDC Features

Each of Pixels and PDC features has two models with different structures, both are used in an ANN. Appropriate models with Pixels, PDC features have the same structure for hidden layers and similar parameter counts (see Table 1).

4.2 Structures of the Models with Conv Feature

Three different architectures of CNNs have been examined. These structures differs in the number of filters on the first and second layers and thereby in the number of parameters. In Table 2 first row is traditional LeNet-5 architecture of Yann LeCun [10]. Two different pooling choices—maximum and average for subsampling used in pooling layers of each CNN model.

Table 1 Structures of ANN models

N	Model	Number of layers	Number of neurons on layers	Number of parameters
1	Pixels_400_50_28	3	400-50-28	21,478
2	Pixels_400_50_50_28	4	400-50-50-28	24,028
3	PDC_256_50_28	3	256-50-28	14,278
4	PDC_256_50_50_28	4	256-50-50-28	16,828

Table 2 Structures of CNN models

N	Model	Number of feature maps on the 1st layer	Number of feature maps on the 2nd layer	Number of neurons on fully connected layers	Number of parameters
1	LeNet5-filters-6-16	6	16	120-84-28	63,236
2	LeNet5-filters-6-8	6	8	120-84-28	38,028
3	LeNet5-filters-6-4	6	4	120-84-28	25,424

5 Numerical Experiments

5.1 Realization of Experiments

Totally 40 different models examined:

- ANN: {quantity of models} × {quantity of DB} = 4 × 4 = 16;
- CNN: {quantity of models} × {quantity of DB} × {quantity of pooling methods} = 3 × 4 × 2 = 24.

In the training process of each model cross-entropy [11] was used as the objective/cost function. On the last layers of networks activation function was Softmax, on the other layers ReLU was used [12]. Optimization was done with Adam method [13].

Implementation of ANN and CNN was done with Keras library that works with Tensorflow framework [14, 15].

Experiments were carried out with 5 different initial points. Results of experiments were given in the Sect. 5.2 according to the highest accuracy among these five initial points.

5.2 Results

Recognition results are given below (Tables 3 and 4).

Charts represent followings:

- Comparison of recognition results of CNN models using average and max pooling as subsampling layers (see Fig. 3),
- Recognition results of all models with respect to training datasets with different sizes (see Fig. 4),
- Comparative analysis of recognition accuracy of one CNN and one ANN models with similar number of parameters (see Fig. 5),
- Comparison of precision of PDC and Pixels models with appropriate structures (see Fig. 6).

Table 3 Recognition results for all neural networks

N	Model	Train data	Num. of params	Recog. on train data (%)	Recog. on test data (%)
1	LeNet5-filters-6-16	14,000	63,236	99,64	89,32
2	LeNet5-filters-6-16	70,000	63,236	99,58	89,14
3	LeNet5-filters-6-16	42,000	63,236	99,23	88,82
4	LeNet5-filters-6-8	70,000	38,028	99,26	88,75
5	LeNet5-filters-6-8	14,000	38,028	99,57	88,57
6	LeNet5-filters-6-8	28,000	38,028	99,38	88,43
7	PDC-256-50-28	14,000	14,278	99,50	88,21
8	LeNet5-filters-6-4	70,000	25,424	99,02	88,18
9	LeNet5-filters-6-16	28,000	63,236	99,49	88,18
10	LeNet5-filters-6-4	28,000	25,424	99,21	88,07
11	LeNet5-filters-6-4	42,000	25,424	99,22	87,96
12	LeNet5-filters-6-8	42,000	38,028	99,53	87,96
13	PDC-256-50-28	70,000	14,278	97,79	87,86
14	PDC-256-50-50-28	42,000	16,828	98,58	87,54
15	PDC-256-50-50-28	70,000	16,828	97,97	87,50
16	LeNet5-filters-6-4	14,000	25,424	99,31	87,43
17	PDC-256-50-28	42,000	14,278	98,69	87,32
18	PDC-256-50-50-28	14,000	16,828	99,60	87,32
19	PDC-256-50-28	28,000	14,278	98,53	87,29
20	PDC-256-50-50-28	28,000	16,828	98,74	87,29
21	Pixels-400-50-50-28	70,000	24,028	98,30	84,04
22	Pixels-400-50-28	70,000	21,478	98,76	83,96
23	Pixels-400-50-50-28	42,000	24,028	98,91	83,36
24	Pixels-400-50-50-28	14,000	24,028	99,98	83,25
25	Pixels-400-50-28	42,000	21,478	99,20	83,04
26	Pixels-400-50-50-28	28,000	24,028	99,27	82,43
27	Pixels-400-50-28	14,000	21,478	99,94	82,04
28	Pixels-400-50-28	28,000	21,478	99,49	81,46

For a compact view of this table for CNNs, recognition results of models with different subsampling methods are not shown, but the maximum values over all initial points for each model are selected

6 Conclusion

Convolutional Neural Networks are compared to Artificial Neural Networks in recognition of handwritten letters of Azerbaijani alphabet with some image recognition features, multiple structures and the training datasets with different sizes. Convolutional Neural Networks performs better recognition results than Artificial Neural

Table 4 Recognition results for ANNs

N	Model	Train data	Num of params	Recog on train data (%)	Recog. on test data (%)
1	PDC-256-50-28	14,000	14,278	99,50	88,21
2	PDC-256-50-28	70,000	14,278	97,79	87,86
3	PDC-256-50-50-28	42,000	16,828	98,58	87,54
4	PDC-256-50-50-28	70,000	16,828	97,97	87,50
5	PDC-256-50-28	42,000	14,278	98,69	87,32
6	PDC-256-50-50-28	14,000	16,828	99,60	87,32
7	PDC-256-50-28	28,000	14,278	98,53	87,29
8	PDC-256-50-50-28	28,000	16,828	98,74	87,29
9	Pixels-400-50-50-28	70,000	24,028	98,30	84,04
10	Pixels-400-50-28	70,000	21,478	98,76	83,96
11	Pixels-400-50-50-28	42,000	24,028	98,91	83,36
12	Pixels-400-50-50-28	14,000	24,028	99,98	83,25
13	Pixels-400-50-28	42,000	21,478	99,20	83,04
14	Pixels-400-50-50-28	28,000	24,028	99,27	82,43
15	Pixels-400-50-28	14,000	21,478	99,94	82,04
16	Pixels-400-50-28	28,000	21,478	99,49	81,46

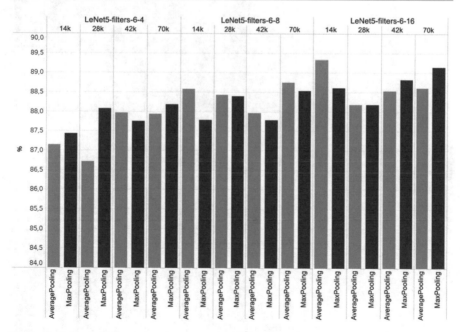

Fig. 3 CNNs with different pooling parameters

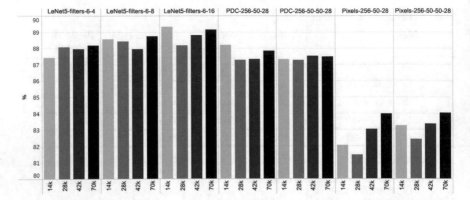

Fig. 4 Recognition results of all model structures depending on the sizes of the training set

Fig. 5 Comparison of
recognition accuracy of
Le-Net5-filters-6-4 (Conv)
and Pixels_400_50_50_28
(NN) with similar number of
parameters—25,424, 24,028
parameters accordingly

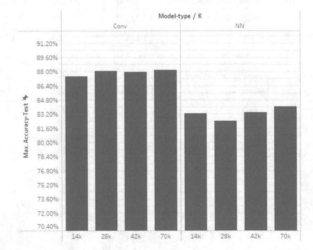

Fig. 6 Comparison of
precision of models—
PDC_256_50_50_28 and
Pixels_400_50_50_28

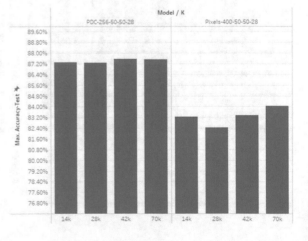

Networks which is examined with two different features. Subsampling layer type— average and max doesn't have tangible effect on accuracy results tested with test set. Among two features used with Artificial Neural Networks PDC have more accuracy than pixels of an image. Increase in training set size didn't show a definite impact on CNNs and ANNs with PDC feature. But in Pixels feature test accuracy almost increased accordingly with the increase of training sets. Traditional Convolutional Neural Network architecture—LeNet-5 with 6, 16 filters in the first and second convolution layers accordingly was the best among all Convolutional Neural Network architectures examined in the study.

References

1. Wang, X.: A Review of Image Recognition Technology (2018)
2. Padmappriya, S., Sumalatha, K.: Digital image processing real time applications. Int. J. Eng. Sci. Invent. (IJESI), 46–51 (2018). ISSN (Online): 2319–6734
3. Saliba, E.: An overview of Pattern Recognition (2013)
4. Jiang, X., Liu, L., Shan, C., Shen, Y., Dong, X., Li, D.: Recognizing Vector Graphics without Rasterization (2021).
5. Hrenatoh Courses. http://www.hrenatoh.net/curso/matematica/uso_cores.pdf
6. Serteep, H., Ibrahim, A.-W.: Transfer color from color image to grayscale image 5 (2017)
7. Aida-zade K.R., Mustafayev, E.E.: Intelligent handwritten form recognition system based on artificial neural networks. In: Proceedings of the International Conference on Modeling and Simulation, 28–30 Aug., pp. 609–613. Konya, Turkey (2006)
8. Mori, M., Wakahara, T., Ogura, K.: Measures for structural and global shape description in handwritten Kanji character recognition (1998). https://doi.org/10.1117/12.304621
9. Wu, X., Wu, M.: A recognition algorithm for Chinese characters in diverse fonts. In: Proceedings/ICIP … International Conference on Image Processing, vol. 3, pp. 981–984 (2002). https://doi.org/10.1109/ICIP.2002.1039139
10. Lecun, Y., Bottou, L., Bengio, Y., Haffner, P.: Gradient-based learning applied to document recognition. Proc. IEEE **86**(11), 2278–2324 (1998). https://doi.org/10.1109/5.726791
11. Shannon, C.E.: A mathematical theory of communication. Bell Syst. Tech. J. **27**(3), 379–423 (1948). https://doi.org/10.1002/j.1538-7305.1948.tb01338.x
12. Nwankpa, C., Ijomah, W.L., Gachagan, A., Marshall, S.: Activation Functions: Comparison of trends in Practice and Research for Deep Learning (2018). arXiv:1811.03378
13. Kingma, D.P., Ba, J.: Adam: A Method for Stochastic Optimization (2015). CoRR:1412.6980
14. Keras Homepage. https://keras.io/
15. Tensorflow Homepage. https://www.tensorflow.org/

Research and Analysis Criterion Quality of Service and Experience Multifractal Traffic Using Fuzzy Logic

Bayram G. Ibrahimov⊚, Sevinc R. Ismaylova⊚, Arif H. Hasanov⊚, and Yalchin S. Isayev⊚

Abstract The paper considers the solution of an urgent scientific problem—the study and analysis criterion quality of service (QoS) and quality of experience (QoE) indicators perception of the multifractal nature traffic differentiated by classes using the theory of fuzzy sets (Fuzzy sets) and fuzzy logic (Fuzzy logic). These tasks are related to the development heterogeneous streaming model of multiservice telecommunication networks, taking into account the property self-similarity useful and service traffic when rendering various classes multimedia services. The multicriteria tasks multiservice telecommunication networks (MTN) are analyzed based on the architectural concept subsequent NGN (Next Generation Network) and future networks FN (Future Networks) using information and communication technologies (ICT), which are based on the theory of fuzzy sets. A new approach to monitoring the quality of service indicator QoS and the quality of experience criterion QoE multimedia traffic using fuzzy logic apparatus is proposed. On the basis of the new approach, a streaming model and a system fuzzy inference are proposed, membership functions and rules fuzzy products are defined taking into account the numerous requirements QoS and QoE multimedia traffic and the quality of state CS (Current Service) communication channel. In this paper, we investigate the formulation of the problem and propose new approaches to its solution.

Keywords Multifractal nature traffic · Fuzzy logic · QoS · Membership function · QoE · Multiservice telecommunication networks · CS · Hurst coefficient

B. G. Ibrahimov (✉) · S. R. Ismaylova
Azerbaijan Technical University, H. Javid Ave. 25, AZ 1073 Baku, Azerbaijan
e-mail: i.bayram@mail.ru

A. H. Hasanov · Y. S. Isayev
Military Academy of the Republic of Azerbaijan, Qizil Sherg, 13, AZ1042 Baku, Azerbaijan

1 Introduction

Currently, the intensive development infrastructure digital economy, the formation strategic plans for a single information space and multioperator environment require new principles and innovative approaches to building highly efficient MTNs based on the architectural concept of the next NGN (Next Generation Network) and future FN (Future Networks) networks with increased throughput ability [1–4].

In order to build highly efficient multiservice telecommunication networks with wide capabilities multimedia services, it is necessary to use the apparatus theory of fuzzy sets, which are generalizations of the classical theory of fuzzy sets, fuzzy-neural networks and fuzzy logic [5–7]. Moreover, recent mathematical theories are generalizations classical set theory and classical formal logic [8–15]. The investigated packet-switched MTNs—perform multicriteria tasks and support a wide range multimedia services and applications with the implementation ICT technology, which provide QoS and QoE indicators multifractal traffic nature [3, 4].

It is known [2–4] that the emergence global information exchange networks, primarily the Internet, has become a qualitative leap in the development multiservice communication networks. With the development of the Internet, 5G/IMT-2020 and IoT (Internet of Think), using innovative technologies, research and analysis QoS and QoE indicators multiservice traffic has intensified, which determine the dynamics traffic in various switching nodes and communication channels [5–8].

The use promising technologies helps to accelerate the launch new multimedia services and applications, as well as to reduce the overall cost of their implementation in communication systems and requires the use innovative ICT technologies such as SDN (Software Defined Networking), 5G/IMT-2020, NFV (Network Functions Virtualization), LTE (Long Term Evolution), IMS (Internet Protocol Multimedia Subsystem), so and NR (New Radio). The latter provide the following indicators such as the requested quality of service of the multifractal nature traffic, resource management in communication networks, sufficient flexibility classifier management, and the required QoS and QoE criteria using a fuzzy logic apparatus [5, 9, 12].

It was found [2, 9, 12] that a special role in MTN belongs to the provision required level QoS and QoE multimedia traffic. To adequately describe the dynamics traffic, as well as to determine the required level QoS and QoE in a multiservice network, it is necessary to create adequate systems and approaches to assess the efficiency network in conditions self-similarity traffic, which is an important task. Indeed, studies of the properties multimedia traffic in modern packet-switched networks show that the traffic data transmission networks, in contrast to the classical representation traffic by a Poisson stream, has the property self-similarity [12].

Currently, research continues in the field mathematical modeling multiservice traffic, which makes it possible to explain the physical reasons for the phenomenon self-similarity in traffic and assess its impact on network efficiency. In this case, an important parameter self-similar processes is the Hurst exponent. This indicator is used to assess the degree self-similarity of a random process generated by multimedia services and applications [3, 12].

When studying communication networks based on a new approach using fuzzy logic, the following criteria are taken into account [3, 12, 14]:

- parameter self-similar traffic taking into account the Hurst coefficient;
- introduction new services and applications such as "Triple Play services" & "Bandwidth on Demand";
- effective management physical, informational and channel resources;
- use limited network bandwidth resources;
- the capacity buffer memory when exposed to DoS (Denial of Service) and DDoS (Distributed DoS) attacks;
- information protection system from authorized access;
- indicators quality of services QoS and information protection system from authorized access;
- reliability of the network and indicators quality of perception QoE.

Considering the above, the task preliminary assessment numerous indicators multimedia traffic using fuzzy logic in MTN is relevant. On the basis of the study, it was established [2, 7, 8] that an adequate assessment quality of criterion MTN functioning is crucial for the successful creation fuzzy logic inference system.

The purpose this work is to study and analyze indicators, criteria quality of service and indicators the perception quality of multimedia traffic differentiated by classes, taking into account the Hurst parameter when using the mechanisms theory of fuzzy logic.

2 General Statement of the Problem and Description System

Given the importance building multi-service communication networks based on FN [3, 5] with packet switching (ITU-T, Y.3000, ..., Y.3499) for QoS and QoE multimedia traffic, special attention should be paid to complex indicators. In the works, a generalized technique for fuzzy control of the QoS classifier [9] and methods for improving the performance communication networks taking into account the self-similarity multimedia traffic [2] using the fuzzy logic apparatus are studied [15].

The indicators quality of service and quality of experience traffic [3], operational distribution information and network resources in MTNs [3, 5], as well as a method for calculating the Hurst coefficient parameters using fuzzy models to describe [2, 7] and modeling weak formalized multiservice systems and telecommunication processes [9].

The analysis showed [1–6] that the effective load factor communication channel is determined by the quality of the QoS and QoE self-similar traffic control. Based on the study, it was determined [14] that the standard procedure for using the classifier multiservice communication networks for outgoing traffic multimedia service different class does not allow organizing the required (Requested Service, RS) quality of service for the telecommunication infrastructure.

Unified telecommunications infrastructure does not have the following: sufficient flexibility to control the classifier communication networks; limited ability real communication channels; not rational use of the channel resource These shortcomings in a single telecommunications infrastructure and multi-carrier environment can lead to the loss important—useful and service packages.

Given the components vector, the classifier multiservice communication networks using a single telecommunications infrastructure is functionally described by the following relationship [12]:

$$Q_{VK}(H, \lambda_i) = G[R_{SE}, E_i(\lambda_i)], \quad i = \overline{1, k}, \tag{1}$$

where $E_i(\lambda_i)$–function that takes into account the indicators stationary state of the communication channel with when providing i–th multimedia services with intensity λ_i, $i = \overline{1, k}$; R_{SE}–the required indicator quality of service and the quality of perception telecommunications infrastructure with the uneven nature formation of the current state traffic:

$$R_{SE} = W[R_{QoS}, \rho(\lambda_i, H_i), R_{QoE}], \quad i = \overline{1, k}, \tag{2}$$

where $\rho(\lambda_i, H_i)$–coefficient effective loading of the channel MTNs, taking into account the self-similarity traffic H_i in the provision i–th multimedia services, $i = \overline{1, k}$. As the degree self-similarity increases, the Hurst parameter takes values from 0.5 to 1.

The obtained expressions (1) and (2) are a complex indicator of the classifier multiservice communication networks, taking into account the quality of multimedia traffic management QoS and QoE and indicators the quality of state communication channel of the telecommunication infrastructure. Thus, in this paper, the formulation of the problem is investigated and new approaches to its solution are proposed.

3 Block Diagram of the Communication Service Quality of Control System

To implement the algorithms operation and control quality of service classifier systems in dynamics, a fuzzy control system is proposed [7, 9, 12, 14]. Figure 1 shows the dynamics generalized structural diagram fuzzy control system for multiservice communication networks.

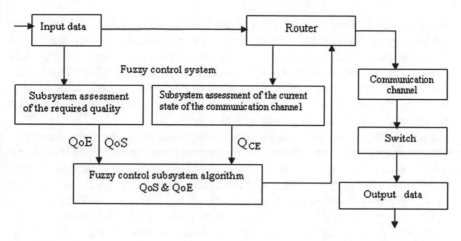

Fig. 1 Generalized block diagram fuzzy control system in multiservice telecommunication networks

The proposed scheme consists input and output data, a fuzzy control system, a router, a communication channel, and a switch. In this case, the fuzzy control system combines the following functional blocks:

- subsystem for assessing the required quality QoS and QoE;
- subsystem for assessing the current state of the communication channel R_{CS};
- subsystem fuzzy control of support algorithms R_{SE}.

Thus, the study effective loading communication channels and analysis of the structural diagram shows that when using the fuzzy logic apparatus, the parameters required quality of service are clearly higher than when using standart parameters communication network.

4 Description of the Fuzzy Inference System Taking into Account the Self-Similarity Property Multimedia Traffic

Based on the study, it was established [3, 12] that an important parameter in (1) and (2) is the effective load factor communication channels $\rho(\lambda_i, H_i)$ in a fuzzy control system, which changes the control parameter multiservice communication networks.

Given the self-similarity multimedia traffic $\rho(\lambda_i, H_i)$, the objective functions is to maximize the effective loading communication channels of multiservice communication networks as follows:

$$\rho(\lambda_i, H_i) = \frac{\lambda_i \cdot L_{i.p}}{C_{i.\max} \cdot N_m} \cdot f(H_i) \leq 1, \quad i = \overline{1, k}, \tag{3}$$

where λ_i–the flow rate i–th of the multimedia service packet; $C_{i.\,max}$–maximum service capacity multimedia service traffic packet streams; N_m–the number of communication channels for the services served at the access point to multimedia services, i.e. the number channel resource units multiservice networks; $f(H_i)$–a function that takes into account the self-similarity of the incoming load and is equal to $f(H_i) = 2H_i$; H_i–Hurst coefficient for i–th packet flow; $L_{i.p}$–length transmitted streams of the i-th multimedia service packet, $i = \overline{1,\ k}$.

Expression (3) is the objective function that determines the effective use communication channels using the fuzzy logic apparatus based on the required QoS and QoE parameters, taking into account the dynamics changes in the Hurst coefficient. However, as the degree self-similarity increases, the Hurst parameter significantly affects the required QoS and QoE parameters and the quality of network Q_{CS}.

Based on the proposed approach to monitoring the QoS and QoE indicator multimedia self-similar traffic and the quality of network, we consider the method calculating the Hurst coefficient using the mathematical apparatus of fuzzy sets. Here, an important characteristic of a fuzzy set is the Membership Function.

It should be noted that when using the apparatus in question, the expected value of the coefficient H_i from area $H_{э}$ according to the experimental results for various methods can be determined using the fuzzy integral [3]. In this case, it is a priori necessary to determine the distribution of the fuzzy density of the weights these values $g(H_i)$, $H_i \in H_{å}$. Here $H_i \in H_{å}$ means the aggregate of the expected value of the Hurst coefficient in the management QoS and QoE traffic.

Given the processes that occur using fuzzy logic, it is advisable to use the membership function $\mu_W(H_i)$ of an element H_i to a fuzzy set W. The analysis systems and decision-making under fuzzy conditions is the subject of the fundamental work Zade and Bellman [10, 11, 14].

On the basis of [9, 12], which allows changing the QoS and QoE parameters using the fuzzy logic apparatus from the domain magnitude W and H_e there is a set of ordered pairs and is described as follows:

$$W = \{\rho(\lambda_i,\ H_i),\ \mu_W(H_i)\}, \quad H_i \in H_e, \tag{4}$$

where $\mu_W(H_i)$–function that determines the degree membership of an element H_i to a fuzzy set W.

Given the algorithm for calculating the Hurst coefficient using the fuzzy logic apparatus, the membership function is expressed as follows:

$$\mu_W(H) = \sum_i \mu_W(H_i)/(H_i), \quad H_i \in H_{å}, \tag{5}$$

As a result of the study different method for calculating indicators self-similar traffic, the following values Hurst coefficient are obtained: $H_i = \{(i = 1, ..., 4)$: $H_1 = 0,68, H_2 = 0,75; H_3 = 0,86; H_4 = 0,95\}$.

Now we can determine the value of the function membership function ordered by decreasing degrees as follows:

$$Y(H_i) = \sum_{i=1}^{4} H_i \cdot \mu_W(H_i), \quad i = \overline{1,\,4}, \tag{6}$$

Then, according to the calculated data, numerical values $Y(H_i)$ are found as $Y(H_i) : h(H_i) = 0{,}43$, $Y(H_2) = 0{,}52$, $Y(H_3) = 0{,}90$, $Y(H_4) = 0{,}98$.

Based on the fuzzy control technique, the numerical values of the membership function of the process under study can be determined as follows:

$$\mu_W(H_i) = 0{,}43H_1^{-1} + 0{,}52H_2^{-1} + 0{,}90H_3^{-1} + 0{,}98H_4^{-1}, \tag{7}$$

Given (7) under given conditions, fuzzy measures take the following values:

$$g(F_1) = 0{,}52, \quad g(F_2) = 0{,}81, \quad g(F_3) = 0{,}95, \quad g(F_4) = 1{,}0, \tag{8}$$

Based on the proposed approach and calculation, the QoS and QoE indices are considered both the average service time and the probability packet loss when using the predicted value H_i compared to determining the listed indicators by the average value of the Hurst coefficient $H_i^{av} = 0{,}70$, where $0 < H_i^{av} \le 1$.

5 Construction and Verification of the Degree Adequacy Apparatus Fuzzy Model

It is worth noting that when constructing and checking the degree of adequacy of the fuzzy model apparatus, the Fuzzy Logic Toolbox software package Matlab was used [2–4, 7]. Figure 2 shows the function belonging $\mu_W(H_i)$ to the input variables of the coefficient effective loading communication channels in multiservice networks at $H_i^{av} = 0{,}70$ and $C_{i.\,max} = 155$ Mbps. From the graphical dependence it follows that an increase in the effective loading communication channels leads to an increase in the share lost packets, there by reducing the required QoS and QoE parameters.

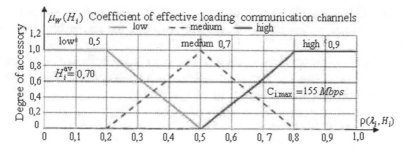

Fig. 2 Graphs membership function $\mu_W(H_i)$ input variables of the coefficient

For various services, its noticeable change begins with the value $\rho(\lambda_i, H_i) >$ 0.55, ..., 0.65 at $H_i^{av} = 0,70$ and $C_{i.\max} = 155$ Mbps.

To set membership functions in the Fuzzy Logic Toolbox, it was found that membership functions become non-linear. And this greatly complicates further research effective loading communication channels.

6 Estimation Probability-Time Characteristics Network Based on the Model Fuzzy Sets

Therefore, it was decided to build mathematical models for assessing the quality of transmitted traffic based on the fuzzy logic apparatus. In view (1)–(4) and (5), the formal description of the model is represented by the expression:

$$Q_{VK} = \langle R, \mu_W(H_i), T \rangle, \quad i = \overline{1, k}, \tag{9}$$

where R–many indicators reflecting the required parameters R_{SE} and R_{CS} network, $R = [R_{SE}, R_{CE}]$; T–many indicators describing the probabilistic-time characteristics of the network, $T = TD(\lambda_i, H_i)$; $\mu_W(H_i)$–membership function.

The criterion quality of service and perception QoS and QoE multimedia self-similar traffic using fuzzy logic apparatus is a probability-time characteristic [12]:

$$T[Mark_{\text{mod}}[TD(\lambda_i, H_i)] \to^i \min, \quad i = \overline{1, k}, \tag{10}$$

where $Mark_{\text{mod}}[TD(\lambda_i, H_i)]$–model assessment of the probability-time characteristics when providing the required multimedia services; $TD(\lambda_i, H_i)$–delays in transmitting packet streams based on parameters λ_i and H_i, $i = \overline{1, k}$.

Given (9) and (10), the criteria for the quality indicator of the state communication channel for representing a variety of services is defined as follows:

$$K_{opt}[Mark_{\text{mod}}(R)] \to K_{opt}[Mark_{\delta\hat{a}\vartheta\acute{a}}(R)], \quad i = \overline{1, k}, \tag{11}$$

where $K_{opt}[Mark_{\delta\hat{a}\vartheta\acute{a}}(R)]$–quality of service and experience R_{SE}, and quality of condition communication channel R_{CE} when providing the service at the required level.

Based on (9), (10) and (11), the criteria or rules for removing the partial uncertainty quality of state communication channel when providing a service that describes the operation fuzzy logic are represented by the expression:

$$\mu_W = \{0 \ \ddot{\imath}\vartheta\grave{e} \ R \notin W, \ \mu_W(R) \ \ddot{\imath}\vartheta\grave{e} \ R \in W, \ \mu_W(R) \in (0, 1], \tag{12}$$

where μ_W–membership function that associates each element fuzzy set W with a certain number, which is interpreted as the degree of belonging of the element R to a fuzzy set W.

Expression (7)–(12) describes a new approach to constructing a mathematical model for assessing the quality of traffic and the state of the communication channel in communication networks based on the fuzzy logic apparatus.

7 Analysis of the State Communication Channel Based on the Fuzzy Control System

System-technical analysis [3, 12, 14] showed that having information about the current state of the system and the required services at the access point to multimedia services, it is possible to dynamically control the parameters communication channels. This mechanism allows you to increase the quality indicators communication channel (Current Service, R_{CS}) in the network $Q_i^{CS}(\lambda)$. The indicators of the quality state of the communication channel in communication networks based on a fuzzy control system are functionally described by the following relationship:

$$Q_i^{CS}(\lambda) = F[K_i(\lambda), E_i(\lambda)], \quad \lambda = \sum_{i=1}^{k} \lambda_i, \quad i = \overline{1, k}, \tag{13}$$

where $K_i(\lambda)$–a coefficient characterizing the intensity multimedia traffic and taking into account the number units of the channel resource N_m multiservice communication networks is determined by the expression:

$$K_i(\lambda) = N_m \cdot [\lambda_i^s / \sum_{i=1}^{N_m} \lambda_i^s] \cdot f(H_i), \quad i = \overline{1, k}, \tag{14}$$

where λ_i^s–traffic arrival rate i–th service on one of the N_m terminal devices; $E_i(\lambda)$–total non-delivery rate i–th service packet streams.

Given the loss coefficients packets i–th traffic $K_{i.nn}(\lambda)$ during transmission over communicat- ion channels, the total coefficient non-delivery packets i–th traffic is found as follows:

$$E_i(\lambda) = K_{i.nn}(\lambda) + K_{i.BER}(\lambda), \quad i = \overline{1, k}, \tag{15}$$

where $K_{i.BER}(\lambda)$–the coefficient erroneous reception packet flows i–th traffic when transmitted over communication channels.

Expressions (13), (14) and (15) characterize the indicators quality of state communication channel in the network and describe the algorithms of the odd logic system.

8 Conclusion

1. As a result of the study, criteria for the quality MTN work were selected using the mechanisms theory of fuzzy sets, a new approach to the analysis of the QoS and QoE indicator multimedia traffic differentiated by classes was proposed under the influence the self-similarity effect, taking into account the prediction of the Hurst coefficient.
2. On the basis flow model, a function of the element's membership H_i in a set using the apparatus fuzzy logic is proposed W, a graph of the function $\mu_W(H_i)$ input variables coefficient effective load communication channels is presented, and a generalized block diagram fuzzy control system in MTN is constructed.
3. The conducted research and analysis show that the quality of service and quality of perception self-similar multimedia traffic based on the fuzzy control methodology, allow you to manage information resources in MTN using the apparatus fuzzy logic.
4. The proposed approach and system fuzzy logical control of the MTN classifier most fully takes into account the change in the current state multimedia service in the telecommunications infrastructure and makes it possible to increase the quality of traffic transmission with limited channel resources.

References

1. Zadeh, L.A.: Fuzzy sets. Inf. Control **8**, 338–353 (1965)
2. Borisov, V.V., Kruglov, V.V., Fedoulov, A.S.: Fuzzy models and networks. Hot line-Telecom (2012)
3. Ibrahimov, B.G., Alieva, A.A.: An approach to analysis of useful quality service indicator and traffic service with fuzzy logic. In: 10-th International Conference on Theory and Application of Soft Computing. Computing with Words andPerceptions–ICSCCW-2019. Advances in Intelligent Systems and Computing, vol. 1095, pp. 495–504. Springer, Cham (2019)
4. Sokolov, D.A.: Fuzzy quality assessment system. Technol. Means Commun. **4**, 26–28 (2009)
5. Ibrahimov, B.G., Humbatov, R.T., Ibrahimov, R.F.: Analysis performance multiservice telecommunication networks with using architectural concept future networks. T-Comm. **12**(12), 84–88 (2018)
6. Gaidamaka, Yu.V., Zaripova, E.R., Vikhrova, O.G.: Approximate method of session initiation delay performance evaluation in IP multimedia subsystem. T-Comm. **8**, 17–23 (2014)
7. Islamov, I.J., Hunbataliyev, E.Z., Zulfugarli, A.E.: Numerical simulation characteristics of propagation symmetric waves in microwave circular shielded waveguide with a radially inhomogeneous dielectric filling, Cambridge University Press. Int. J. Microw. Wirel. Technol. **9**, 1–7 (2021)
8. Bellman, R., Giertz, M.: On the analytic formalism on the theory of fuzzy sets. Inf. Sci. **5**, 149–157 (1974)
9. Ibrahimov, B.G., Ismaylova, S.R.: The effectiveness NGN/IMS networks in the establishment of a multimedia session. Am. J. Netw. Commun. **7**(1), 1–5 (2018)
10. Belman, R.E., Zadeh, L.A.: Decision-making in a fuzzy environment. Manag. Sci. **17**, 141–164 (1970)

11. Zadeh, L.A.: The role soft computing and fuzzy logic in understanding, designing and developing information, Intelligent systems. In: Artificial Intelligence News, No. 2–3, pp. 156–164 (2001)
12. Ibrahimov, B.G., Alieva, A.A.: Research and analysis indicators of the quality of service multimedia traffic using fuzzy logic. In: 14-th International Conference on Theory and Application of Fuzzy Systems and Soft Computing. Advances in Intelligent Systems and Computing, vol. 1306, pp. 773–780. Springer, Cham (2021)
13. Drutskoy, D., Keller, E., Rexford, J.: Scalable network virtualization in software-defined networks. IEEE Internet Comput. **17**(2), 20–27 (2013)
14. Yusifbayli, N., Nasibov, V.: Trend in Azerbaijan is electricity securite short-tem periods. In: 14-th International Conference on Theory and Application of Fuzzy Systems and Soft Computing–ICAFS-2020. Advances in Intelligent Systems and Computing, vol 1306, pp. 565–571. Springer, Cham (2021)
15. Zadeh, L.A.: The concept of a linguistic variable and its application to approximate reasomng. Part I-III, Information Sciences, No 8, 199-249; No. 8, 301–357; 1975. No 9. 43–80 (1975)

Author Index

Printed in the United States
by Baker & Taylor Publisher Services